Structured Decision Making

Structured Decision Making

A Practical Guide to Environmental Management Choices

R. Gregory, L. Failing,
M. Harstone, G. Long, T. McDaniels, and D. Ohlson

WILEY-BLACKWELL

A John Wiley & Sons, Ltd., Publication

This edition first published 2012
© 2012 by R. Gregory, L. Failing, M. Harstone, G. Long, T. McDaniels, and D. Ohlson

Blackwell Publishing was acquired by John Wiley & Sons in February 2007. Blackwell's publishing program has been merged with Wiley's global Scientific, Technical and Medical business to form Wiley-Blackwell.

Registered Office
John Wiley & Sons, Ltd, The Atrium, Southern Gate, Chichester, West Sussex, PO19 8SQ, UK

Editorial Offices
9600 Garsington Road, Oxford, OX4 2DQ, UK
The Atrium, Southern Gate, Chichester, West Sussex, PO19 8SQ, UK
111 River Street, Hoboken, NJ 07030-5774, USA

For details of our global editorial offices, for customer services and for information about how to apply for permission to reuse the copyright material in this book please see our website at www.wiley.com/wiley-blackwell.

Library of Congress Cataloging-in-Publication Data

Structured decision making : a practical guide to environmental management choices / R. Gregory ... [et al.].
      p.   cm.
   Includes index.
   ISBN 978-1-4443-3341-1 (hardback) – ISBN 978-1-4443-3342-8 (paper)
1. Environmental management–Decision making.   I. Gregory, Robin (Robin Scott)
   GE300.S7834 2011
   333.7–dc23

2011038034

A catalogue record for this book is available from the British Library.

Set in 10/12pt Photina by SPi Publisher Services, Pondicherry, India

1   2012

Contents

COMPANION WEBSITE

This book has a companion website:

www.wiley.com/go/gregory/sdm

with Figures and Tables from the book for downloading

Foreword

This book is about the creative and messy process of making environmental-management decisions. The approach we describe is called Structured Decision Making, a distinctly pragmatic label given to ways for helping individuals and groups think through tough multidimensional choices characterized by uncertain science, diverse stakeholders, and difficult trade-offs. This is the everyday reality of environmental management, yet many important decisions currently are made on an ad hoc basis which lacks a solid value-based foundation, ignores key information, and results in selection of an inferior alternative. Making progress – in a way that is rigorous, inclusive, defensible, and transparent – requires combining analytical methods drawn from the decision sciences and applied ecology with deliberative insights from cognitive psychology, facilitation, and negotiation. We review key methods and discuss case-study examples based in our experiences in communities, boardrooms, and stakeholder meetings. Our goal is to lay out a compelling guide that will change how you think about making environmental decisions.

We gratefully acknowledge partial funding support from the Decision, Risk and Management Science program of the US National Science Foundation (Award No. 0725025). We also acknowledge intellectual support from many colleagues and clients over the years who – through a mix of interest, curiosity, and frustration with conventional methods – have seen fit to work with us on a wide range of environmental-management problems, including Joe Arvai, Trent Berry, Cheryl Brooks, Mark Burgman, Bob Clemen, Jean Cochrane, Alec Dale, Nate Dieckmann, Daryl Fields, Baruch Fischhoff, Bill Green, Nicole Gregory, Paul Higgins, Dan Kahneman, Jack Knetsch, Howard Kunreuther, Sarah Lichtenstein, Lynn Maguire, Steven Morey, Ellen Peters, Mike Runge, Terre Satterfield, Basil Stumborg, William Trousdale, Nancy Turner, Terry Walshe, Carl Walters, Tim Wilson, and Kristin Worsley. Many of you (along with Rachel Flood and Leisha Wharfield) also helped by commenting on draft chapters of this book, for which we (and our readers) send thanks. We also thank the editorial staff at Wiley-Blackwell, which has been helpful, professional, and efficient. We take full responsibility for any errors or confusion in how these ideas and case studies are presented, but we recognize and appreciate the many contributions of your clear thinking and practical insights.

This book is dedicated to Ralph Keeney and Paul Slovic. Your enthusiasm for improving how society addresses environmental and risk-management decisions, and your continuing excitement over each new opportunity to learn a little more about how people join emotions with reason to make tough choices, informs and inspires us. Without your friendship and guidance, this book would not exist.

Preface

This book is about decisions. More specifically, it is about making decisions concerning the management of environmental resources. By 'decision' we don't mean a routine i-dotting and form-filling exercise to fulfill an administrative requirement. We mean clear and considered decisions generated through working as part of a team to develop a resource-management plan, prepare an environmental assessment, design a policy framework, propose an infrastructure project, or engage in a host of other activities that may significantly influence the use or protection of valued natural resources.

Our goal in writing this book is to reset the guideposts – even, we might boldly hope, to propose a standard – for what constitutes defensible decision making for the management of environmental resources. In so doing, we are divulging a well-kept secret: there are 'best practices' in decision making. Every decision context is different. Yet there is an emerging consensus about what constitutes a responsible and defensible approach to difficult decision making about environmental resources. This book sets out a guide for people who want to lead or be part of such decision processes – whether as managers, facilitators, technical experts, decision makers, community residents, resource users, or as members of non-governmental organizations or interest groups.

The approach we describe is called *Structured Decision Making* (hereafter SDM). It is a distinctly pragmatic label that we believe captures the essence of the approach. We could equally have called it 'a useful way to think about addressing tough environmental choices', but this abbreviation would be unwieldy. An SDM approach combines analytical methods from the decision sciences and applied ecology with insights drawn from cognitive psychology and the experience of facilitators and negotiators. People use SDM methods to organize complex issues in ways that help individuals and groups build common understanding, identify relevant information, and find innovative solutions to difficult environmental-management problems.

SDM is particularly helpful for *groups* of people working together on *solutions* in a way that is *rigorous, inclusive, defensible and transparent*. This framing profoundly changes how decisions are approached: who is involved, how the purpose is defined, how attention and resources are applied across different issues, and how success is gauged.

One of the distinguishing features of SDM is that it gives attention to both the values of people affected and factual information concerning the potential consequences of actions. This dual focus is the foundation of defensible decisions: explicit value-based choices based on the best available information. SDM neither should nor can 'make' decisions, but by linking values and facts it provides enormous insight to those charged with this task. To this end, we will introduce methods developed in the decision sciences, but tested and implemented in communities, boardrooms, and stakeholder meetings over the past decades in Canada, the United States, and other parts of the world. We emphasize how the fundamental goal of SDM processes – providing clear insights about possible courses of action to those charged with making a decision – can still be achieved in the midst of deadlines, resource constraints, limited attention, and overworked people with diverse perspectives.

Although we have a range of research and academic credentials, we are primarily practitioners and our experience is in applying SDM to environmental problems in the real world. Thus our approach in this book is unapologetically practical. Our goal is to bridge the gap between theory and practice in environmental decision making. We've tried to synthesize an enormous amount of thinking by minds greater than ours into a volume that will be understandable and useful in the real world. To keep us grounded, we've used an abundance of examples to illustrate both the pros and cons of suggested methods. We include success stories from our own work and the work of others, as well as some of our favorite mistakes – from which we like to think we've learned a great deal.

Who should read this book? Our comments are addressed to a broad audience of resource managers, risk analysts, land-use planners, industry leaders, environmental NGOs, facilitators or negotiators, and government decision makers as well as concerned resource users and the residents of potentially affected communities. In our experience, all of these people can and should have the opportunity to participate in meaningful and productive decision-making processes about environmental resources that lead to the implementation of recommended actions. This book also is intended to be appropriate as supplementary reading for university courses in environmental decision making or for courses in planning, ecology, resource economics, and environmental management, at either advanced undergraduate or graduate levels.

Despite the emphasis on methods, this is not primarily a how-to text. It's meant to introduce you to a new way of thinking about problem solving and accountability in decisions that involve public resources and multiple interests. You don't need to master any fancy mathematics or complex technical tools to benefit from SDM. You may need to change how you think about choices. We hope this book provides a guide to this new way of thinking and gives you a set of tools that are useful for finding solutions to tough environmental-management problems, particularly those that involve multiple stakeholders and choices over diverse and conflicting interests.

Structured Decision Making contains 12 chapters. The first documents why a new approach to environmental decision making is needed. The second outlines the philosophical and theoretical underpinnings of SDM, just enough to thank our predecessors and mentors and to allay any fears that we've made it all up. Chapter 3, on decision sketching, introduces the idea of structuring your resource management problem as a decision from the start. Chapters 4 through 9 cover core methods, emphasizing the role of behavioral considerations as well as analysis and deliberation. Chapter 10 discusses critical concepts of iteration, learning, and monitoring – keys to responsive decision making under uncertainty. Chapter 11 provides a reality check, meant to keep our feet on the ground and focus on what is possible in terms of people, institutions, legal constraints, information, and politics. The final chapter briefly summarizes key points. At the end of most chapters we also list key readings that provide additional details on concepts and methods.

Case studies are included in every chapter. These demonstrate SDM methods in practice and provide ideas, based on our collective experience, about which analytical techniques will work under what circumstances and how to deal with decision makers or other participants who may be hostile, bored, inflexible, or looking for miracles. Despite our emphasis on environmental-management problems, SDM is fundamentally about working with people. If a process to aid resource decision making and environmental management is going to be effective in this modern world of participatory democracy and stakeholder involvement, then it must be understandable and responsive to the needs of the participants.

The bottom line is that SDM, done properly, can help to clarify values, build a common understanding of the best available information, improve the quality of deliberations, and help to identify and implement better management alternatives. It can also save time and money, and – to the extent that the application of SDM methods helps resource management agencies to avoid litigation, lengthy negotiations, or costly remedial efforts to overcome management failures – these savings can be substantial. An SDM process requires knowledge of a mix of analytical and deliberative techniques, along with a willingness to learn and to be both frustrated and delighted by how people address tough environmental and societal choices. It also requires that the minds and hearts of participants (including SDM analysts and their clients) be open to restructuring environmental issues in ways that can lead to new and broadly acceptable management opportunities.

1

Structuring Environmental Management Choices

In late 1999, 24 people representing a variety of interests, perspectives, and agencies signed off on a consensus agreement that fundamentally changed water flows on a disputed stretch of a managed river in British Columbia, Canada. Up until then, hydroelectric facilities on the river had been operated primarily for power production, with limited consideration given to effects on fisheries, wildlife, recreation, and local communities. Relationships among the diverse stakeholders (BC Hydro, which produces electricity from the dam, the federal Department of Fisheries and Oceans, the provincial Ministry of Environment, and community members) were strained. Court actions were threatened by both the local aboriginal community and the federal Fisheries regulator. From the utility's perspective, water management options were complicated by an unclear regulatory environment that offered little guidance about how to involve other stakeholders, how to address trade-offs affecting water flows, or how to adapt management practices to public values that had changed over time.

Conventional thinking suggested a choice between negotiation and litigation. Instead, the utility, along with provincial and federal regulators, collaboratively developed and adopted a structured decision making (SDM) approach. In addition to achieving consensus agreements at all but one of 23 facilities, the SDM process produced a common understanding among key stakeholders about what could and couldn't be achieved with different management alternatives, about which trade-offs were acceptable, and about which uncertainties were most important. By focusing on mutual learning, it built trust and stronger working relationships among key stakeholders, and institutionalized a commitment to improving the information available for decision making over time. The process won a range of international awards for sustainability.

Although there are many reasons for the remarkable success of water-use planning in British Columbia, one key factor was the use of SDM methods to guide both analysis and deliberations[1]. Over 10 years later, SDM continues to play a prominent role in framing important environmental management decisions in the province. The provincial government regularly requests the use of SDM to help guide environmental assessment and project or program planning efforts. BC Hydro, the government regulated provincial energy utility, uses SDM approaches to assess its electricity generation options and incorporates SDM in its triple bottom-line approach to corporate purchasing policies. In both BC and the adjacent province of Alberta, several indigenous communities are using SDM as part of environmental management and ecosystem restoration initiatives[2]; one has adopted

Structured Decision Making: A Practical Guide to Environmental Management Choices, First Edition. R. Gregory, L. Failing, M. Harstone, G. Long, T. McDaniels, and D. Ohlson.
© 2012 R. Gregory, L. Failing, M. Harstone, G. Long, T. McDaniels, and D. Ohlson. Published 2012 by Blackwell Publishing Ltd.

its own SDM guidelines as the overarching framework for planning and negotiations in its territory with the provincial and federal governments[3]. The federal fisheries regulator wrote SDM-based procedures into its Wild Salmon Policy[4] and has used the process to produce an interim agreement on the management of a widely recognized threatened species[5]. Forestry practitioners in western Canada recently used SDM to help assess climate-change vulnerabilities and adaption options for sustainable forest management[6].

In the United States, the Fish and Wildlife Service (USFWS) has adopted SDM as a standard of practice and is using SDM methods in a variety of environmental management contexts. In its technical guide for the conduct of adaptive management, the US Department of the Interior (USDOI) states that 'Adaptive management is framed within the context of structured decision making, with an emphasis on uncertainty about resource responses to management actions . . . '[7].

This interest in SDM is not limited to North America. In Australia, the Department of Agriculture, Fisheries and Forests has a community of practice in SDM and is using it to develop an approach to the management of agricultural pests and invasive species. SDM approaches have been used in New Zealand to aid recovery of the endangered Hector's dolphin[8] and the country is debating use of an SDM approach to help develop a risk framework for management of genetic organisms, with a special emphasis on ways to integrate concerns of the aboriginal Maori culture alongside concerns developed through western scientific studies.

Why all the interest? What's different about SDM? Fundamentally, SDM reframes management challenges as choices; not science projects, not economic valuation exercises, not consultation processes or relationship builders. You have a decision (or a sequence of decisions) to make. The context is fuzzy. The science is uncertain. Stakeholders are emotional and values are entrenched. Yet you – or someone you are advising – has to make a choice. This decision will be controversial. It needs to be informed, defensible, and transparent. This is the reality of environmental management. It has been said that reality is what we deal with when there are no other options. We think that SDM is a useful way to deal with the realities of everyday environmental management.

1.1 Three typical approaches to environmental decision making

Let's look first at three dominant paradigms that guide how environmental management decisions are conventionally made: science-based decision making, consensus-based decision making, and analyses based in economics or multi-criteria decision techniques.

1.1.1 Science-based decision making

A file arrives on the desk of a resource manager working in government. A biologist by training, she learns that a species recovery plan for the recently listed split-toed frog will need to be in place in 18 months. This is a priority issue and she has been given lead responsibility. She pulls together an inventory of all the science on split-toed frogs and launches a science review and planning team. Within a few months, work is underway to produce a comprehensive risk assessment and a state of the art habitat model. The modeling is completed within 15 months, an extraordinary accomplishment. Our scientist heaves a big sigh of relief and settles down to develop a plan. Two months later, however, she is disillusioned, attacked by angry environmental activists who reject the recommended captive breeding options on ethical grounds, by local tourism operators who claim that the proposed road closures will ruin their businesses, and by frustrated recreationists who demand that recovery funding be used instead for the protection of more visible species. Faced with the impending deadline, her embattled boss pushes through a band-aid solution that slightly soothes stakeholders' ruffled feathers but will never protect the frog. Everyone is frustrated and disillusioned with a world where, once again, politics trumps science.

This scenario is (perhaps) a little exaggerated, but it illustrates the deep-rooted reliance on science of many decision makers and resource managers and their desire to produce 'science-based' decisions. When a solution supported by scientific experts fails to receive wide support, environmental managers often throw up their hands and decry the vagaries of 'irrational' social values and power politics. The problem is not with the science: sound science must underlie good environmental management decisions. The problem is with society's tendency to ask too much of science in making decisions and to leave out too many of the other things that matter to people. First of all, science is not the only credible or relevant source of knowledge for many environmental management decisions. Secondly, social considerations and ethics and the quality of dialogue play important roles in shaping environmental management choices. Most importantly, rarely is there a single objectively right answer and science provides no basis for dealing with moral or value-based choices. The biologist Jane Lubchenco, in her Presidential Address to the American Association of the Advancement of Science, reminded the audience (in the context of environmental planning) that 'Many of the choices facing society are moral and ethical ones, and scientific information can inform them. Science does not provide the solutions . . .'[9].

There is increasing recognition that when management choices are characterized by a high degree of stakeholder controversy and conflict, the decision process must address the values held by key participants[10]. Unfortunately, most resource management agencies have little knowledge about how to deal constructively with value-based questions. Nor is it generally recognized that many so-called environmental initiatives also will have implications for economic, social, and other considerations. If a narrow, environmentally focused agency mandate means that these related concerns have not been identified carefully, then progress in implementation may be blocked. The frequent result is an 11th-hour, behind closed-doors, largely ad-hoc capitulation

to vaguely defined 'social values' and 'political' pressures – as in our scientist scenario.

One of the things we want to do in this book is to help scientists and scientifically trained managers figure out how to contribute usefully to public policy decisions that are as much about values as about science. Making good choices requires the thoughtful integration of science and values – the technical assessment of the consequences of proposed actions and the importance we place on the consequences and our preferences for different kinds of consequences – as part of a transparent approach to examining a range of policy options. While credible environmental management relies on carefully prepared technical analysis, it also relies on creating a deliberative environment in which thoughtful people can express their views in a collaborative yet disciplined way. Science alone will not make good environmental policy choices. But a values free-for-all will not get us there either.

For some types of problems the objectives are simple and clear, the range of alternatives is well understood, and the evaluation of them involves few and relatively uncontroversial value judgments. For example, if a policy decision has been made to reduce waste or emissions by 30%, then the task of deciding how to achieve that target might be quite technical, largely driven by cost-effectiveness or least-cost analysis (implement the lowest cost alternatives up to the point where the target is reached.) Scientific or technical analysis can perhaps provide 'answers' in this constrained decision context, with scientists acting as 'honest brokers'[11]. For other, morally charged questions – regarding genetically modified foods, the hunting of baby seals, or lethal predator control, for example – beliefs are so deeply entrenched that the influence of scientific or technical information on decisions may be small. These choices often end up in the hands of political leaders who will make a value-based choice with little reference to scientific information.

The problem for environmental managers is that the vast majority of environmental decisions fall into a messy middle ground where science plays a bounded but critical role and values and preferences, often strong and initially polarized, are also critical but not fixed. Research in behavioral decision making emphasizes that, particularly in less-familiar evaluation contexts, preferences are often 'constructed' based on information gained during a process[12], rather than uncovered or revealed as fixed pre-existing constructs. Factual information will never, by itself, make a decision, but it informs and shapes values, which do determine choices.

This clearly implies that what is needed is a framework for making environmental management choices that deals effectively with both science and values. Yet when managers and scientists – and most other people as well – talk about values, they find themselves tip-toeing around, more than a little uncertain how to proceed. Most often, efforts to resolve value-based conflicts focus on bargaining and negotiation or on consensus building. Unfortunately, an overemphasis on process, dialogue, and consensus can create its own problems.

1.1.2   Consensus-based decision making

As the name suggests, consensus-based decision-making processes are those that focus on the endpoint of bringing a group to a consensus agreement.

What could possibly be wrong with this? As an outcome, nothing; we're fans of consensus, just as we're fans of laughter or happiness. Our criticism arises whenever consensus is a *goal* of group deliberations, because we've often seen an emphasis on consensus take environmental management processes in the wrong direction. The biggest problem is that the group will often push too soon, too hard toward convergence, at the expense of a full exploration of minority views and creative solutions. An approach based on building consensus presumes that people have a good idea at the start of what they want to see happen, and that this reflects a good understanding of what the various alternatives

will deliver. When addressing tough environmental management problems, this is rarely the case. Whenever decisions are characterized by multiple and conflicting objectives and a complex array of alternatives with uncertain outcomes – a nearly universal situation in environmental management – people are likely to enter into a decision-making process with plenty of emotions and strong positions but a poor understanding of relationships between actions and consequences. And as we discuss more fully in Chapter 2, it is naïve and misleading to assume that working with people in a group is a simple cure for the shortcomings of individual decision makers.

In addition, insufficient attention typically is given to dealing with uncertainty in the anticipated consequences of actions and to what this means for establishing an effective and robust management strategy. Although in some cases significant reductions in uncertainty are possible, at other times key sources of uncertainty will be irreducible, at least with available resources and within the time scale of management concern. Reaching agreement in these cases necessarily involves tackling directly the thorny issue of risk tolerance – how much risk people are willing to accept and to which of the things they value. Recovery plans often bring these issues into the fore: with the split-toed frog, it's likely that some stakeholders will be highly risk averse ('we must guarantee long-term survival') and others will be more risk neutral ('we need to improve chances of survival'). Bargaining and negotiation frameworks offer little that will help groups work through these issues in a constructive and collaborative manner.

Finally, because of the emphasis on consensus as such, it is tempting for both participants and facilitators to ignore difficult trade-offs and to favor vaguely defined or relatively safe solutions so long as everyone agrees to them[13]. Questioning the motives or aspirations of the group, reminding them of the larger problem context, or introducing participants to demanding – albeit appropriate and insightful – analytical methods, is rarely attempted because the fragile consensus might well be upset. Little is done to combat insidious 'decision traps' that (as we'll discuss later) have been shown to foil the judgments of even sophisticated decision makers[14]. From a decision-making perspective, however, the goal is to reach beyond the least common denominator of a universally supported plan and, instead, to deliver one that is creative and demonstrably effective, that will survive further scrutiny from a wider audience, and that is likely to prove robust (to changing values, circumstances, and politics) over time. This requires that conflicting views be viewed not as problems to be hushed or appeased but as opportunities to clarify the reasons behind apparent differences in values and the various interpretations given to factual information.

1.1.3 Economics and multi-criteria analysis

Imagine if the split-toed frog project had landed on the desk of an economist rather than a scientist. An economist might immediately see the need for a quantitative analysis that will yield a summary calculation showing the ratio of costs to benefits of the alternative courses of action. He is likely to take the list of initial management alternatives he's been given, calculate the expected values of key effects, and begin the process of monetizing a long list of ecological and social impacts. Knowing that this is a complex and controversial task, he is likely to allocate his 18 months to conducting benefit transfer studies, or perhaps to initiate a travel-cost study or contingent valuation survey[15] – tools that help to assign monetary values to non-monetary effects. There will be little constructive debate about the science and the uncertainties underlying estimates of ecological effects, as discussions are dominated by defense of the controversial monetization techniques. The final results are subject to wide-ranging criticism, as various participants either disagree with the monetization efforts or protest that important values

have been left out of the analysis because they are too difficult to quantify. In the meantime, no alternative solutions have been generated.

This scenario demonstrates a more technocratic approach to decision making. The focus is on finding a formula that will calculate a summary answer: the analyst wants to do the right thing, but above all seeks a method that will yield a number (e.g. a net present value or a benefit-cost ratio) and provide the required answer. For the economist, the primary techniques are monetization, benefit transfer studies, and cost-benefit analysis, informed by a variety of specialized non-market valuation methods. This scenario could equally feature a decision analyst; the tools would be multi-attribute utility functions, normalization, weighting, and related techniques. Yet the effect would be the same – a formula-based score that identifies the preferred solution.

What is lost with these technocratic approaches is the focus on making sound decisions. If you're a manager, you need solutions – creative solutions – that are directly responsive to stakeholders' perceptions and concerns and that are developed with their collaboration and support. Instead, a technocratic approach reduces the management task to a project valuation and selection exercise. The essence of good decision making lies in understanding the problem, gaining insight into what matters to people, and then generating responsive alternatives. In a cost-benefit process, there is little room for these tasks. The emphasis is on analyzing one preferred solution: rarely are alternatives compared explicitly or broken down into their components in hopes of combining elements to develop a new, better (i.e. more effective or cheaper or quicker or more widely supported) management alternative. As we discuss further in Chapter 2, economic and multi-criteria approaches might produce a decision, but it may not be one that addresses the real problem at hand and, without the involvement of key parties in a creative problem-solving process, it'sunlikely to enjoy broad-based support. Of course there are experienced practitioners in both economics and multi-criteria decision analysis who emphasize the need for good problem structuring, creative thinking and mutual learning. But in their conventional applications, both cost benefit analysis and multi-criteria methods lack the structuring and deliberative aspects of SDM and, to the extent that they represent expert-driven processes, are unlikely to generate broad-based community or stakeholder support.

1.2   Structured decision making

1.2.1   What is structured decision making and where does it come from?

We define SDM as the collaborative and facilitated application of multiple objective decision making and group deliberation methods to environmental management and public policy problems. It combines analytical methods drawn from decision analysis and applied ecology with insights into human judgment and behavior from cognitive psychology, group dynamics, and negotiation theory and practice. The primary purpose of an SDM process is to aid and inform decision makers, rather than to prescribe a preferred solution.

In more everyday terms, we think of SDM as an organized, inclusive, and transparent approach to understanding complex problems and generating and evaluating creative alternatives. It's founded on the idea that good decisions are based on an in-depth understanding of both values (what's important) and consequences (what's likely to happen if an alternative is implemented). Designed with groups in mind, it pays special attention to the challenges and pitfalls that can trap people working together on emotionally charged and technically intensive problems – mental shortcuts and biases, groupthink, positioning, and a host of difficult group dynamics and communication challenges. Because it has decisions about public resources in mind, it emphasizes decision structuring approaches that

can contribute to consistency, transparency, and defensibility, particularly in the face of technical and value-based controversy.

Although SDM approaches could be applied to a range of public policy and management applications, our focus in this book is on problems involving environmental management and policy choices[16]. The examples we discuss span challenges related to the management of competing water uses, air quality, climate change, species at risk, pest outbreaks, cumulative effects, wildfire risks, parks and recreation, fish and wildlife harvest policies, oil and gas development, mining, water supply options and infrastructure investment. An SDM process can't guarantee great outcomes – both politics and uncertainty will influence what takes place – but it provides a sensible decision-making process for multi-dimensional choices characterized by uncertain science, diverse stakeholders, and difficult trade-offs.

Decision-making methods are often grouped into three categories[17]. 'Normative' methods define how decisions should be made, based on the theory of rational choice. The problem, of course, is that only rarely are people truly rational; instead, decisions usually reflect a mix of cognitive and intuitive or experiential responses. 'Descriptive' methods describe how people actually make decisions. They provide helpful insights about how and when decision-making processes need to be modified in light of how people typically form and express judgments. 'Prescriptive' approaches, such as SDM, suggest ways to help individuals or groups to make better decisions, based on decision theory but adapted for the practical needs and constraints facing real decision makers operating in the real world. This emphasis on practical, real-world solutions – to ensure concepts are understood, or analyses are undertaken promptly, or recommendations are implemented rather than stalled or ignored – is a theme that will recur throughout the book.

Although there are different types of environmental management decisions and different deliberation contexts, the use of an SDM approach usually requires that each of the following questions is addressed:

1 What is the context for (scope and bounds of) the decision?
2 What objectives and performance measures will be used to identify and evaluate the alternatives?
3 What are the alternative actions or strategies under consideration?
4 What are the expected consequences of these actions or strategies?
5 What are the important uncertainties and how do they affect management choices?
6 What are the key trade-offs among consequences?
7 How can the decision be implemented in a way that promotes learning over time and provides opportunities to revise management actions based on what is learned?

None of this should look surprising – you may recognize these as the most basic of steps in developing or evaluating almost any significant choice: from buying a home, to choosing a name for a company, to developing effective public policy. The difference with an SDM approach is that each of these steps is undertaken formally, openly, and in a way that supports collaborative learning and defensible decision making.

In SDM, these core steps are used to structure and guide thinking about complex choices. Sometimes, the steps are used quite literally as a guide: an explicit step-by-step process that a group agrees to follow. This has the benefit of ensuring that everyone knows where they are and what comes next. At other times, they are used just to inform constructive thinking about complicated management problems – an individual manager or stakeholder with these steps in mind, learns to ask 'what are our objectives?', and so on.

The steps are supported by structuring tools and techniques that have been developed in the decision sciences over the past 50 years. These structuring tools are designed to help individuals and groups deal with technically complex decisions and difficult group dynamics. Key SDM structuring tools that are almost universally applicable (and discussed later in this book) include objectives hierarchies, means-ends diagrams, influence diagrams, consequence tables, and elicitation protocols. Other analytical tools, such as Bayesian belief networks, strategy tables, decision trees, value of information calculations, or multi-attribute trade-off techniques are critical to some decisions but typically see more limited application. Economic methods (such as cost–benefit analysis), technical models (such as life cycle analysis or ecological risk assessment and modeling), or statistical uncertainty techniques (such as Monte Carlo simulation or sensitivity analysis), can all play a role in informing a decision, but do not of themselves constitute the decision-making framework.

What exactly is done at each step of an SDM process, to what level of rigor and complexity, will depend on the nature of the decision, the stakes and the resources, and timeline available. In some cases, the appropriate analysis may involve complex modeling or data collection spanning months or years. In others, it will involve elicitations of experts' judgments conducted over several days or weeks. At other times, a half-day workshop to carefully structure objectives and alternatives may be all that is needed to clarify thinking about a particular decision. A key point is that structured methods do not have to be time consuming. Even very basic structuring tools and methods can help to clarify thinking, minimize judgmental biases, and improve the consistency of decision-making processes and the quality of their outcomes.

There are no 'right' decisions. The goal of SDM approaches is to clarify possible actions and their implications across a range of relevant concerns. In the context of environmental management, clarity is achieved by sound technical analysis, informed and transparent value judgments, and a process that engages people in recognizable best practices with respect to decision making.

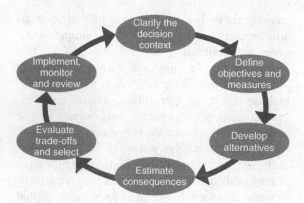

Figure 1.1 Steps in structured decision making.

1.2.2　What does a structured decision making process look like?

There are six core steps in SDM that usefully can be applied to nearly any environmental management decision, as shown in Figure 1.1. What exactly is done at each step will vary depending on the decision context. One of the most important lessons we have learned is the value of iteration. Get started, keep moving, and come back to refine previous work when you need to.

To help to familiarize you with the terminology we will use when discussing these steps, we'll refer to the simple and familiar situation of buying a flight ticket. We know this isn't really an environmental management question, but real problems quickly get complicated and our purpose here is primarily to establish a basic working vocabulary.

Clarify the decision context

This step involves defining what question or problem is being addressed and establishing the scope and bounds for the management decisions. At this early stage, it is important to answer three initial questions:

1　What is the decision (or series of decisions) to be made, by whom and when?
2　What is the range of alternatives and objectives that can be considered? – not the details

at this stage, just the general range and bounds: what's in and what's out.
3　What kind of decision is it and how could it usefully be structured? What kinds of analytical tools will be needed? What level and kind of consultation is appropriate?

This initial problem-structuring phase of SDM lays out a road map for the deliberations that follow so that all parties understand what will happen and when. A key technique at this point is 'decision sketching' – working quickly through the first steps of SDM at a scoping level to clarify what the decision is about and what will be required to make an informed choice. This is the focus of Chapter 3.

Define objectives and performance measures

Objectives concisely define 'what matters' about the decision; performance measures are specific metrics for assessing and reporting how well an

---

Text box 1.1:　Example: buying a flight ticket

Suppose that, for whatever reason, you need to purchase a flight ticket from Vancouver to New York.

There is a surprising amount of additional information you need to know about the context of the trip before running to Expedia: *When is the trip? How flexible are the dates? Who's paying? Is there a travel budget?* You could also challenge constraints: *Do I really need to go – perhaps I could teleconference instead? Maybe we could meet halfway?* All of these details (and many more) need to be addressed in the 'clarify the decision context' phase of the process. Neglect these and other important questions, and your entire analysis could end up being useless.

For now, let's assume it's a personal trip – so you're paying, dates are somewhat flexible, you don't need to consult other family members, and teleconferencing isn't an option.

alternative performs with respect to an objective. Together, objectives and performance measures do two critical things: they drive the search for creative alternatives and they form a consistent and transparent framework for comparing them. They define key concerns related to ecological, social, cultural, economic, or health and safety considerations, to the extent that these might be affected by alternatives under consideration. They include important but hard-to-quantify values and outcomes as well as those more easily quantifiable.

Objectives and their corresponding performance measures must be carefully defined and accepted by key stakeholders as the basis for evaluating management alternatives. It's understood that different parties will attach different importance to different objectives, but weighting is not addressed at this stage; it will be addressed if and only if it's found to be necessary and useful as part of a later trade-off analysis. The goal of this initial stage is for participants to agree on what things matter and need to be assessed in order effectively to compare alternatives.

Of course, the word 'objective' is in wide use, and everyone thinks they have their objectives pretty well figured out. But to be useful for decision making, objectives need to follow a few simple but important rules. We provide some guidance and useful tools in Chapter 4. We devote an entire chapter (Chapter 5) to the development of performance measures, a reflection of how critical they are to a successful decision process.

Develop alternatives

In some decision contexts (like our flight ticket for example), identifying alternatives is a fairly passive affair – the alternatives are discrete things that are 'out there' waiting to be picked from a shelf. In many environmental management contexts, however, things are not so straightforward. Alternatives are usually complex sets of actions that need to be created rather than discovered. That act of creation is what SDM is all about – the

---

> **Text box 1.2:   Example: buying a flight ticket**
>
> Still not ready for Expedia. Ask yourself: 'what do I care about in a flight ticket?' – perhaps it's one that leaves at a convenient time, flies directly, is inexpensive and comfortable. Maybe you'd avoid an airline with a dodgy safety record, or are interested in cancellation flexibility or reward miles? You could probably list dozens of items. But after reflection, suppose these boil down to three fundamental objectives that you'd base your decision on:
>
> 1 Minimize cost.
> 2 Minimize travel time.
> 3 Maximize comfort.
>
> Now for performance measures: of your three objectives, cost should be straightforward to estimate (in dollars), as should travel time (in hours). Comfort is a bit trickier, so you'd need to come up with a scale that incorporated aspects of comfort that concerned you (e.g. legroom, in-flight entertainment options, etc.). Let's say we come up with a scale of 1 to 5, where 1 is the least comfortable, most basic seat you can imagine and 5 is the most luxurious surroundings you could expect for your budget and for the airlines flying this route.

---

development of creative alternatives that are responsive to the defined objectives. In contrast to economic or scientific approaches that focus primarily on valuation or risk, the focus in SDM is on identifying, comparing and iteratively refining alternatives. Alternatives should reflect substantially different approaches to a problem based on different priorities across objectives, and should present decision makers with real choices. In most environmental management contexts, it is important to search for alternatives that are robust to key uncertainties or that are likely to reduce them over time.

OK, now to the Internet. Let's suppose your
search yields three candidate flights that could
work within the dates that you want to travel.

With the alternative flights sorted out, we
now have a start on a 'consequence table',
which compares the alternatives in terms of
the specified objectives and performance
measures.

Objectives	Performance measures	Flight 1	Flight 2	Flight 3
Minimize cost	$			
Minimize time	Hours			
Maximize comfort	Scale (1 = uncomfortable, 5 = luxury)			

In Chapter 7, we describe some methods
for developing alternatives in the context of
multi-stakeholder, multi-issue environmental
management problems. In these situations, the
number and diversity of alternatives can be
overwhelming. And for decision makers some of
them are scary – there is a fear of unrealistically
raising both expectations and the costs
associated with evaluating a large number of
alternatives. SDM includes methods for both
helping groups come up with a range of creative
alternatives, organizing them to facilitate effective evaluation, and then iteratively simplifying
and improving them.

Estimate consequences

At this step, the consequences of the alternatives
on the performance measures are estimated. This
is a technical task, undertaken by 'experts'.
Who's an expert depends on the nature of the
task, but for any given decision may include a
combination of scientists (which we define
broadly here to include natural and social scientists, economists and engineers) and local or
traditional knowledge holders. The presence of
uncertainty complicates the assessment of
consequences, as does the presence of multiple
legitimate forms of expertise. Groups involved in
an SDM process will need to be prepared to learn,
to explore competing hypotheses, and to build a
common understanding of what constitutes the
best available information for assessing consequences. An honest and accurate representation
of uncertainty based on a diversity of expertise
will be essential. Consistent with the project
goals, timeline and resources, attempts to reduce
uncertainty may include collecting additional
information, developing predictive modeling
tools, or eliciting judgments about the range of
potential outcomes from experts. Particular
emphasis is placed on building an understanding
of uncertainty as it affects the evaluation of alternatives. (Are some alternatives more uncertain or
less well-understood than others? If so, how
might these sources of uncertainty be addressed?
Are management actions sufficiently flexible to
incorporate what is learned over time?)

A useful tool for summarizing consequences is
a 'consequence table'. Variously termed a decision table, objectives by alternatives matrix, or
facts table, this deceptively simple tool presents
the consequences of policy or management
options in a matrix with objectives and performance measures on one axis and the alternatives
on the other. In each cell of the matrix is an
estimated value or impact score for the measure –
the impact or distribution of impacts that is
predicted to occur under that policy alternative –
not unlike a consumer report comparison of
stoves or vacuum cleaners. The level of rigor
involved in estimating consequences can vary
considerably depending on the nature of the
decision. Some major decision processes may
have months or even years of supporting analyses; others may require nothing more sophisticated than a one-to-five point scoring system
sketched on a flipchart.

We discuss uncertainty in detail in Chapter 6,
and key considerations in estimating consequences in Chapter 8. This book is not meant to

provide detailed technical guidelines on estimation methods; rather, we focus on approaches to developing a decision-relevant information base. In the context of environmental management problems, expert judgments will almost certainly play an important role. We provide guidance on how to elicit judgments from experts using best practices that minimize the effects of known biases, aggregate the judgments of multiple experts, and improve the consistency and transparency of judgments. We explore some ways to present probabilistic information and explore decision makers' risk tolerances. We'll touch on issues related to 'best available information', including practical ways to address calls for the democratization of science. We'll also discuss some of the difficulties that arise when seeking to evaluate consequences using information from both scientists and community (including aboriginal) resource users, and we review considerations for leveling the playing field between these two important sources of knowledge. Here again, the treatment of facts versus values plays a central role.

Evaluate trade-offs and select

The goal of this fifth step is to choose an alternative (or set of alternatives) that achieves an acceptable balance across multiple objectives. Although an SDM process often delivers win–wins and synergies, most decisions will nevertheless involve trade-offs. Choosing a preferred alternative will involve value-based judgments about which reasonable people may disagree: are the incremental health benefits from cleaning up a contaminated site worth the incremental costs?; do the greenhouse gas reductions from a 'green' energy project outweigh its noise and visual impacts?; does improved protection from wildfires in a rural community offset losses in forest biodiversity that would result from proposed wildfire management actions? SDM promotes exposing and facilitating an open dialogue about these trade-offs.

Methods for making choices should allow participants to state their preferences for different alternatives based on credible technical information about the estimated consequences. There are a range of structured methods for this, some quantitative and some less so. Where the potential outcomes of choices are significant and controversial, formal multi-attribute methods can be used to bring clarity, consistency, and transparency to decision making. However, for most environmental management choices, decision makers will retain the discretion and responsibility for making difficult choices; although quantitative trade-off methods can be helpful aids, they should not serve as formulas to prescribe an answer. Further, for a large proportion of environmental management decisions, thoughtful structuring of the problem (in terms of the fundamental objectives and alternatives), sound estimates of consequences and associated

---

Text box 1.4:   Example: buying a flight ticket

Describing consequences requires researching the best available information. The ticket prices should be easy enough, but what about the flight times? Do we believe the airlines' published figures? Is there some uncertainty (as shown for Flight 3) about the actual flight times? Perhaps we could do some digging around in travel forums to see how reliable they are. The comfort scale might also be a bit tricky to populate; can we find a suitable expert (e.g. an experienced traveler on this same Vancouver – NYC routing) to help us out? Even so, the comfort ratings might also be subject to some uncertainty.

Objectives	Performance measures	Flight 1	Flight 2	Flight 3
Minimize cost	$	2000	1200	600
Minimize time	Hours	6	5	10–12
Maximize comfort	Scale (1 = uncomfortable, 5 = luxury)	3	2	1–3

uncertainties, and structured deliberations about key trade-offs will result in wise, well-informed choices without explicit weighting or formal quantitative trade-off analysis. And because SDM clarifies areas of agreement and disagreement among stakeholders and the reasons for these disagreements, the results of an SDM process are useful to decision makers whether a consensus is reached or not.

In Chapter 9 we discuss different approaches to making trade-offs and provide examples showing how groups have successfully used both quantitative and qualitative methods to aid their understanding. We also do a little myth-busting

---

Text box 1.5:   Example: buying a flight ticket

Although we know that comfort is important to us, the consequence table shows that because of uncertainty it's not possible to know if there's any difference in the performance of alternatives. Unless we can refine this information, comfort can't help us decide anything. On the remaining two objectives, Flight 2 outperforms (dominates) Flight 1 because it's both quicker and cheaper. So, unless there are any other factors we're missing – and, at this point, it's always sensible to pause and ask yourself this question, whether something vital has been omitted from the analysis – we should eliminate Flight 1. Flight 3 takes between five and seven hours longer than Flight 2, but is $600 cheaper. So the real choice comes down to this: is avoiding five to seven hours in an airplane worth $600 to you?

This, of course, is a value question that will depend on your personal preferences and finances. Some people will prefer Flight 2, others Flight 3: the use of SDM approaches won't 'make the choice' for you, but they have successfully identified the key trade-off and laid out a consistent, defensible process for making this decision.

---

in this chapter as we show that there are rigorous ways to deal with values, and that you can – respectfully – ask people to justify, reflect on, and (at times) revise their expressed preferences.

Implement, monitor, and learn

A structured decision process should promote learning and build management capacity to make better decisions in the future. This learning may be related to technical understanding (for example, reducing uncertainty in the estimation of consequences), human resources (for example, training local community members in monitoring methods) or institutional capacity (for example, building trust and partnerships and/or developing systems for tracking and storing data).

Making decisions about things we care about can be hard at the best of times, and it's even more difficult if we're very uncertain about the outcomes. In such cases, an emphasis on learning over time, accompanied by a formal commitment to review the decision when new information is available, can be the key to reaching agreement on a controversial management strategy.

Particularly when resources are limited, it's necessary to focus on the most important sources of uncertainty, those for which reductions would be of greatest value. As discussed in Chapter 10, in SDM, monitoring programs are designed to address those uncertainties that are thought to be the greatest barrier to making an informed choice. To ensure their relevance to the choices facing decision makers, the results of learning should be closely linked to the objectives and performance measures used to evaluate management alternatives.

Learning in SDM doesn't just occur at Step 6. We believe that the ability of an SDM approach to facilitate mutual learning throughout the process – about both facts and values – is one of the key factors in its success. In Chapter 10 we emphasize the central role of learning throughout the SDM process and introduce some ideas for encouraging learning as part of the application of SDM to applied problems.

Text box 1.6:   Example: buying a flight ticket

As noted, there is insufficient information to know which of the alternatives might be better for comfort, our third objective. If this was a one-off flight, there might not be much we could do about it. However, if we regularly traveled on the same route, we might want to purchase tickets on a different airline once in a while in order to learn more about the relative comfort ratings for each of the three flight alternatives. With uncertainty reduced, comfort might well come back into play and change which flight we prefer to take from then on. But would the price we pay for this additional learning really be worth it?

1.2.3   Who 'does' structured decision making?

As the case studies presented in later chapters will demonstrate, there is no standard formula for deciding who gets together to 'do SDM' or how many sessions should be held. SDM is a way of thinking and talking your way through complex choices and can be applied informally, with a few colleagues and a flipchart, or formally in high stakes decisions using advanced computer software and state-of-the-art technical analysis.

SDM approaches most often are used to structure decisions involving a group of 10–20 people who agree to work iteratively and collaboratively through the SDM steps in a sequence of perhaps three to eight meetings. In some cases, all of the participants are highly trained and may have worked together for years. In other cases, groups involve a mix of experts and non-experts who may or may not have met each other prior to the start of the process. There might be additional technical working groups or expert panels who provide input to the main group. There might be auxiliary public

---

Text box 1.7:   What's an 'SDM practitioner'?

Often in this book, we refer to 'SDM practitioners'. What we mean is anyone who wants to organize and lead a sound public decision process, participate meaningfully in it, or simply know whether it's being conducted according to a reasonable standard of care. Often these people will be environmental managers or other decision makers, but the reference also covers any other government, city or industry employee, local resident, resource user, or interested party who is either directly involved in the environmental management process or concerned about its outcome and potential effects. There's no single right way to conduct a decision process, and every process is limited by resources and timelines. But the methods in this book describe a sensible approach that can be used (and, as the case studies will attest, currently is being used) under a wide variety of conditions by people with varying degrees of experience and interests and training.

---

meetings or sessions with aboriginal groups or senior elected officials that take place in parallel. But there is an assumption that a core group of people is getting together repeatedly to work through a complex problem. Who are these people and how long does the sequence of meetings last? It depends. It could take a few weeks, with meetings attended by individuals within a single agency, corporation, or NGO. Or it could be an inter-agency group or a full multistakeholder contingent, with members coming from several regions or countries for meetings that span several months or even years. The key is that we envision a group of people who agree to work together, who seek to build a common understanding of a complex issue, and whose goal is to develop good solutions that can be implemented.

Text box 1.8:   The language of environmental something

Sometimes the language commonly used in the world of resource management can inadvertently be off-putting people who have come to participate in a decision. Here are a few examples, based on our own experiences when words that we introduced with confidence were met, to our surprise, with frustration or anger:

1 *Resources.* To some, this word denotes a commodification of the natural world. Trees, water, and wildlife are not resources, they are the essence of life.

2 *Management.* To some, this word denotes failure[18]. The only reason we need management is to try new approaches to make up for past mistakes, but why should our current attempts be any more successful? Planning can have a similar connotation.

3 *Choices.* To some, a choice is very limiting; decisions can be wide open, but once the language of choice is introduced it means that someone already has limited the options or presented an ultimatum.

4 *Process.* To some, saying SDM is a process means that we're introducing yet another long-winded, self-serving, convoluted approach that might work elsewhere but won't work here. As we've been told by stakeholders on several occasions at the start of our work, 'we already have been processed to death – what we want now are results'.

5 *Analysis.* To some, introducing the word analysis means that decision-making power will be taken out of their hands. Analysis is seen as another way that outsiders seek to control resource decisions and to be secretive about what is really taking place.

6 *Trade-off.* To some, the word trade-off doesn't mean careful value-based balancing but, instead, is synonymous with an imposed loss or sacrifice. It means something of value is about to be stolen from individuals, their community, or the natural environment in which they live.

Anyone who has worked with multi-interest stakeholder groups on environmental issues could add to this list. Our intent is not to be comprehensive; instead, we emphasize the need to be sensitive to the prior experience of participants and ready to be flexible in the terminology that's used.

1.3   Case study: using structured decision making to develop and evaluate creative water-use strategies

In the beginning of this chapter we described the water-use planning process in British Columbia, Canada. Twenty-three water-use plans were developed at large and small facilities throughout the province, all using an SDM process. Each one progressed a little differently. Some were developed over the course of three or four meetings in four to six months. Others required a dozen or more meetings over one to two years. Here we describe one typical example, with the goal of demonstrating how the steps of SDM just outlined are applied in a situation much more complex than buying an air ticket. It's an example involving a great deal of technical analysis of the consequences of alternatives. For an example of a more qualitative application of SDM, see the case study at the end of Chapter 11, where SDM is used to help communities in a rural African community to choose drinking water filters. Other case studies demonstrating particular steps of SDM are included throughout the book.

1.3.1 Decision context

Consider a large hydroelectric facility located in a rural part of British Columbia, Canada[19]. A dam forms a large reservoir, and power generation operations affect both the water level of the reservoir and water flows in the river below the dam. There are longstanding conflicts about how water should be managed at the facility. Competing interests are related to fisheries (in the reservoir and the river), wildlife (especially wildlife that uses the riparian areas along the river and reservoir), revenues from power production (which flow through the crown-owned utility to the provincial government), cultural sites located in the drawdown zone of the reservoir, and opportunities for recreational use and quality (boating, bird watching, picnicking, etc.) at sites along the river or reservoir.

Facing public controversy and impending court challenges about how facilities are operated, the utility invites a diverse range of people to participate in a SDM process to explore and make recommendations about alternative ways to manage water at the facility, with the ultimate goal of finding a better balance among competing interests. Participants include the energy utility, the provincial and federal regulators for fish and wildlife, the provincial treasury board, local communities and the First Nation[20] on whose territory the facilities lie. A set of 'water-use planning guidelines' prepared by the utility in partnership with the provincial and federal governments establishes the ground rules for this consultative process, outlines the SDM process that will be used in a multi-stakeholder context to develop recommendations, and describes how those recommendations will feed into formal regulatory approval and water licensing processes[21]. The mandate is to identify and evaluate 'water-use' alternatives at the hydroelectric facility – that is, alternatives related to the elevation of water in the reservoir and the timing and magnitude of flow releases from the dam. Alternatives related to structural changes to the facilities or upstream watershed management

activities (such as land use or forestry practices) are therefore out of scope. At minimum, the multi-stakeholder committee is required to evaluate and address the impacts of proposed flow changes on electricity revenues, fisheries, flooding and First Nations communities; other items can be considered at the discretion of participants, but these four items are required. A budget and timeline for the process is established. An SDM facilitator and analyst is selected who is responsible for ensuring that all the essential elements for an informed decision are addressed, with emphasis on the effective integration of technical analysis into the deliberative multi-stakeholder process.

1.3.2 Objectives and performance measures

Participants arrive at the first consultative committee meeting with a messy and emotionally charged list of issues and concerns. One of the first tasks is to turn these into a set of fundamental objectives and associated performance measures that can be accepted by all participants as the basis for identifying and evaluating alternatives.

By emphasizing careful structuring, especially the separation of 'means' and 'ends', and the development of a common vocabulary so that people can communicate effectively, the group relatively quickly reaches agreement on a set of fundamental objectives. While they almost certainly don't agree on the relative importance of these objectives, they don't need to. They all agree that these are important considerations when choosing among water-management alternatives – establishing this common ground helps to create a more collaborative environment.

The discussion turns to defining performance measures – specific metrics for assessing and reporting the effects of alternatives on the objectives. This is a complex task, requiring both technical and value-based judgments, and takes substantially more time. Upon conclusion, the group has a clear set of objectives and performance

Table 1.1 A clear set of objectives and performance measures to serve as the basis for identifying and evaluating alternatives.

Objectives	Measured by
Maximize the abundance and diversity of fish in the reservoir	Primary productivity, in tonnes of carbon per year
Maximize abundance and diversity of fish in the river	Spawning habitat, in square meters
	Frequency of stranding events (days per year)
Maximize the quality of riparian habitat for wildlife	Inundation duration of riparian zone in days per year (during the growing season only)
Minimize losses in the value of power	Annual revenue from electricity sales in levelized dollars per year
Maximize access to culturally significant sites	Frequency of access, in days per year
Maximize quality and quantity of recreational opportunities	Quality-weighted annual recreation days (where quality weights were assigned by recreational users as a function of reservoir levels)
Minimize flood damage to infrastructure	Return period of a 'major' flood in years (where major is a flood that will affect infrastructure in a nearby community)

measures that all agree will serve as the basis for identifying and evaluating alternatives (Table 1.1 above).

1.3.3 Alternatives and consequences

To generate alternatives thought to be worth considering, the consultative committee breaks into subgroups. They use an approach called value-focused thinking to generate 'bookend' alternatives. The subgroups include people with diverse backgrounds and world views – fish and wildlife biologists, power engineers, local residents, traditional knowledge holders. Working through the objectives one at a time, they generate alternatives that would be good for that objective alone – without consideration of the trade-offs that might be required to balance with other objectives. Initially, this generates some alternatives that are quite polarized – a 'fish friendly' alternative for example and a 'power friendly' one. While these aren't very good candidates for a balanced solution as-is, they generate creative ideas, especially when combined with the cross-disciplinary thinking as a result of participants having a range of technical backgrounds. Some of these creative solutions ultimately form the basis for

unexpected win–win opportunities – water use changes that simultaneously benefit *both* power and fish, for example[22].

The generation of alternatives progresses through several 'rounds'. Elements of the initial bookend alternatives are combined with other ideas to create hybrid alternatives that seek a balance among all the competing alternatives. Models are developed to predict the consequences of each alternative with respect to each performance measure and the results are summarized in a consequence table. Alternatives are iteratively refined, eliminated and re-combined by the committee to form new and better alternatives. In some cases this process results in eliminating apparent trade-offs and replacing them with win–wins; in other cases, it exposes irreducible trade-offs. As inferior alternatives are eliminated, some of the performance measures initially identified become irrelevant – they are no longer affected by the refined alternatives and can be eliminated from subsequent analyses. This allows participants to focus in on key trade-offs. Flooding for example in this case is eliminated early as none of the proposed alternatives affect it. Wildlife concerns and recreational opportunities were initially very useful for helping to design good

Table 1.2 Consequence table exposing trade-offs among economic, ecological, and cultural resources.

Objective	Performance measure	Base case	Alternative 4	Alternative 5
Fish	Primary productivity (tonnes carbon per year)	1900–3300	2600–4600	2500–4200
Power	Levelized annual revenue ($/year)	$2 000 000	$2 500 000	$1 800 000
Cultural Sites	Frequency of access (median no. of days/year)	22	0	40

alternatives, but became irrelevant when inferior alternatives were eliminated. It turns out that industrial concerns and downstream fisheries concerns, the latter probably the single most important controversy at the start of the process, are addressed by creative win–win solutions.

1.3.4   Trade-offs and uncertainty

After several rounds of analysis and dialogue about alternatives, the final decision is narrowed down to a value-based choice between two management alternatives and three objectives. At this point, the estimates of fisheries benefits in the reservoir are playing a key role in the decision, so they are refined. Experts are consulted to provide judgments about uncertain quantities and relationships and a statistical analysis is performed (Monte Carlo) to produce a defensible estimate of the range of likely outcomes.

At the core of this decision now are value-based trade-offs between the treatment of heritage sites with great cultural and spiritual value to the local First Nation, the potential but uncertain benefits to fisheries, and the changes in revenues associated with power production. Early in the process, First Nation participants had rejected the notion of placing a 'value' on cultural sites and making 'trade-offs' about a resource of such fundamental spiritual value. However, they agreed to use performance measures (frequency of access per year) to allow comparisons of the degree of access provided by alternative operating strategies. In the end, the provision of access to cultural sites turns out to be a key driver in the selection of

an alternative. The use of explicit performance measures for this spiritual value helps to ensure that it is evaluated on equal footing as part of the decision process, along with power, fisheries, and other objectives. In contrast, any attempt to quantify in monetary terms the value of these resources would have alienated the First Nations community and quickly terminated their participation.

With two alternatives and three objectives (see Table 1.2), there is no need for fancy quantitative trade-off tools to support decision making. What's needed is dialogue – an opportunity to talk about the significance of these impacts to the people affected. Participants learn a great deal about things they knew little about before the process. This influences their opinions about which alternative provides the best balance across interests. At the end, participants are asked to identify which alternatives they enthusiastically support, oppose, or can live with and why. Consensus is reached on Alternative 5, largely influenced by (a) frank dialogue among participants about the spiritual significance of the First Nation's heritage sites and the importance of access to them, and (b) an agreement to monitor and learn about the influence of reservoir operations on fish productivity.

1.3.5   Monitoring and learning

There are important uncertainties, and agreement doesn't come easily. Are the estimates of primary productivity under the new operating regime accurate? Will they lead to the anticipated increases in fish abundance? Will the First Nation make use of the access to the culturally significant sites? Will they continue to be important over time? Will the win–win actions identified early in the process to address longstanding fisheries conflicts turn out to be effective? The final agreement includes a formal commitment to monitoring for confirming the effect of water use on fisheries productivity in the reservoir and river. It also establishes an interagency monitoring committee, and a commitment to review the water-use decision upon completion of the monitoring program.

1.4   The art and science of decision making

The methods that underlie the SDM approach are drawn from the decision sciences, specifically multi-attribute utility analysis and behavioral decision research, and from applied research by ecologists into the choice of management alternatives under uncertainty. Each of these is a well-developed and carefully documented area of theory, research, and application. We discuss this fundamental work in more detail in the next chapters.

Yet it is also important, at this early stage in our description of SDM practices, to emphasize that helping government agencies, industry, indigenous resource managers, or community advisory groups to make better decisions is also an art. It calls on emotions as well as reason, creativity as well as discipline. Knowing what to do, from a technical standpoint, is often not sufficient: the practitioner of SDM, often working as both facilitator and analyst, must also know how to introduce different topics or ideas, when to do so, with what emphasis, and using what language. As with great stories, paintings, or music, great decisions are not achieved by applying formulas or following procedures. But neither are they pulled out of the sky. Jazz musicians are not simply making it up as they go along: knowledge of theory and fundamentals are the tools of composition and improvization. Players rely on their informed intuition and experience to judge whether the music is taking them in an interesting and inspiring direction. They also feed off each other to create new and creative musical ideas. If something isn't working, they move off in different directions – but not randomly – they draw on an established toolkit and it is their knowledge of the fundamentals that suggests which ideas and techniques are likely to lead to interesting musical results.

So why bother with structure? There is a narrative to every decision process, and as with a good story this requires structure – a beginning, a middle, and an end. In a decision process, a good solution will not be recognized and supported as such without a properly defined context, a clear set of objectives, and confidence that an appropriate range of alternatives has been identified and evaluated. Structure is particularly critical in the context of group decisions. When playing music in a solo setting, you have complete freedom to change the tempo or feel of the music at any time; if you do the same thing in a group setting, you end up with noise (and frustrate both colleagues and audience). When people come together to make complex decisions as a group, it is essential that everyone works through the same steps, uses the same vocabulary, and has access to similar tools; only then can effective analysis and constructive dialogue take place.

So both science and art contribute to good decisions, and that's part of what makes decision making fun – but there's more than that. Over the course of our years of working with groups in structured decision making settings we've been inspired by what a motivated group of people can do. We've become addicted to the feeling that accompanies that magical 'a-ha' moment when a group that's been working hard with a tough problem finally reaches agreement on a way to move forward. Well before that there are a whole series of smaller 'a-ha' moments, when there is finally clarity about previously opaque objectives, when impact models or expert judgments deliver some critical insights about likely consequences, when the key trade-offs are finally crystallized, and when an alternative emerges that might be supported by everyone. And we've come to know that anyone can participate meaningfully in a decision that affects them, no matter how technically complex the decision and no matter what their background. There's no reason or excuse for leaving environmental decision making in the hands of a select few; that can only impoverish the solution. In our view there are only three criteria for participating in SDM: a genuine concern about the outcome, a willingness to learn, and a willingness to work collaboratively toward solutions that address a broad range of interests.

The successful manager will give as careful attention to people and the decision context as to analytical techniques and modeling results. Our hope is that by reframing environmental management problems as decisions, and by unpacking the objectives, uncertainties and trade-offs associated with proposed alternatives, new ways of thinking might be fostered and more effective actions taken. The examples we introduce in the following chapters have been selected to illustrate how SDM approaches can help to deal with these diverse challenges of environmental decisions.

1.5 Key messages

SDM is the facilitated and collaborative application of multi-objective decision making and group deliberation methods to environmental management and public policy problems. It's designed to help managers working with groups first build a common understanding of a complex problem and then identify and evaluate innovative management alternatives. Although closely related to decision analysis, SDM is distinct in its emphasis on addressing the social and political needs of public planning efforts as well as the cognitive and behavioral challenges commonly faced by people when discussing novel and multidimensional choices under conditions of uncertainty. It's also unique in its emphasis on developing better alternatives rather than simply evaluating them and facilitating mutual learning.

It's particularly useful for *groups* of people working together on controversial environmental management problems in a way that's *rigorous, inclusive, defensible, and transparent.* The goal is clarity and insights for those responsible for making a decision or for developing recommendations about a difficult choice.

1.6 Suggested reading

Hammond, J.S., Keeney, R.L. & Raiffa, H. (1999) *Smart Choices: A Practical Guide to Making Better Decisions.* Harvard Business School Press, Cambridge, MA. Few books on decision making become best sellers; *Smart Choices* takes the steps and techniques of decision analysis and applies them to everyday choices. By the end of this book, placing renewed emphasis on improving personal and societal decision-making methods appears to be both essential and enjoyable.

US National Research Council (1996) *Understanding Risk: Informing Decisions in a Democratic Society.* National Academy Press, Washington, DC. This lucid report was one of the first to articulate the need to link rigorous analysis with structured deliberations as a pre-requisite for informed societal risk management decision-making processes. By making a compelling argument, the NRC report has influenced many of the initiatives subsequently undertaken by environmental management agencies in North America and Europe to meaningfully involve potentially affected parties in risk decision making processes.

1.7 References and notes

1 Gregory. R. & Failing, L. (2002) Using decision analysis to encourage sound deliberation: water use planning in British Columbia, Canada. *Journal of Policy Analysis and Management,* **21**, 492–499. Reviews the use of structured decision-aiding methods as part of a comprehensive stakeholder-based review of operating plans at all major hydroelectric facilities in British Columbia.

2 McDaniels, T. & Trousdale, W. (2005) Resource compensation and negotiation support in an aboriginal context: using community-based multiattribute analysis to evaluate non-market losses. *Ecological Economics,* **55**, 173–186.

3   Failing, L., Gregory, R. & Higgins, P. (2011) *Leveling the playing field: structuring consultations to improve decision making*. Manuscript submitted for publication.

4   Department of Fisheries and Oceans (DFO) (2005). *Canada's Policy for Conservation of Wild Pacific Salmon*. Fisheries and Oceans Canada, Vancouver, BC.

5   Gregory, R. & Long, G. (2009) Using structured decision making to help implement a precautionary approach to endangered species management. *Risk Analysis*, **29**, 518–532.

6   Ohlson, D.W., McKinnon, G.A. & Hirsch, K.G. (2005) A structured decision-making approach to climate change adaptation in the forest sector. *The Forestry Chronicle*, **81**, 97–103.

7   Williams, B.K., Szaro, R.C. & Shapiro, C.D. (2007) *Adaptive Management: The US Department of the Interior Technical Guide*. US Department of the Interior Washington, DC. Retrieved from http://www.doi.gov/initiatives/AdaptiveManagement/TechGuide.pdf

8   Conroy, M.J., Barker, R.J., Dillingham, P.W., Fletcher, D., Gormley, A.M. & Westbrooke, I.M. (2008) Application of decision theory to conservation management: recovery of Hector's dolphin. *Wildlife Research*, **35**, 93–102.

9   Lubchenco, J. (1998) Entering the century of the environment: a new social contract for science. *Science*, **279**, 491–497.

10 Renn, O. (2006) Participatory processes for designing environmental policies. *Land Use Policy*, **23**, 34–43

11 Pielke, R., Jr (2007) *The Honest Broker: Making Sense of Science in Policy and Politics*. Cambridge University Press, New York.

12 Slovic, P. (1995) The construction of preference. *American Psychologist*, **50**, 364–371.

13 Gregory, R., McDaniels, T. & Fields, D. (2001) Decision aiding, not dispute resolution: creating insights through structured environmental decisions. *Journal of Policy Analysis and Management*, **20**, 415–432.

14 Hammond, J., Keeney, R.L., & Raiffa, H. (1998) Even swaps: a rational method for making trade-offs. *Harvard Business Review, March–April*.

15 Freeman, A.M., III (2003) *The Measurement of Environmental and Resource Values: Theory and Methods*. Resources for the Future, Washington, DC.

16 In the context of child-welfare applications in the United States, the term 'Structured Decision Making' also has been used to refer to a specific process trademarked by the National Council on Crime and Delinquency.

17 Bell, D., Raiffa, H. & Tversky, A. (1988). *Decision Making: Descriptive, Normative, and Prescriptive Interactions*. Cambridge University Press, New York.

18 In 2001 Donald Ludwig, a mathematical ecologist, wrote a compelling article titled 'The Era of Management is Over', in which he argued that standard management paradigms fail when experts are no longer trusted and objectives are not clearly defined (Ludwig, D. [2001] The era of management is over. *Ecosystems*, **4**, 758–764). More recently, Dean Bavington (2010) has called into question what he refers to as 'the holy grail of manageability' in the context of the collapse of the Newfoundland cod fishery (Bavington, D. [2002] *Managed Annihilation: An Unnatural History of The Newfoundland Cod Collapse*. UBC Press, Vancouver).

19 This case draws heavily on the Stave River water-use plan, but omits some details and modifies others to make the story more concise. As a result, it is a hypothetical but typical example of the methods and insights from 23 water-use plans developed under the program.

20 First Nations and First People are terms used to describe Canada's aboriginal people, analogous to 'Native Americans' in the United States or to the terms 'Indigenous' or 'Tribal' peoples used in other countries.

21 These can be found at http://www.env.gov.bc.ca/wsd/plan_protect_sustain/water_use_planning/cabinet/wup.pdf

22 This occurred on the Stave system because through the collaborative work of power engineers and fisheries biologists, it was found that previously imposed restrictions on power operations could be relaxed in particular seasons without negative impacts on fish.

2

Foundations of Structured Decision Making

This chapter covers the eclectic mix of ideas that underpin the practice of SDM. As noted in Chapter 1, SDM is fundamentally about helping resource managers make better decisions. Because decision making is an activity that occurs largely at the borders between disciplines, in this chapter we draw on selected concepts and methods drawn from psychology, economics, decision analysis, biology and ecological science, engineering, management science, facilitation, and negotiation analysis.

We first review nine foundational concepts that we find particularly useful; these provide much of the underlying philosophy and analytical basis that helps to distinguish SDM methods. We then describe some of the other common approaches that often are used by environmental managers to help generate, select, and evaluate resource-management options. Although our coverage of their principles and methods – both complementary and competing with SDM – is admittedly brief, our hope is that the comparisons with SDM will help to highlight the essential, distinguishing elements of a SDM approach to environmental management.

2.1 Underlying concepts

SDM is not a unified theory or specified technique. Instead, it is both a way of thinking and a bundle of approaches aimed at providing insight

to decision makers about difficult decisions. These decisions may be difficult because they affect a wide range of concerns, or because of the high levels of uncertainty that characterize outcomes, or because the different stakeholders potentially affected by the decision hold very different views about what might happen, or about how important these consequences will turn out to be.

The nine concepts we introduce in this chapter (summarized in Table 2.1) will reappear throughout the book in various forms because they fundamentally influence the choice of methods to provide insights for decision making in the context of environmental management – affecting everything from setting of objectives to consequence estimation and trade-off analysis.

2.1.1 Multi-attribute utility theory

Multi-attribute utility theory (MAUT) describes a consistent yet varied set of approaches for making decisions with multiple, competing objectives (which succinctly describes most problems faced by resource managers). The approach was carefully outlined in the late 1960s in Howard Raiffa's *Decision Analysis*, published in 1968[1], followed by Ralph Keeney and Raiffa's *Decisions with Multiple Objectives*, first published in 1976[2]. Through its applied offspring of decision analysis, MAUT is now widely recognized as a rigorous, practical, and accessible set of principles for helping people

Structured Decision Making: A Practical Guide to Environmental Management Choices, First Edition. R. Gregory, L. Failing, M. Harstone, G. Long, T. McDaniels, and D. Ohlson.
© 2012 R. Gregory, L. Failing, M. Harstone, G. Long, T. McDaniels, and D. Ohlson. Published 2012 by Blackwell Publishing Ltd.

Table 2.1 Conceptual foundations of structured decision management.

Concept	What it is
Multi-attribute utility theory and analysis (MAUT)	MAUT is a prescriptive approach to multi-objective decision making, designed to help people make better decisions under uncertainty. It provides the analytical rigour that underlies SDM.
The integration of analysis and deliberation	Defensible environmental management and evaluation require 'an analytic-deliberative process' – the effective and ongoing integration of systematic analysis and deliberation. Both require careful attention to best practice.
Constructed preferences	People often do not have fixed preferences for environmental-management actions. Instead, their preferences are constructed in response to information about both relevant facts (the consequences of proposed actions) and the values (preferences or priorities) held by themselves and others.
Separation of values and facts	It is possible and useful to distinguish, in practical terms, between values, which define what matters to people, and facts, which describe the likely effects of management actions. Both are fundamentally important to decision making, but they need to be treated differently in the decision process.
Value-focused thinking	Good decisions start by fully exploring what we want (our values or objectives) rather than what is typically available to do (alternatives). This shift has profound implications for how a decision process unfolds.
Two systems of thinking	People draw on two modes of thinking when they make choices: analytical (which is slower and thoughtful) and intuitive (which is fast and often automatic). Both have an important role to play in making good decisions.
Mental short cuts and biases	All people, expert and non-expert, are influenced by predictable judgmental biases and cognitive short cuts that can adversely affect decision quality. There are well-developed methods that can minimize the negative effects of these influences on the quality of environmental management choices.
Best available information	The information needed to understand likely consequences of actions comes from many sources, including but not limited to science. Information relevance and quality can only be determined in the context of the decision at hand, based on both analysis and deliberation.
Group wisdom and dysfunction	Groups are the source of both important insights and debilitating biases. Best practice involves an interplay between eliciting individual judgments and facilitating group dialogue.

make better decisions. Although initial applications focused on problems involving a single decision maker, MAUT practitioners soon emphasized ways to involve diverse stakeholders[3] and to apply simplified multi-attribute methods[4] to a wide range of important public policy problems.

SDM approaches are based fundamentally on the theory of MAUT and the applied practice and methods of decision analysis (DA). Both MAUT and DA methods decompose a management problem to identify the relevant objectives, assign performance scores to the alternatives under consideration, and then develop weighted 'utility' or preference scores for each alternative using relative importance weights calculated for the objectives in light of the anticipated consequences of actions. This information can then be provided to decision makers, and provides a basis for making a choice. A variety of different approaches or models are used for summarizing the performance of different alternatives. SDM is primarily focused on a simple but very robust approach – the linear value model. A linear value model calculates preference scores using a weighted sum (see Text box 2.1). It is easy to use and produces results decision makers can readily understand. If the objectives have been well structured (as we outline in Chapter 4), then it can be used with confidence as the basis for informing decisions[5–7].

Text box 2.1: Linear value modeling

A linear additive model calculates an overall performance score for each alternative, based on a weighted sum of its consequences. In other words, multiply an alternative's score on an attribute by the relative importance weight assigned to the attribute. Do that for all the attributes and sum the products to get the overall performance score. Repeat the process for the other alternatives under consideration, and the alternative with the highest overall score is preferred. Mathematically, a linear value model represents an alternative's performance score as:

Overall score $= W_1X_1 + W_2X_2 + W_3X_3 + \cdots$

where X_1 is the score assigned to attribute 1, W_1 is the weight or importance assigned to attribute 1, X_2 is the score assigned to attribute 2, W_2 is the weight of attribute 2, etc. In order to calculate a meaningful overall score when the individual performance scores are recorded in different units (e.g. dollars, hectares, jobs, etc.), the individual performance scores must first be 'normalized', and then weighted. Despite the complicated name, linear additive value models are commonly employed: summary ratings of automobile performance (e.g. incorporating fuel efficiency, driver comfort, handling, and the like) are calculated using this basic model, as are widely cited reviews of colleges (which rate college performance in terms of variables such as reputation, financial resources, and student retention). Chapters 6 and 9 will provide more detailed environmental management examples.

Good decision making doesn't require the use of a quantitative value model. However, the thinking about a decision can be more disciplined and more transparent if it's structured in ways consistent with sound decision theory. In SDM, care is taken to carefully define the objectives, alternatives and their consequences using recognized best practices. These are often summarized in a consequence table. These are the essential elements of a value model; the only thing missing is the weights. When a preferred alternative is chosen based on a review of the performance scores in a consequence table, participants are either implicitly or explicitly assigning weights to each objective and then aggregating across the objectives as they consider the trade-offs. If a preferred alternative isn't apparent – either because individual participants are unsure, or a group can't reach agreement, then sometimes further work to assign explicit weights and perform quantitative modeling can help.

Of course, when decision makers make choices among alternatives with multiple objectives, they are assigning importance weights – they're just doing it implicitly. When decision makers assign weights implicitly, there is likely to be a loss of rigor and transparency (because others have no idea how or why they are doing it), consistency (because the decision makers themselves do not fully understand how or why they are doing it), and inclusiveness (because the views and logic of others – whether similar or different – are hidden). Structured methods seek to increase the quality of importance judgments by making the weighting explicit. As discussed in Chapter 9, dealing with value trade-offs and weighting requires special attention in multi-stakeholder applications to ensure that participants understand and fully engage in the process.

It's worth remembering that, in practice, no value function is ever fully complete or fully right. As in ecological modeling, or any other impact modeling, the role of value modeling is to provide insight, not answers. The primary benefits of quantitative value models used in decision making are the same as those in ecological or other impact modeling – to improve the structure and discipline of thinking about relationships, interdependencies and possible outcomes, to permit peer review, and to provide insights about the range of possible consequences that might result from management decisions[8].

The decomposition required to build value models serves as an important cognitive aid to resource managers: it helps them to think through for themselves, and to communicate to others, both the science basis for a decision (what are the estimated consequences?) and the values basis (what is the relative importance of different kinds of consequences?). In a public decision-making context, especially one involving multiple stakeholders, decomposition also provides a kind of audit trail that facilitates peer and public review. Although value judgments are not subject to criticism in the same way as are technical judgments, it is not unreasonable to expect that participants in a decision process be asked to make their values explicit, to provide a rationale for their judgments, to demonstrate that they have thought about their own choices in light of others' expressed views, and to explain or resolve apparent inconsistencies. In a public decision-making context, this expectation is not just reasonable but essential.

To summarize the relationships among MAUT, DA and SDM:

1 MAUT is the theory behind sound decision making. It is broad and encompassing, and sufficiently robust to address virtually any multi-attribute problem and any decision model structure with a high degree of technical rigor. More complex model forms can only be applied by a very small set of experienced practitioners.

2 DA is the application of the theory and includes very practical methods that can be used to support decision makers. Ralph Keeney[9] (p. 806) neatly refers to decision analysis as 'a formalization of common sense for decision problems that are too complex for informal use of common sense'. DA methods have been widely used in a diverse range of settings, including environmental management questions as well as numerous energy,

agriculture, medical, national defense, and business contexts[10].

3 SDM is a framework for making decisions that is founded on the theory of MAUT and the practice of DA, but includes theory and practices from other fields (e.g. psychology, ecology, facilitation) that also influence decision quality. Methods related to multi-attribute evaluation and to quantitative decision modeling are thus a subset of methods used in SDM. In most applications, SDM approaches place a relatively greater emphasis on group interactions and the role of focused deliberations in generating (in contrast to evaluating) management alternatives that are responsive to participants' expressed values.

2.1.2   *The integration of analysis and deliberation*

In 1996, the United States National Research Council (NRC) published an influential report called *Understanding Risk: Informing Decisions in a Democratic Society*[11]. A central conclusion is that defensible environmental management and evaluation require 'an analytic-deliberative process' – the effective and ongoing integration of systematic analysis and deliberation.

Our approach to SDM was developed largely in response to this call. It is probably the single most defining feature of SDM, the core idea that distinguishes it from other approaches. A defensible SDM process — whether taking place over a day, a week, or a year – always will seek to foster deliberation informed by analysis. The sophistication of the analysis will vary depending on the situation – the goal is that it should be 'good enough' for the decision at hand, in line with the urgency and complexity of the problems and the preferences of the stakeholders[12]. In some cases, it might be back of the envelope. In others, it will be the state of the art in engineering, ecological or social science, with potentially sophisticated treatment of interactions and uncertainties. Yet an SDM process recognizes that it is people – usually working together – who make decisions. It puts equal emphasis on deliberation – ensuring that participants in the planning or decision-making process thoughtfully identify and consider their relevant values, the alternatives and their consequences, and key trade-offs before making a choice or taking action. When dealing with the management of public resources, thoughtful consideration necessarily involves talking with others – a deliberative exchange of both facts and values in an environment designed to peel away biases and traps, and allow for reasoned exchanges.

A good SDM process seeks to translate complex technical ideas into language and decision-relevant messages that allow people without technical expertise to meaningfully consider technical information, and it seeks to express people's full range of potentially affected values and concerns in a way that decision makers will understand and notice. It enables and demands that participants think, feel, talk and learn on their way to a decision. You wouldn't want to make simple decisions in your personal life – from naming a child to buying an apartment to choosing a career – without these four elements. SDM applies these same basic elements, in an explicit and transparent manner, to public policy decisions such as the management of valued environmental resources.

2.1.3 Constructed preferences

Many participants in environmental management decisions come to the process with strong and passionate views and a firm position about a preferred solution. For this reason, it's easy to assume they have firmly entrenched preferences. However, the reality is that their views and opinions are shaped – sometimes consciously and sometimes unconsciously – by the context within which the deliberation takes place and by the specific elements of a proposed plan. The bottom line from several decades of research is that 'true' preferences for many environmental choices simply do not pre-exist; instead, they tend to be constructed – rather than revealed – as part of the deliberation and elicitation process[13].

The construction of preferences is particularly strong whenever participants are faced with novel choices or a new evaluation setting. Evidence shows that when people are asked to make decisions involving unfamiliar outcomes or options, their choices can easily shift in response to a wide variety of cues[14,15]. This lack of stability in expressed preferences is often present when the decision context involves negative consequences (e.g. potential injuries or deaths for valued species, including humans) or when difficult trade-offs need to be made across very different dimensions of a problem[16]. Sometimes this shifting of preferences is good – for example, people learn from exchanges with others, both about technical considerations (the consequences of alternatives) and about value considerations (how other people perceive the impacts and trade-offs). At other times the construction process is less obviously productive. For example, preferences can change in response to relatively arbitrary changes in how information is worded, how outcome states are framed (e.g. as a gain vs the restoration of a prior loss), and ordering effects that involve the sequencing of questions or judgment tasks[17].

Sometimes people resist the idea that preferences or values change – after all, if values define who we are, then how can they change? To a large degree this question reflects differences in terminology. What tends to be fixed is a fundamental belief or a worldview – a tendency to value the environment highly, or to place a strong value on social justice or economic efficiency or decentralized governance[18]. Our preferences, defined in terms of the trade-offs we are willing to make in a specific situation, are definitely not fixed – as any politician or journalist or advertising executive knows well. There are many reasons why: we struggle to make sense of disparate or confusing information, our computational abilities are limited, and – as part of another very endearing human characteristic – we learn (although not always very quickly). SDM processes are designed to support the ability of both individuals and groups to make sense of the information that is available and to learn about their own values.

Recognizing that preferences are constructed through a decision process has important implications:

1 People need an opportunity to talk and to learn together, about both the facts basis and the values basis for choices. Preferences that seem fixed at the beginning of a process may change as a result. In this sense people are more like architects than archeologists, building their values from the information and cues that are at hand, rather than simply uncovering them.

2 People need time, to think and to feel, in order to come up with a relatively stable understanding of their own preferences. Participants in a decision process should be allowed to reflect on choices outside of the meeting room environment.

3 Seemingly arbitrary choices by analysts and technical experts about how information is presented can inadvertently influence preferences. In some cases, multiple representations of the same information may help people reach a more robust interpretation, and consequently a more stable expression of preference.

2.1.4 Separation of facts and values

Making good choices requires appropriate information about values as well as facts – identifying, understanding, and acting on the set of values which best represents the interests and concerns of the individuals and groups likely to be affected by a decision. If you're like many environmental scientists and managers, the mere mention of the word 'values' may cause you to break into a cold sweat. In this day and age, 'values' has become an inflammatory term and an entry-point for controversy and disagreement: everyone knows the importance of values, generally in society and specifically in environmental decision making, but relatively few know how to address values constructively.

An SDM process treats values or value judgments as statements about what we want or what we think is important in the context of making a choice. Values may encompass beliefs about the relevant concerns (what matters? what things should be considered?), priorities (what is their relative importance?), preferences for different outcomes (what trade-offs are acceptable?), tolerance for risk (how acceptable are alternatives that have a small but non-zero probability of an extreme outcome?), or time preference (e.g. what is the relative importance of immediate versus longer term benefits?). Facts or technical judgments, in contrast, are descriptive statements about the condition of systems or processes (what is the best estimate of salmon abundance?), relationships between elements of the system (how is salmon abundance affected by increases in water temperature?), or predicted consequences of management actions (how will salmon abundance change if we modify the volume of water flowing in the river?). Facts are (often) stated probabilistically, reflecting the degree of uncertainty about the truth or accuracy of a statement.

From a decision-making perspective, it is possible and useful to distinguish, in practical terms, between facts and values. Both are fundamentally important to decision making, but they are treated differently. We acknowledge that this distinction between facts and values is not without controversy, and much has been made from a philosophical perspective about the difficulty in completely distinguishing between facts and values[20,21]. It's true that much factual information is colored by the values of those presenting it, and that value-based choices are profoundly affected by our attitudes toward uncertainty and risk. However, our inability to cleanly separate facts and values does not mean they are one and the same thing. From a practical perspective, most people would agree that in the course of conducting an evaluation of resource management alternatives some judgments will be

---

Text box 2.2: Facts and values

In a widely cited paper, Maguire[19] explores the use of decision-aiding methods for integrating values- and facts-based information as a way to improve managers' decisions about invasive species. She is concerned, in general, that risk assessments often either ignore the values of different stakeholders or confound the social values at stake in decisions with the factual knowledge necessary to predict the likely impacts of management actions. Maguire illustrates this issue with the case of feral pigs in Hawaii. Scientific analyses had emphasized predictions of how many pigs would be at specified locations and their likely impacts on native plants, non-native competitors, and the like. The more difficult, and largely neglected, questions – critical to the choice of generation of options and the successful implementation of management actions – had to do with stakeholder preferences over a range of predicted management options and outcomes. This neglect was significant because several of the participating groups, such as Native Hawaiian hunters and conservancy land managers, differed dramatically in their perspectives and preferences. By using decision-aiding methods to separate factual questions (and uncertainty concerning the outcomes of management actions) from values considerations, managers are able to develop more responsive and broadly supported management alternatives.

primarily fact based (e.g. how will migratory birds respond to a new wind farm?) and others will be primarily value based or prescriptive (e.g. how much are we willing to spend to protect migratory birds, or what priority should we put on the birds relative to production of electricity?). We propose that, from a practical perspective, how these two types of judgments are treated in decisions and who has legitimacy to make them should differ.

2.1.5 Value-focused thinking

In 1992, Ralph Keeney – one of the major contributors to decision analysis – published a book called *Value-Focused Thinking: A Path to Creative Decisionmaking*[8]. In simple terms, Keeney reminds us that 'values are what we care about' and 'values should be the driving force for our decisionmaking'. In spite of the common sense appeal of this notion, most resource management choices continue to reflect what can instead be termed 'alternatives-focused thinking': anchoring on the readily identified management actions and choosing the best of a potentially mediocre lot. Value focused thinking in contrast, involves defining what we really want, and then working to achieve it. Simple as it sounds, value-focused thinking profoundly changes how people approach decision problems:

1 Do not rely only on the first few objectives that come to mind. Successful solutions depend critically on developing a complete understanding of what matters and structuring these things into explicit and useful objectives.
2 Include things that matter, whether they are conventional or easy to measure or not.
3 Distinguish between means and ends. Ends are the things we fundamentally care about. Means are the things we can do to achieve them. Both are important but have different roles in decision making. Means are typically identified as alternatives in SDM; ends are often termed fundamental objectives. Means need to be evaluated based on how they perform, relative to the ends. This emphasis on what we care about guides both technical analysis (how we report the consequences of alternatives) and the quality of deliberation (because it demands that alternatives be justified on the basis of their performance).
4 Develop alternatives specifically to meet multiple objectives – don't just tweak the status quo or respond to the pet projects of vocal participants. Instead, invite people to develop

alternatives that would best serve the thing(s) that matter, temporarily ignoring constraints in order to generate creative ideas.

2.1.6 Two systems of thinking

Most resource managers would readily acknowledge that tough choices among management actions will force people to both think and feel. The conventional interpretation is that emotions lead us astray, whereas reason brings us back to more logical decisions. Recent research suggests that it may be more accurate to view reason and emotion as inextricably linked during decision making[22], so much so that logical decision making, particularly in social situations, might be impossible or at least seriously compromised without a good injection of feelings.

In his 2002 lecture accepting the Nobel Prize in Economics, the psychologist Daniel Kahneman distinguished two generic modes or 'systems' of thinking that characterize how all of us – whether scientists, engineers, elected officials, aboriginal elders, or rock stars – operate in the world. System 1 is an experiential system that involves intuition and feelings. Fast, automatic, and effortless, System 1 generates impressions. These impressions are not always explicit, nor are they necessarily intentional. System 2 involves reasoning and is based on slower, more rational and cognitively effortful ways of comprehending reality. Often called our analytic system, System 2 generates judgments. Judgments are usually explicit and always intentional. There's plenty of theoretical debate about whether these two ways of thinking are really 'systems'[23], but for our purposes, it's useful to think of them that way.

These two systems work together to inform our choices. Rational, cognitive systems form the basis for most training received by resource managers. Yet emotional and intuition-based responses also play a major role in our evaluations of different alternatives. They often lead us to an immediate perception of a thing as good or bad: psychologists call this 'affect'. As analysts or managers, we can't presume to know what

stimuli will be perceived affectively as good or bad by participants in a decision process. Although some stimuli are almost universally received as good (e.g. smiling babies) or bad (e.g. oil-fouled beaches), most are subject and context specific – different people will perceive them differently, based on their prior experiences and reactions to the specific wording that is employed (numerous examples come to mind: people responding quite differently to actions labeled as restoration vs enhancement, or to trade-offs vs balancing among objectives). These positive and negative affectively based images serve as mental short cuts or guides, sometime conscious and sometimes not, to judgments and choices[24]. Affective responses thus function as powerful but often invisible determinants of choice.

As individuals, we are typically unaware of these influences and of the ways in which the two systems interact as information is processed[25]. Yet Kahneman notes that the 'highly accessible impressions produced by System 1 control judgment and preferences, unless modified or overridden by the deliberate operations of System 2'. System 1 is in part informed by our experience – a good chess player can quickly 'see' moves that a less-experienced player would not. But it is also strongly affected by factors that determine the 'accessibility' of different features of a situation – how naturally, quickly and easily these features come to mind and stay there. Highly accessible features will influence decisions while less-accessible features will be largely ignored. Unfortunately, there is no reason to believe that the most accessible features are also the most relevant to a good decision[26].

2.1.7 Judgmental short cuts and biases

The existence of these two modes of thinking has generated considerable interest in recent years because they help to explain some aspects of human behavior and judgments under uncertainty that have puzzled researchers for decades. One implication, confirmed by much experimental work, is that people are not the rational, individual maximizers of utility portrayed by much of modern economics or the natural sciences[27]. Instead, people routinely make use of simplifying rules of thumb – which psychologists call judgmental heuristics – as a means of coping with the complex cognitive demands that are placed on them[28]. Sometimes these rules of thumb help us – they offer a degree of efficiency that allows us to make the many dozens of small decisions we face every day. But sometimes they become 'traps', leading us astray. Below some common examples.

Representativeness

Our understanding of an event or condition frequently is based on its apparent similarity to something else. This ability to recognize that a person or event is representative of a larger class can be helpful to the extent that we are quickly able to make sense of something new by relating it to what already is known. A stakeholder we've just met for the first time is too quickly classified as representing a single perspective, or a new marine policy issue is too quickly assumed to fit into a class of familiar problems that we've dealt with before. But it causes problems when we ignore disconfirming information. Representativeness is also an important source of error when eliciting expert judgments, as many experts (even statistically literate ones) ignore fundamental information about the probability of events or conditions when presented with interesting but only peripherally relevant details. Evidence shows that they routinely overweight irrelevant details and/or misprocess them, causing them to make logical errors in assigning probability. Consider for example the following three scenarios. Which situation is more likely?

1   A forestry road washes out.
2   A forestry road washes out, triggered by a combination of deforestation and unseasonably heavy rains.
3   Forest sector geoscientists and regulators are working together on terrain stability guidelines when a forestry road washes out.

Despite the fact that scenarios 2 and 3 cannot be more probable than scenario 1 because they impose more conditions, studies show that they are likely to be assigned a higher probability, even by experts. The increased detail, even if it's not relevant to the occurrence of the event (as in scenario 3), tends to increase both salience and perceived likelihood[29].

Availability

Our estimate of the likelihood of an event is affected by how readily we can recall similar events/conditions to mind – predictions of increases in recreational use for a site or estimates of the effectiveness of enforcement policies can easily be biased by recent experiences. The strength of this mental short cut is affected by chance (what we happen to have read yesterday), emotions (what frightens or surprises us), imagery (what we can easily visualize) and beliefs (we remember information that confirms our beliefs more than information that contradicts it). The influence of availability is surprisingly strong; experts with years of experience have been known to alter their expressed views substantially based on evidence from a single recent event. Availability bias has obvious implications for fact-based judgments, but it can also affect value-based judgments.

Sunk costs

Individuals tend to justify and protect earlier choices, often long after it's obvious to an external observer that it's time to make a change. As a result, large proportions of agency budgets might be spent so that years 4 and 5 of a planned five-year study can be completed, even though the results already in hand clearly show that the remaining funds should be spent investigating new options. Reasons for the strength of the sunk-cost (or status quo) bias include everything from institutional constraints to psychological and self-esteem factors (who likes to admit that they've made a mistake?). But facing the sunk-cost bias head-on is important for good resource management; as the investor Warren Buffet once said, 'when you find yourself in a hole, the best thing you can do is stop digging'.

Anchoring (with insufficient adjustment)

An initial impression can significantly influence our responses to questions or our interpretation of subsequent information. The results from an initial pre-test of habitat restoration alternatives may overly influence the interpretation of subsequent trials; a stakeholder who is having a bad day may long afterwards be cast as the moody or angry participant. Even if the 'information' in question is clearly irrelevant, responses can still be affected. In a still-widely cited study by Tversky and Kahneman[29], people were asked to provide answers to factual questions. To kick off, they were given a starting number, produced by spinning a wheel of fortune. They were asked to state if the true number was higher or lower than the number produced by the wheel, and then to estimate the true number by moving up or down from the starting point. The responses clearly showed that this entirely irrelevant information (where a needle landed on a wheel of fortune) systematically influenced their judgments[30]. Although the effect – insufficient adjustment from an initial, often poorly chosen anchor – is universal, its problematic influence on judgments is made worse by careless elicitation, unfamiliarity with the context, or questioning that inadvertently introduces an arbitrary anchor.

Overconfidence

We often claim a greater degree of confidence about our judgments than is warranted, even when we already have acknowledged that we are uncertain. This decision bias is widespread and has been the subject of numerous studies particularly related to the elicitation of judgments from experts – who unfortunately can simultaneously be quite confident and also wrong[31]. In these studies, researchers ask people to assign a numerical estimate to an uncertain quantity. Wording varies, but essentially subjects are asked to provide a low and a high estimate, in such a way that they are 90% sure that the correct answer will fall between the two. When asked a series of such questions, unbiased respondents should in theory be right 90% of the time. Inevitably they are not[32]. Cooke[33] and Vose[34], for example, found that experts asked to provide a range of values that contain the true value of a quantity at least 90% of the time provided judgments that were accurate only 40% of the time[34]. What's interesting about this is not that the respondents didn't know the answer, but that they *believed they knew more than they did*. More troubling still, studies show no relationship between confidence in an assessment and accuracy. The degree of overconfidence will vary across experts and problem settings – the likelihood of a novel or rare event is harder to estimate than one that occurs frequently, for example – but the finding is both significant and widely replicated. It implies that many of the unaided estimates provided by experts, much of the time, are prone to systematic error.

Motivation

What we believe is unduly influenced by what we care about, what's good for us, or what we think others will want to hear. This is a frequent cause of disagreements in groups, as information is coloured by the perspective of different individuals and the points of view they represent. This could be based on which industry, government ministry, or department they work for, or in some cases, intellectual ideas (theories or technical methods) that they have a personal interest in. It can also be a cause of false consensus, as people in a workshop environment don't want to be the spoilers of an emerging consensus.

These mental short cuts are useful when they yield quick and easy guides that help navigate through a complex world. However, they can lead to systematic biases that result in judgmental errors. Importantly, research has shown that these errors take place regardless of geography, gender, or training: by-and-large, doctors at Harvard[35], or foresters and ecologists working for a governmental resource management agency, are as prone to make unconscious yet serious errors of judgment – in much the same way, and for many of the same reasons – as are politicians, member of the general public, college students, or professional investors. As described in later chapters, there are ways to anticipate and to minimize – but not eliminate – their influences on our judgments[36].

2.1.8 Best available information

In the past, decision makers often looked to the 'best available science' to deliver 'science-based decisions' – in other words, the right answer based on the right information. We've already talked about the fact that because of the values basis of choices, there is no such thing as a right answer. However, we'll also make the case that there's no 'right' information.

As we discuss more fully in Chapter 8, society's notion of what constitutes best available information is changing. We are used to relying on science to deliver consistent, repeatable, peer reviewed results to guide decision making. But conventional science often has little to say about ecological systems that span large spatial and temporal scales (what practical role is there for replicated laboratory experiments in an ecosystem?). This introduces an element of uncertainty that may be pervasive and largely irreducible. In parallel, there is a growing call for dialogue between experts, stakeholders and decision makers and for a fundamental rethinking of what constitutes 'peer review'. Regulators and the public are increasingly calling for a broader peer review of scientific results that involves both scientists and lay people[37]. This call is largely driven by increasing awareness of the role of values in shaping everything we do – from relatively benign choices about what to measure, to formal risk assessments or cost-benefit studies, to our interpretations of uncertain monitoring results[38].

As a result of these and other trends (e.g. a general trend toward more inclusive, bottom-up planning processes), our understanding of who's an expert has changed markedly in recent years. Academically trained scientists clearly have a role to play in resource management decisions, but so do local communities and traditional knowledge holders. Today there is a large cast of characters who have something legitimate to say about which information should be used for decision making, and often there is no clear line of demarcation between science and non-science[39]. A good decision-making process will need to investigate all relevant knowledge, regardless of its source; the emphasis is on ensuring that all competing knowledge claims or hypotheses receive equitable treatment in the decision process and are evaluated on the basis of their merit and relevance to the decision[40].

The important implication for groups charged with decision making is that participants need to examine – together – the quality and relevance of information for decision making: what's relevant, who's an expert, how reliable is the information, and is it sufficient or 'good enough' to make a decision. The methods in this book attempt to strike a balance between drawing on the specialized knowledge of experts and drawing on the common sense of ordinary people to interpret and use that knowledge in ways appropriate to the decision at hand. We further explore the implications for how we estimate and characterize consequences and judge the quality of information for decision making in Chapter 8.

2.1.9 Group wisdom and dysfunction

Group decision making is unavoidable in decisions about public resources. Groups can be the source of both creative synergy and sometimes crippling dysfunction. We already have suggested that an important benefit of SDM processes is that people learn together: they build a common information base, agree on which information is most relevant, and learn about both the facts and the values basis for making decisions. We continue to believe in this positive contribution of structured deliberations.

However, unstructured group processes can encourage other types of responses. In a classic book from the 1980s, Janis[41] warned against the dangers of 'groupthink', whereby a group prematurely converges on a conclusion with insufficient analysis. Group loyalty, conformity, illusions of vulnerability, excessive risk taking, and censorship of deviations from the apparent group consensus can all adversely affect the quality of group deliberations. Creativity and independent thought are stifled when individuals follow the lead of a dominant voice; more reserved group members, in particular, are at risk of remaining unheard. More recently, Surowiecki[42] has emphasized the importance of independent thought (i.e. ideas not influenced by the ideas of others) and decentralization (ability to draw on local knowledge) in contributing to group 'intelligence'.

These are all important considerations. Given the increase in society's use of social media, we anticipate that both the expectations and capabilities for incorporating a broad range of concerns, information, and alternatives in at least some types of environmental management decisions will continue to grow. But pulling this input together in some meaningful way to make choices that involve biological uncertainties and difficult value trade-offs will require structure and critical thinking, learning and collaboration. Although this book does not provide detailed guidance on group facilitation (for helpful ideas, see Schwarz[43] or Forester[44]), in our experience best practices in working with groups usually will involve:

1 Ensuring the process is led by an impartial leader who ensures the group follows sound decision making practices.
2 Drawing on a range of stakeholders with a diversity of viewpoints.
3 Encouraging the introduction of new evidence and data.
4 Encouraging the expression and exploration of dissenting views.

5 Eliciting independent, individual judgments as well as facilitating group discussions and learning.
6 Encouraging people to go outside the group to share what is being done with others and bring back their questions and insights.
7 Providing a chance to reflect and allow 'second thoughts' to be discussed.

There is an essential interplay between allowing a group's decision process to diverge – exploring new, creative, and potentially peripheral issues and ideas – and to converge – synthesizing ideas in a way that allows a group to move forward toward a solution. The structured interplay between the independence and disciplined thought that goes with individual thinking and the creativity and learning that comes from group interaction is central to SDM. It is also a powerful tool for leveling the playing field. We have found that individuals and groups who tend to feel marginalized in formal public processes (e.g. aboriginal groups), appreciate structured methods as a means of having their voices heard [40,45].

2.2 Comparison with other evaluation approaches

We turn now to examining how SDM is related to some of the other evaluation approaches that commonly contribute to environmental management decisions – cost-benefit analysis (CBA), multi-criteria decision-making methods, negotiations and dispute resolution, risk analysis and adaptive management. The bottom-line is that SDM is a decision-focused organizing framework within which many of the insights and methods of other evaluation tools can be incorporated. Many of these other methods share the goal of maximizing net benefits to society, but both their approach to getting there and – importantly, from both conceptual and practical perspectives – the relative emphasis placed on essential analytical and deliberative tasks varies greatly.

2.2.1 Structured decision making compared with cost-benefit analysis

SDM is based on the same core goal that drives economic approaches to CBA – maximizing net benefits or utility – and shares many common conceptual roots. A primary difference is in how net benefits are identified and estimated and whose interests are being addressed. With its decision analysis roots, SDM evaluates costs and benefits in terms of multiple dimensions of value. Impacts are reported in natural units, which may be quantitative (e.g. number of lost hectares of rangeland) or qualitative (e.g. significance of the lost critical wildlife habitat). In contrast, a CBA assigns a monetary value to impacts based on some combination of financial costs (e.g. the cost of additional feed that would substitute for the lost acres of rangeland) and either individuals' 'willingness to pay', to gain benefits or to avoid losses, or their 'willingness to accept compensation' in return for agreeing to a negative outcome such as losing critical wildlife habitat.

Cost-benefit analyses compare the costs and benefits of different alternatives and provide decision makers with a summary indicator of their relative merits for a reference group or society as a whole[46]. If the dollar-based ratio of benefits to costs exceeds 1 then a project said to be worth doing, and it follows that projects with higher ratios are preferred. Cost-benefit analyses account for both goods sold in established markets, such as timber or fisheries, and environmental services for which markets do not exist, such as contributions made by ecosystems to water and air quality, or wildlife habitat, or to the creation of interface fire hazards. Environmental economists have developed a variety of ways to convert these non-market streams of environmental services into monetary units of value, which then can be incorporated directly into CBA; good summaries of techniques are readily available (see Freeman[47] or Champ, Boyle and Brown[48]).

There are aspects of SDM and CBA that are complementary, but the differences – both conceptual and practical – between the approaches are significant. Five fundamental differences are noted below.

1 *Decision context*. At the end of an evaluation, CBA seeks to provide decision makers with a single monetary estimate of the net costs and benefits of the alternatives under consideration. SDM on the other hand, seeks to provide decision makers with clarity about what to do, based on a good understanding of the available alternatives, the key trade-offs and uncertainties, and the preferences of stakeholders. A CBA process needs to run start to finish, in order to provide the final number or ratio. By focusing instead on *structuring* the alternatives evaluation process, SDM often has done its job by the end of step 2 or 3 – once objectives are clearly defined, it's often easy for decision makers to recognize and agree that one of the possible alternatives clearly is preferred, across every dimension, to any other alternative.

2 *Monetization vs multiple measures*. CBA approaches assign dollar-based measures to all environmental services and goods. A multi-attribute approach does not preclude the monetization of some impacts, and many analyses will use a combination of the two to bring insight to complex decisions: in SDM, whenever monetizing impacts will bring insight to decision makers, we do it. However, all other potential impacts of an action – anticipated effects on environmental and ecosystem, social, cultural, health and safety, or process considerations – are expressed in relevant natural units or, if these don't exist, then using proxies or scales that express health or cultural or ecological concerns which are specifically defined for this context (as described in Chapter 6). The final numbers provided by CBA are expressed in monetary units, presumably based

on individuals' revealed or expressed willingness to pay. The trade-offs and recommendations provided by SDM are expressed in natural units and reflect preferences or attitudes.

3 *Generation of alternatives.* CBA typically evaluates a *given* set of alternatives. SDM uses well defined objectives and performance measures to *generate* a creative suite of alternatives. Alternatives are then iteratively refined and improved. Dominated alternatives[56] are eliminated (before costly valuation exercises) and promising ones are refined, seeking joint gains. The priority is on finding creative solutions to complicated problems.

4 *Treatment of trade-offs.* In CBA, trade-offs are expressed to decision makers indirectly, through comparisons of dollar-denominated values. In SDM, trade-offs are expressed directly, using either quantitative or qualitative terms depending on the nature of the values involved. The exploration of trade-offs leads to learning and the development of creative solutions.

5 *Role of decision makers.* A final and very significant difference is related to the emphasis placed on the decision-making team. In SDM, it is decision makers, informed by what they know about what matters to stakeholders, who identify the values to be considered, the weights assigned to them, and the acceptability of trade-offs. In CBA, the values of external groups of people (e.g. those participating in an activity or survey) are central, with their preferences used to assign a value to the decision at hand. It can be argued that CBA studies canvas a broader and more representative sample of society, whereas SDM and other multi-criteria methods that typically work with smaller groups of people are less democratic or can be overly influenced by special interest groups. Both are fair comments: SDM gives up some 'breadth' and representativeness in favor of 'depth' in analysis and deliberation.

In practice then, we view SDM as the organizing framework for decision making. When monetization makes sense for some objectives, we use those methods (see Chapter 5, section 5.3.3). If decision makers want to (or have to) use CBA to make or justify their decisions, then we encourage them to use an SDM approach first to clarify objectives and identify good alternatives, and then to use CBA to look only at short-listed alternatives. And we caution them that many of the costs and benefits of environmental management alternatives may not be well represented (and may entirely be omitted) by monetary measures.

2.2.2 Structured decision making compared with other multi-criteria methods

In Section 2.1 we described SDM as rooted in MAUT and its applied offspring, decision analysis. However, a variety of other multi-attribute methods have been developed as offshoots of MAUT and are widely used to address environmental management problems, particularly in Europe and Asia. These include multi-criteria decision making (MCDM) approaches as well as analytical hierarchy methods (AHP[57]). There are important differences between these methods and SDM. Consider the generic steps in these approaches:

1 Define the decision context (establish criteria, classify by type) and picture in terms of a decision tree, with branches denoting the primary choices[58].

2 Assign weights (i.e. scores or preferences) to each branch that reflects its expected importance in determining outcomes.

3 Discuss scores (e.g. among a group of participants) and normalize after any revisions have been made.

4 Aggregate the performance of each option across all criteria.

5 Conduct a sensitivity analysis by varying scores and weights.

Text box 2.3:   What's not to like about contingent valuation approaches?

Over the past 30 years, contingent valuation methods (CVM) have been widely used by economists to derive estimates of the value of environmental goods and services that are not priced in conventional markets[49]. Hundreds of CVM studies have been conducted for estimating the benefits associated with proposed environmental initiatives (e.g. improving water quality) or the damages associated with natural resource losses (e.g. an oil spill); the approach has been embraced by federal agencies (including the US Environmental Protection Agency), recognized by the courts, and widely promoted by consultants. At its core is the idea that individuals can express the value they place on a current or future environmental good or service by stating their maximum willingness to pay if a hypothetical market for the good were to be created.

From a decision-making perspective, a fundamental problem with this approach is that because the affected goods, services or resources are not exchanged on an actual market, people may be unable to think clearly about the associated value in monetary terms[50]. The high correlations that have been observed in studies that compare economic (dollar-based) and psychological (attitudinal) measures of value strongly support this interpretation, which is directly contrary to CVMs market analogy assumptions[51]. A useful measure of economic or psychological value should be able to demonstrate that people are sensitive to theoretically-relevant aspects of the good, and insensitive to theoretically irrelevant information; CVM methods repeatedly fail both tests[52,53]. Numerous studies demonstrate anchoring and 'range' effects, whereby participants' expressed willingness to pay for actions is insensitive to dramatic (e.g. hundred-fold) changes in the numbers of potentially affected individuals – surely something one would expect to be relevant.

From the standpoint of a resource manager faced with helping a group of stakeholders to make a recommendation on a tough policy choice, a related issue is that all the effort spent on development of a market analogy and a dollar-denominated ratio largely misses the point[54]. There are many cases where monetization (even if it were successful) simply doesn't help – estimating a willingness to pay for the preservation of cultural sites or a genetically distinct salmon run might produce a number, but this will do little or nothing in terms of identifying key concerns, creating novel management strategies, or encouraging deliberations and agreement among diverse participants.

The most fundamental problems with CVM, however, arise from moral and ethical rather than methodological concerns. Consider a situation such as restoring wildlife habitat on a Native American reserve damaged by years of contamination from mining operations. In this (illustrative) case, assume that losses to the community have been substantial and include changes in culture and lifestyle (because people can no longer engage in traditional hunting activities), a loss of self or community identity, adverse impacts on health (related to changes in diet), emotional concerns (related to the uncertainty associated with eating possibly contaminated foods), and losses of knowledge regarding traditional practices. To express these activities in natural units (days of hunting, pounds of berries collected) is difficult but often possible. To take the extra step of converting these effects from natural to market-based, dollar units is sufficiently tenuous as to be meaningless and, depending on the context, easily might be viewed as unethical or immoral.

In recent years, behaviorally minded economists have begun to work in tandem with psychologists and decision scientists to address these issues in the hope of developing new methods of environmental evaluation; examples include work on stated preference and conjoint evaluation methods[55]. These new approaches are much closer in spirit to SDM approaches, particularly with respect to their emphasis on the careful definition of objectives and attributes.

These steps are similar to the middle portions of SDM practice, once (and only if) it has been determined that a quantitative approach to trade-offs will be useful. A key difference is the emphasis of SDM on broader participation in decision making and in helping participants to think critically about the decision. The tendency with MCDM or AHP approaches is to underemphasize the up-front problem structuring process; insufficient attention to defining the issue and working with stakeholders to identify objectives and performance measures can lead to biased or misleading results and to a less well-structured problem (e.g. with confusion between means and ends, or with improperly specified and weighted performance measures). AHP and MCDM approaches also tend to be more linear and therefore give less attention to the later stages of iteration, monitoring, and incorporating learning into revised management plans. Our experience also suggests that adoption of an AHP or MCDM approach often can result in needless expenses: SDM practitioners find that only a small percentage of decision problems ever require a formal quantitative trade-off analysis, because after careful structuring the general outlines of a preferred management plan and its likely outcomes are often clear.

2.2.3 Structured decision making compared with negotiations and dispute resolution

A structured decision process is based on an integration of analysis and deliberation, more so than on negotiation. It is designed to inform a decision by a responsible authority, rather than to serve as a bargaining or dispute resolution framework among adversaries. That said, the approach clearly supports negotiated solutions and there has been much thoughtful theory and case-study application showing how decision analysis can contribute to negotiated solutions[59,60]. One of the most widely cited (and best-selling) guidebooks on negotiations, *Getting to Yes*[61], promotes essentially the same steps, albeit with objectives called interests. Many more recent negotiation texts

(e.g. Thompson[62]) reflect insights from behavioral decision research, with an emphasis on the role played by emotions, ethics, and efforts to build trust through structured discussions.

SDM can be used as a pre-negotiation process[63]. In this context, parties who need to reach agreement are brought together to clarify their interests (objectives), generate potential solutions (alternatives), examine the implications of the potential solutions for different parties, clarify the trade-offs and uncertainties, and discuss areas of agreement and disagreement. If SDM produces a consensus choice, which it often does, then it can replace formal negotiations or eliminate the need for dispute resolution. If it does not – and sometimes it does not or is not designed to – then at minimum it lays the ground for more informed and streamlined negotiations. Because it is not a formal negotiation, parties are free to explore creative alternatives. Because participants indicate their support or reservations about various alternatives, the parties leave the process with a clear idea about where mutually agreeable solutions may lie – and they thoroughly understand how the interests of others are affected.

All this brings us back to the concept of consensus. The explicit search for consensus, we believe, can be problematic, at least from a management and policy development perspective. Consensus itself, whether among decision makers or stakeholders, can of course be a helpful result of a well-constructed decision-focused dialogue, and it can strengthen recommendations made to a decision-making authority. The main problem is focusing too early on consensus. When that happens, two problems affect decision quality.

The first is that the quality of judgments suffers. Environmental problems and their potential solutions tend to be technically complex, and participants in them are subject to a variety of errors and biases in judgment. While the structuring techniques of SDM don't eliminate these problems, they do improve the extent to which participants understand the objectives, the alternatives and their consequences, and help to minimize the effects of systematic biases. A second concern is that premature emphasis on seeking consensus may limit the exploration of controversial but creative alternatives[61, 64]. A willingness to explore creative and in some cases extreme alternatives is the source of important mutual learning, which we discuss more in Chapter 7.

Somewhat counter intuitively, experience has shown that consensus can be achieved as the product of good decision practices – appropriate analysis, informed deliberation and a creative exploration of alternatives – even when (or perhaps particularly when) consensus is not required[65]. Support and trust therefore ultimately depend critically on collaboratively creating a

basis for informed judgments: neither consensus nor a good feeling among participants are to be avoided, but for SDM processes a sensitivity to group dynamics is a means to providing insights for decision makers that can lead to lasting environmental management solutions.

2.2.4 Structured decision making compared with risk assessment and management

A probabilistic risk assessment estimates the risk(s) of an activity based on the probability of events whose occurrence is expected to lead to undesirable consequences[66]. A conventional risk-based approach to the assessment of resource development or allocation options involves four principal steps: define the problem, identify hazards (i.e. activities with a potential for producing undesirable outcomes) and endpoints, assess exposure pathways and potential effects, and characterize how probable it is that the undesirable consequences will occur. This basic probability and consequence assessment, which sometimes is combined with analyses of socio-economic and political factors, then establishes an information basis for selecting risk management actions. Suter[67] describes numerous applications in the context of ecological risk assessments.

Risk assessment has served an important and highly useful role for environmental decision makers over the past four decades. It is a useful tool for one part of an SDM process — estimating and characterizing uncertain consequences. Two shifts in thinking about risk assessment are particularly important with respect to how it is used in decision making. Both concern the role of values.

The first concerns the conventional separation of the 'science' of risk assessment from the 'policy' of management and decision making. Today, there is increasing acknowledgement that this separation can be problematic: it forces analysts to make assumptions about what managers or decision makers need or want, and many of these assumptions involve important value judgments.

The selection of environmental objectives for example, the measures that will be used to assess them, and in some cases even the specific analytical methods to be used, all involve value judgments, and should be made with the involvement of managers or decision makers who have legitimacy to make them.

A second shift involves increasing recognition of the role of values in defining and making judgments about risk. 'Risk' is a multi-attribute concept. It's widely recognized that factors such as the familiarity or dread, catastrophic potential and the voluntary nature of exposure affect perceptions and ranking of risks as much as frequency and magnitude estimates. As a result, questions about what constitutes a risk and what should be the relative ranking of risks are value based as much as science based. Risk assessment and ranking processes that ignore the values side will necessarily be flawed. Further problems arise when risk assessors are asked to determine what constitutes an acceptable level of risk. What is acceptable depends on the available alternatives, a realistic depiction of trade-offs, and a clear understanding of participants' values[14], none of which are explored by risk assessment.

Environmental managers need tools and approaches for *managing* risks, not just analyzing them. This involves balancing the costs and benefits of the available management *alternatives*, not just characterizing the hazards, stressors, or pathways associated with baseline conditions[68]. Risk management is much more similar to SDM than risk assessment, as it explicitly seeks to balance costs and benefits. There are three challenges with using 'risk management' to frame environmental management problems:

1 It frames everything as a risk – a bad thing to be avoided. While environmental management often involves risks, there are usually important benefits from proposed actions – improvements to habitat for example, or increases in jobs. It is unfortunate to use an overall framing for a decision that inherently emphasizes the negative.

2 Risk management frameworks tend to focus on one (or perhaps a few) endpoints. A typical approach identifies alternatives for managing this endpoint 'while minimizing impacts on social values'. Putting all other interests into this secondary category marginalizes them and opens the decision up to criticism later – recall the 'scientist' scenario from Chapter 1.

3 Perhaps most importantly, risk management frameworks simply provide little actual guidance about how to make decisions. There are many frameworks (and some are better than others), but inevitably they present a variety of boxes and arrows leading in one way or another to a box called 'manage risks' or 'decide'. In other words, ecological risk assessment + socio-economic factors + political factors = decision. That's all OK, but how? A great deal of technical analysis may be done without a clear idea of how it will be used to make decisions. In contrast, SDM starts with this management or decision focus and builds an analysis around it. As a decision-focused approach, it provides the broader framework within which more specific risk analyses might take place.

Consider a northern region facing development pressure that may put polar bears at risk. A risk-based approach tends to start with a species-at-risk and asks (in this case) 'what is the risk to polar bears'? In setting objectives, it will therefore identify a range of stressors to polar bears and set objectives to minimize them. In contrast, a SDM approach focuses on the management decision and thus typically includes a broader set of objectives, including economic and social or cultural as well as environmental considerations. Objectives might include 'maximize polar bear population', 'enhance traditional ways of life', and 'increase employment opportunities'. An ecological risk analysis therefore may be used to inform the environmental portions of an SDM analysis of alternatives, or to provide guidance about management alternatives that could be considered.

Ultimately, we see SDM as a broader decision-making framework within which ecological (and other types of) risk assessment may play a role in examining and describing the range of uncertain consequences that could result from an action. Risk management by definition involves making decisions to balance risks, but risk management frameworks provide little guidance for making multi-objective choices that involve value-based trade-offs. In short, risk assessment and management are important parts of decision making, but don't provide guidance to the whole decision making process. Paul Slovic[69] sums up the issue well:

> Risk is a complex and controversial concept that typically has no direct implications for decision making. Assessing a risk as 'high' does not necessarily mean we should act to reduce it. Similarly, assigning a risk as 'low' does not mean we should ignore it. Risk management decisions depend on the balancing of options, benefits, and other costs – not just risk. In this sense, we need to look beyond measurements of something called 'risk' to make effective risk management decisions. In particular we may need to embed risk decisions more strongly in techniques for sound social and individual decision making. . . .The use of a decision analysis framework for addressing risk management problems allows the tools of modern risk analysis to be used as part of a broader context, where the emphasis is on creating a sound structure for [the] decision maker rather than addressing the narrow concept of risk as some kind of loss (p. 73[69])'

2.2.5 Structured decision making compared with adaptive management

Adaptive management (AM) is a way to deal with uncertainty and surprise. More formally, it's a systematic approach for improving resource management by learning from management outcomes[72–74]. It involves exploring alternative ways to meet management objectives, predicting the outcomes of alternatives based on the current state of knowledge, implementing one or more of these alternatives, monitoring to learn about the

---

Text box 2.5: Using structured decision making to guide a watershed-scale ecological risk assessment

In the late 1990s, the Greater Vancouver Water District set out to develop a science-based watershed management plan based on the principles of ecological risk assessment (ERA)[70]. The primary goal of the plan was to provide source water protection within their three watersheds that provide drinking water to the nearly two million residents of the Greater Vancouver Region. The plan was being developed against a backdrop of increasing stakeholder opposition to a sustained-yield forestry program that had been in place for several decades, with criticism aimed at both the risks that logging posed to water quality and the loss of old forests. Viewed through an ERA lens, the primary assessment endpoints were water quality and forest ecosystem health. The introduction of SDM as the organizing framework for the decision-making process resulted in the addition of visual quality and management costs as assessment endpoints[71]. Key activities of the multi-year process included:

1 Developing assessment endpoints for the full suite of management objectives.
2 Developing distinct management alternatives that prescribed varying levels of forest management, erosion control, etc.
3 Assessing the consequences of the alternatives on the assessment endpoints.
4 Explicitly examining value-based trade-offs among the endpoints – including water quality, environmental, financial and social endpoints.

---

impacts of management actions, and then using the results to update knowledge and adjust management actions. Intuitively appealing, AM has become a central pillar of many ecological management initiatives and is widely seen as a means to manage responsibly under conditions of uncertainty. It's explicitly called for in many management and regulatory contexts – this in spite of the acknowledgement that there remain significant challenges in implementation, for which a variety of reasons have been identified[75,76].

One of the barriers to effective implementation of AM has been the difficulty linking the results of management experiments to decision making. While effective integration of science and management has always been a central tenet of adaptive approaches, early representations of the AM cycle tended to emphasize the scientific aspects – the design of experiments, sampling protocols, and so on. What AM failed to adequately recognize historically was that every management action has multiple objectives. Experiments were set up to learn about the response of a particular environmental endpoint to a management action, with the implicit assumption that this would lead to clarity in decision making. In all but a few very scope-constrained cases (e.g. comparisons of alternative silviculture treatments on experimental plots) it did not live up to expectations. The problem is that AM, as it has historically been approached, does not provide a framework for actually making management choices once the new information is received. This is because there are inevitably multiple objectives that must be balanced. Further, even deciding to monitor or experiment is itself a multi-objective decision – longer experiments may deliver higher quality information but they defer potential on-ground benefits, some experimental designs impose more stress on resources than others, and the selection of which alternatives make the short list for experimentation is itself a value-based choice[77].

More recent literature demonstrates a shift toward ever greater integration with the decision-making process. The US Department of the Interior's technical guidelines for Adaptive Management[76] advise that adaptive management must be conducted within the context of a structured decision making process (see Chapter 10 for more discussion). This is consistent with our view, which highlights four common principles at the heart of both AM and SDM:

1 Acknowledge pervasive and sometimes irreducible uncertainty.
2 Commit to formal learning – make explicit hypotheses and design strong monitoring programs or experiments to test them.
3 Identify creative alternatives to promote learning and deal with uncertainty.
4 Build in review cycles to institutionalize the capacity to learn and to respond to new information about facts and about values.

In practical terms, SDM thus becomes the broader organizing framework within which scientifically rigorous approaches to learning can be implemented. Framing it this way helps to address many of the implementation problems that typically are associated with AM. The explicit and *a priori* emphasis in SDM on defining how subsequent decisions will be made – who will make them, using what objectives and measures as evaluation criteria, and through what analytical and deliberative processes – can enhance the likelihood of successful implementation of monitoring and experimental approaches and subsequent uptake of the results.

2.3 Key messages

A core set of foundational concepts underlie the philosophy and methods used in SDM approaches. We discuss MAUT and analysis; the concept of constructed preferences; the practical need for separation of values and facts; the notion of value-focused thinking; implications of recognizing two systems of thinking; judgmental heuristics and biases; what constitutes the best available knowledge; and methods for addressing group wisdom and dysfunction. Key messages include:

1 SDM approaches are based fundamentally on the insights and methods of MAUT, which provides a technically defensible foundation

and a framework for thinking clearly about the decision. It lays the groundwork for formal, quantitative trade-off analysis, but does not require it.

2 Preferences are constructed through the course of a decision process, as a result of people learning about both the facts basis and the values basis for choices. There are often significant and surprising shifts in preferences over the course of a decision process that allow participants new ways to think, feel, talk and learn.

3 Separating questions of fact (what is?) from questions of value (what matters?) is essential to good decision making.

4 Value-focused thinking involves the discipline of thinking carefully about what's important to people about a decision context before making choices. Key decision steps, including defining objectives, identifying alternatives, estimating consequences, and making trade-offs are all built on the foundation of individuals' expressed values.

5 People draw on two systems of thinking: an emotional (or intuitive) System 1 and a reasoned (or analytic) System 2, to make choices. Both are simultaneously powerful and prone to error, and both have a central role to play in decision making about environmental issues.

6 The cognitive short cuts used by all people, both expert and non-expert, can lead to useful insights or to cognitive biases that adversely affect the quality of judgments. There are methods for anticipating and for minimizing, but not eliminating, these biases.

7 Knowledge is constructed and agreed to by participants in a decision process. This doesn't mean that anything goes with respect to what constitutes relevant information. It acknowledges that uncertainty is pervasive, that there are many legitimate sources of knowledge and no clear line between science and non-science, and that the quality and relevance of information for decision making is determined with reference to the decision at hand.

8 Both individual and group judgments contribute to decisions. The selection of methods and deliberative processes for clarifying the environmental management problem and opportunities, identifying objectives, creating alternatives, and addressing trade-offs must accommodate and encourage informed input from both sources.

9 CBA, multi-attribute methods, risk management, AM and negotiations can all play useful roles in developing, evaluating or reaching agreement on a decision. SDM is a broader organizing framework for the environmental management decision process, within which these tools may be used to help inform a decision.

2.4 Suggested reading (loosely grouped by the authors' primary discipline)

Psychology

Ariely, D. (2008) *Predictably Irrational: The Hidden Forces that Shape our Decisions.* HarperCollins: New York. Entertaining and clearly written account of judgmental factors that underlie our decisions, written by a solid researcher who knows how to tell a good story as well as design a clever experiment.

Kahneman, D. & Tversky, A. (Eds) (2000) *Choices, Values, and Frames.* Cambridge University Press, Cambridge. There is no more complete collection of papers on the psychology of how people make decisions; topics include alternative conceptions of value, preference weighting, choices over time, and applications to public policy.

Plous, S. (1993) *The psychology of Judgment and Decision Making.* McGraw-Hill, New York. A highly readable overview of the psychology of judgment and decision making, emphasizing the role of judgmental biases, social influences, and group decisions.

Slovic, P. (1995) The construction of preference. *American Psychologist,* **50**, 364–371. A clear summary of the behavioral research underlying the concept of constructed preferences, including implications for decision making and public policy.

Decision analysis

Clemen, R.T. (2004) *Making Hard Decisions: An Introduction to Decision Analysis* (4th edn). Duxbury, Belmont. Since the first edition came out in 1996, this accessible book has become the favored text for learning about the techniques and practice of decision analysis.

Fischhoff, B., Lichtenstein, S., Slovic, P., Derby, S.L. & Keeney, R.L. (1981) *Acceptable Risk*. Cambridge University Press, New York. This short and entertaining book, still relevant after 30 years, clearly introduces the concept of acceptable risk and discusses why it's often so difficult to resolve acceptable risk problems.

Keeney, R.L. (1982) Decision analysis: an overview. *Operations Research*, **30**, 803–838. Written for an audience unfamiliar with decision analysis, this concise article provides a great overview of what a decision analytic approach does and does not attempt to do. The discussion also sets decision analysis practice within the context of earlier work in decision making.

von Winterfeldt, D. & Edwards, W. (1986) *Decision Analysis and Behavioral Research*. Cambridge University Press, New York. The authors, among the first to combine behaviorally oriented approaches with techniques of decision analysis, provide clear and detailed descriptions of many of the fundamental methods used in SDM.

Biology and ecological science

Burgman, M. (2005) *Risks and Decisions for Conservation and Environmental Management*. Cambridge University Press, Cambridge. A comprehensive and clearly written introduction to the philosophy, methods, strengths and shortcomings of environmental risk assessment, with an emphasis on the use of decision analytic methods in conservation biology to improve experts' judgments under uncertainty.

Gunderson, L.H., Holling, C.S. & Light, S.S. (1995) *Barriers and Bridges to Renewal of Ecosystems and Institutions*. Columbia University Press, New York. Case studies and theory about learning and adaptation when dealing with uncertainty in ecosystem management.

Walters, C. (1986) *Adaptive Management of Renewable Resources*. MacMillan, New York. Thoughtful discussions of the principles and techniques of adaptive management, with an emphasis on practical insights from numerous case studies.

Policy analysis and management

Bazerman, M.H. (2002) *Judgment in Managerial Decision Making*. Wiley, New York. A clear primer on decision making for managers, with case studies from numerous organizational contexts and good discussions of the role of fairness, methods for negotiations and group deliberations, and the role of uncertainty.

Goodwin, P. & Wright, G. (2004) *Decision analysis for management judgment* (3rd edn). Wiley, Chichester. An overview of both simple and demanding decision-making techniques, with clear examples for aiding choices by managers.

Slovic, P. (1999) Trust, emotion, sex, politics, and science: surveying the risk-assessment battlefield. *Risk Analysis*, **19**, 689–701. The title of this article says it all: 'Trust, Emotion, Sex, Politics, and Science: Surveying the Risk-Assessment Battlefield'.

Negotiation and group facilitation

Fisher, R., Ury, W. & Patton, B. (1991) *Getting to Yes* (2nd edn). Penguin Books, Harmondsworth. It's unusual for a book covering the fundamentals of negotiation analysis to sell millions of copies, but it's no surprise given the entertaining examples, clear thinking, and practical focus of this decision-analysis inspired approach to negotiations.

Forester, J. (1999) *The Deliberative Practitioner: Encouraging Participatory Planning Processes*. MIT Press, Cambridge. Most books on facilitation contain less than common sense; this well-written and experienced author's practical advice reflects lessons drawn from policy analysis, political science, law, and planning.

Renn, O. (2006) Participatory processes for designing environmental policies. *Land Use Policy*, **23**, 34–43. A good overview of principles and advice for working with participatory environmental processes, by a leading European practitioner of decision-analytic methods.

Thompson, L.L. (2009) *The Mind and Heart of the Negotiator*. Pearson Prentice Hall, Upper Saddle River. Introduces effective negotiation techniques, with an emphasis on using insights from behavioral research to acknowledge emotions and build trust among participants as a means to reaching broadly supported agreements.

2.5 References and notes

1 Raiffa, H. (1968) *Decision Analysis: Introductory Lectures on Choices under Uncertainty.* Addison-Wesley, Reading.

2 Keeney, R. & Raiffa, H. (1993) *Decisions with Multiple Objectives: Preferences and Value Tradeoffs.* Cambridge University Press, UK. (Original work published 1976.)

3 Edwards, W. & von Winterfeldt, D. (1987) Public values in risk debates. *Risk Analysis*, **7**, 141–158.

4 Edwards, W. (1977) How to use multiattribute utility measurement for social decisionmaking. *IEEE Transactions on Systems, Man, and Cybernetics*, SMC-7, 326–340.

5 von Winterfeldt, D. & Edwards, W. (1986) *Decision Analysis and Behavioral Research.* Cambridge University Press, New York.

6 Keeney, R.L. (2007) Developing objectives and attributes. In: *Advances in Decision Analysis: From Foundations to Applications.* (eds W. Edwards, R.F. Miles, Jr & D. von Winterfeldt). pp. 104–128. Cambridge University Press, New York.

7 Theoretically, a 'utility function' is the appropriate model in cases where uncertainties are large and are explicitly incorporated into the depiction of consequences (Keeney, R. L. & von Winterfeldt, D. (2007) Practical value models. In: *Advances in Decision Analysis: From Foundations to Applications.* (eds W. Edwards, R.F. Miles, Jr & D. von Winterfeldt). pp. 104–128. Cambridge University Press, New York). However, from a practical standpoint, the use of measurable value models is less demanding in terms of the information that decision makers need to provide, and often more readily understood. Provided objectives have been well-structured, and given the complexities involved in estimating the consequences of the alternatives in many environmental management decisions, the use of a linear value model is unlikely to be the weak link in the analysis.

8 Keeney, R.L. (1992) *Value-Focused Thinking: A Path to Creative Decisionmaking.* Harvard University, Cambridge.

9 Keeney, R.L. (1982) Decision analysis: an overview. *Operations Research*, **30**, 803–838.

10 Keefer, D.L., Kirkwood, C.W. & Corner, J.L. (2004) Perspective on decision analysis applications, 1990–2001. *Decision Analysis*, **4**, 4–22.

11 National Research Council (NRC) Committee on Risk Characterization (1996) *Understanding Risk: Informing Decisions in a Democratic Society.* National Academy Press, Washington, DC.

12 Phillips, L.D. (1984) A theory of requisite decision models. *Acta Psychologica*, **56**, 29–48.

13 Lichtenstein, S. & Slovic, P. (eds) (2006). *The Construction of Preference.* Cambridge University Press, New York.

14 Fischhoff, B., Lichtenstein, S., Slovic, P., Derby, S.L. & Keeney, R.L. (1981) *Acceptable Risk.* Cambridge University Press, New York.

15 Poulton, E.C. (1994) *Behavioral Decision Making.* Lawrence Erlbaum, Hillsdale.

16 A well-known example comes from an experiment conducted by Daniel Kahneman and Amos Tversky in the early 1970s (see Tversky, A. & Kahneman, D. (1981) The framing of decisions and the psychology of choice. *Science*, **211**, 453–458). They first described a straightforward health-case issue involving 600 people and then asked one group of subjects to choose between two treatment alternatives, knowing that 200 people would be saved; a second group was shown the same description and again asked to select a preferred treatment, in the case after being told that 400 people would die. Although the presentation of the alternatives was formally equivalent, the difference in the treatment choice was significant due to differences between the 'lives saved' and 'lives lost' decision frames.

17 Payne, J.W., Bettman, J.R. & Johnson, E.J. (1993) *The Adaptive Decision Maker.* Cambridge University Press, New York.

18 Dake, K. (1991) Orienting dispositions in the perception of risk: an analysis of contemporary worldviews and cultural biases. *Journal of Cross-Cultural Psychology*, **22**, 61–82.

19 Maguire, L. (2004) What can decision analysis do for invasive species management? *Risk Analysis*, **24**, 859–868.

20 Funtowicz, S.O. & Ravetz, J.R. (1992) Three types of risk assessment and the emergence of post-normal science. In: *Social Theories of Risk* (eds S. Krimsky & D. Golding). pp. 251–273. Praeger, Westport.

21 Jasanoff, S. (1990) *The Fifth Branch: Science Advisers as Policymakers.* Harvard University Press, Cambridge.

22 Finucane, M.L., Peters, E. & Slovic, P. (2003) Judgment and decision making: the dance of affect and reason. In: *Emerging Perspectives on Judgment and Decision Research* (eds S. L. Schneider & J. Shanteau). pp. 327–364. Cambridge University Press, Cambridge.

23 Keren, G. (ed.) (2011) *Perspectives on Framing.* Psychology Press New York.

24 Slovic, P., Finucane, M.L., Peters, E. & MacGregor, D.G. (2002) The affect heuristic. In: *Heuristics and Biases: The Psychology of Intuitive Judgment* (eds T. Gilovich, D. Griffin & D. Kahneman). pp. 397–420. Cambridge University Press, New York.

25 Loewenstein, G., Weber, E.U., Hsee, C.K. & Welch, E.S. (2001) Risk as feelings. *Psychological Bulletin,* **127,** 267–286.

26 Kahneman, D. (2002) *Maps of Bounded Rationality: A Perspective on Intuitive Judgment and Choice.* Retrieved from http://nobelprize.org/nobel_prizes/economics/laureates/2002/kahnemann-lecture.pdf

27 This body of work lately has been the subject of several best-selling books; examples include *Nudge* (Thaler, R.H., & Sunstein, C.R. (2008). *Nudge: Improving Decisions About Health, Wealth, and Happiness.* Yale University Press, New Haven), *Predictably Irrational* (Ariely, D. (2008) *Predictably Irrational: The Hidden Forces that Shape our Decisions.* HarperCollins, New York), and *Blink* (Gladwell, M. (2005). *Blink: The Power of Thinking Without Thinking.* Litttle, Brown, New York).

28 Tversky, A. & Kahneman, D. (1974) Judgment under uncertainty: heuristics and biases. *Science,* **185,** 1124–1131.

29 See the extensive literature on 'ignoring base rates' (Bazerman, M.H. (2002) *Judgment in Managerial Decision Making.* Wiley, New York). In an often cited experiment, Kahneman and Tversky (1973) (On the psychology of prediction. *Psychological Review,* **80,** 237–251) provided subjects with information that a group of 100 people contained 30 engineers and 70 lawyers. They then asked subjects to assign a probability that one of them ('Jack') was an engineer. When no other information was given, people correctly assigned a probability of 30%. When additional details were provided (interesting but largely irrelevant details such as – 'Jack spends his free time on carpentry and math puzzles'), subjects tended to ignore the base rate information and assigned a much higher probability to Jack being an engineer.

30 For example, in response to a question about the number of African nations that were members of the United Nations, the median estimate of the group that received '10' as a starting point was 25, while the median estimate of the group receiving '65' as a starting point was 45.

31 Alpert, M. & Raiffa, H. (1982) A progress report on the training of probability assessors. In: *Judgment under Uncertainty: Heuristics and Biases* (eds D. Kahneman, P. Slovic & A. Tversky). pp. 294–305. Cambridge University Press, New York.

32 You can do this yourself. Test yourself on ten questions, such as 'what is the diameter of the moon?' Provide and upper and lower value such that you think there is a 90% chance the true value falls between them. Now go to Google – if you are well calibrated, nine out of 10 of your answers will encompass the true value.

33 Cooke, R.M. (1991) *Experts in Uncertainty: Opinion and Subjective Uncertainty in Science.* Oxford University Press, New York.

34 Vose, D. (1996) *Risk Analysis: A Quantitative Guide to Monte Carlo Simulation Modelling.* John Wiley & Sons, New York.

35 McNeil, B.J., Pauker, S.G., Sox, H.C., Jr & Tversky, A. (1982) On the elicitation of preferences for alternative therapies. *New England Journal of Medicine,* **306,** 1259–1262.

36 Fischoff, B. (1982) Debiasing. In: *Judgment under Uncertainty: Heuristics and Biases* (eds D. Kahneman, P. Slovic & A. Tversky). New York, Cambridge University Press.

37 Ravetz, J.R. (1999) What is post-normal science? *Futures,* **31,** 647–653.

38 Slovic, P. (1999) Trust, emotion, sex, politics, and science: surveying the risk-assessment battlefield. *Risk Analysis,* **19,** 689–701.

39 Lubchenco, J. (1998) Entering the century of the environment: a new social contract for science. *Science,* **279,** 491–497.

40 Failing, L., Gregory, R. & Harstone, M. (2007) Integrating science and local knowledge in environmental science and local knowledge in environmental risk management: a decision-

focused approach. *Ecological Economics*, **64**, 47–60.

41 Janis, I.L. (1982) *Groupthink: Psychological Studies of Policy Decisions and Fiascoes*. Houghton Mifflin, Boston.

42 Surowiecki, J. (2004) *The Wisdom of Crowds: Why the Many are Smarter than the Few*. Random House, New York.

43 Schwartz, B. (2004) *The Paradox of Choice: Why More is Less*. HarperCollins, New York.

44 Forester, J. (1999) *The Deliberative Practitioner: Encouraging Participatory Planning Processes*. MIT Press, Cambridge.

45 Gregory, R.S. & Keeney, R.L. (2002) Making smarter environmental management decisions. *Journal of the American Water Resources Association*, **38**, 1601–1612.

46 Zerbe, R.O. & Dively, D.D. (1994) *Benefit-Cost Analysis: In Theory and Practice*. HarperCollins College, New York.

47 Freeman, A.M., III (2003) *The Measurement of Environmental and Resource Values: Theory and Methods*, 2nd edn. Resources for the Future, Washington, DC.

48 Champ, P.A., Boyle, K.J. & Brown, T.C. (2003) *A Primer on Nonmarket Valuation*. Kluwer Academic Press, Boston.

49 Mitchell, R. (2002) On designing constructed markets in valuation surveys. *Environmental & Resource Economics*, **22**, 297–321.

50 Kahneman, D. & Knetsch, J.L. (1992) Valuing public goods: the purchase of moral satisfaction. *Journal of Environmental Economics and Management*, **22**, 57–70.

51 Kahneman, D., Ritov, I. & Schkade, D. (1999) Economic preferences or attitude expressions?: an analysis of dollar responses to public issues. *Journal of Risk and Uncertainty*, **19**, 203–235.

52 Fischhoff, B. (1991) Value elicitation: is there anything in there? *American Psychologist*, **46**, 835–847.

53 Pidgeon, N. & Gregory, R. (2004) Judgment, decision making, and public policy. In: *Blackwell Handbook of Judgment and Decision Making* (eds D.J. Koehler & N. Harvey). pp. 604–623. Blackwell, Oxford.

54 Gregory, R., Lichtenstein, S. & Slovic, P. (1993) Valuing environmental resources: a constructive approach. *Journal of Risk and Uncertainty*, **7**, 177–197.

55 Adamowicz, W.L. (2004) What's it worth? An examination of historical trends and future directions in environmental valuation. *Australian Journal of Agricultural and Resource Economics*, **48**, 419–443.

56 A dominated alternative is one that performs less well, or no better, than another on all performance measures (see Chapter 9 for additional discussion).

57 Mustajoki, J., Hämäläinen, R.P. & Marttunen, M. (2004) Participatory multicriteria decision analysis with Web-HIPRE: a case of lake regulation policy. *Environmental Modelling & Software*, **19**, 537–547.

58 Goodwin, P. & Wright, G. (2004) *Decision Analysis for Management Judgment*, 3rd edn. Wiley, Chichester.

59 Raiffa, H. (1982) *The Art and Science of Negotiation*. Harvard University Press, Cambridge.

60 Bana e Costa, C.A. (2001) The use of multi-criteria decision analysis to support the search for less conflicting policy options in a multi-actor context: case study. *Journal of Multi-Criteria Decision Analysis*, **10**, 111–125.

61 Fisher, R., Ury, W. & Patton, B. (1991) *Getting to Yes*, 2nd edn. Penguin Books, London.

62 Thompson, L.L. (2009) *The Mind and Heart of the Negotiator*. Pearson Prentice Hall, Upper Saddle River.

63 Bana e Costa, C.A., Nunes da Silva, F. & Vansnick, J.-C. (2001) Conflict dissolution in the public sector: a case-study. *European Journal of Operational Research*, **130**, 388–401.

64 Gregory, R., McDaniels, T. & Fields, D. (2001) Decision aiding, not dispute resolution: creating insights through structured environmental decisions. *Journal of Policy Analysis and Management*, **20**, 415–432.

65 The Water-Use Plan multi-stakeholder process used in British Columbia, Canada, which relied on deliberations based on SDM guidelines, explicitly did not require consensus among members of the Consultative Committees. Nevertheless, consensus was delivered at 22 of the 23 hydroelectric facilities undergoing water allocation reviews.

66 Morgan, G.M. (1993, July) Risk analysis and management. *Scientific American*, pp. 32–41.

67 Suter, G.W.I. (1993) *Ecological Risk Assessment*. Lewis, Boca Raton.

68 Derby, S. L. & Keeney, R.L. (1981) Risk analysis: understanding 'how safe is safe enough?' *Risk Analysis*, **1**, 217–224.

69 US National Science Foundation (2002, July) Integrated research in risk analysis and decision making in a democratic society (Workshop report). Arlington, Virginia.

70 US Environmental Protection Agency (USEPA) (1993) *Comparative risk*. Washington, DC.

71 Ohlson, D. & Serveiss, V.(2007) The integration of ecological risk assessment and structured decision making into watershed management. *Integrated Environmental Assessment and Management*, **3**, 118–128.

72 Holling, C.S. (1978) *Adaptive Environmental Assessment and Management*. Wiley, Chichester.

73 Walters, C. (1986) *Adaptive Management of Renewable Resources*. MacMillan, New York.

74 Lee, K. (1993) *Compass and Gyroscope: Integrating Science and Politics for the Environment*. Island Press, Washington, DC.

75 Walters, C. (1997) Challenges in adaptive management of riparian and coastal ecosystems. *Conservation Ecology*, **1**, 1. Retrieved from http://www.ecologyandsociety.org/vol1/iss2/art1/

76 Williams, B.K., Szaro, R.C. & Shapiro, C.D. (2007) *Adaptive Management: The US Department of the Interior Technical Guide*. US Department of the Interior, Washington, DC. Retrieved from http://www.doi.gov/initiatives/AdaptiveManagement/TechGuide.pdf

77 Gregory, R., Failing, L., Ohlson, D. & McDaniels, T. (2006) Some pitfalls of an overemphasis on science in environmental risk management decisions. *Journal of Risk Research*, **9**, 717–735.

3

Decision Sketching

Most of us have been trained in what could be termed a 'study culture': a problem is identified and, in short order, a study is launched to provide additional information. Why the study is needed, or how any new information will contribute to a better choice among management options, is rarely specified. If there is one message you take away from this book, it is that environmental managers have to start thinking like decision makers and shift from a study culture to a 'decision culture'. If you are a manager, or work in an organization that develops management plans or policies, nearly every problem you face is a decision or one that is linked to a decision. When you frame your problem as a decision – a choice with multiple objectives and alternative courses of action – it changes your point of entry into the problem and, consequently, everything else that you do: the make up of the project team, the allocation of resources, the collection of information, the focus of technical analyses, the timing and content of public consultation, and the characterization of the outcome.

Much of what we conventionally think of as problem solving becomes far easier once the problem is well structured. Yet as a society, we don't teach people how to do this. Think about problems you 'solved' in school. You were given the problem and asked to solve it – which you cleverly did, at least most of the time. But school is about the only place in the world where the problems are already structured for you. In the real world, problem structuring is often by far the trickiest part of problem solving.

Usefully structuring a problem (there is no such thing as a 'correctly' structured problem) requires some combination of skill, experience, inspiration, and persistence. Our best advice is to take a trial run at the decision. Then maybe take another, and another. We call this sketching the decision, and it forms the heart of problem structuring in SDM. In the first part of this chapter, we lay out some decision structuring basics: different types of decisions, steps in decision sketching, examples of decision sketching exercises, and a brief introduction to some structuring tools. In the second part of the chapter, we discuss some of the finer points of the practice, including both tips for more effective sketching, as well as considerations for the conduct of the process beyond the sketch itself.

3.1 The basics

3.1.1 Types of decisions

It should come as no surprise that there are different types of environmental management decisions and that these typically require somewhat

Structured Decision Making: A Practical Guide to Environmental Management Choices, First Edition. R. Gregory, L. Failing, M. Harstone, G. Long, T. McDaniels, and D. Ohlson.
© 2012 R. Gregory, L. Failing, M. Harstone, G. Long, T. McDaniels, and D. Ohlson. Published 2012 by Blackwell Publishing Ltd.

Table 3.1 Common types of environmental management decisions.

Type	What is needed	Examples
Choosing a single preferred alternative	An informed, transparent and broadly supported solution to a policy or planning problem	Developing a management plan for an endangered species or airshed
Developing a system for repeated choices	A system for efficient, consistent, and defensible decisions that are likely to be repeated	Setting annual harvest levels or seasonal water allocations
Making linked choices	A way to separate decisions into higher and lower order choices, or those to be made now as opposed to later	Screening analysis followed by detailed evaluation; decisions that might be informed by investment in research
Ranking	A way to put actions or items in order of importance or preference, according to clear criteria	Prioritizing watersheds for restoration efforts; ranking projects to be funded
Routing	Grouping of actions or items into different categories, so they can be evaluated appropriately. This is often a preliminary action to more detailed assessment.	Screening out ineligible projects or proposals; identifying proposals for more detailed evaluation

different types of decision-making approaches. A first challenge is recognizing that what you have on your hands is a decision. In many cases, problems as initially presented don't look much like choices. They are a messy set of issues and opportunities accompanied by a general sense of urgency and usually uncertainty about how to proceed[1] – hence the usual reliance on a preliminary study to 'clear up' the uncertainties. Alternatively, the problem might be stated as a need for a 'plan' or a 'policy', but without any explicit recognition that developing a good plan or policy requires a choice among alternative plans or policies.

In this section we look at five common types of resource-management decisions (Table 3.1):

1   Choosing a single preferred alternative.
2   Developing a system for making choices.
3   Making several linked choices.
4   Ranking things (risks, problems, actions).
5   Routing things so that they receive appropriate treatment.

These five 'types' of decisions are by no means a comprehensive typology. They simply serve to demonstrate that there are different kinds of decisions, that some things that don't look like

decisions really are, and that the decision-analytic steps of SDM are generally applicable to a wide range of resource-management choices. We know of no decision problem that does not benefit from clearly defined objectives and evaluation criteria. And all decisions involve choosing among or ranking some alternatives – whether these are alternative actions you can do or alternative items you must deal with in some way. Making these choices, in all but the most trivial of cases, will involve thinking hard about the relative importance of multiple criteria – in other words, trade-offs.

Choosing a single preferred alternative

In this type of decision problem, what's needed is an informed, transparent, and broadly supported solution to an environmental policy or planning problem. This is the kind of one-off choice that managers face all the time[2]. Typically there is a pressing problem: a species is designated as 'endangered' or 'at risk', an airshed requires a reduction in emissions, or an integrated approach to watershed management is required. The candidate solutions will have important consequences on multiple objectives (well beyond the biological considerations for the species of concern or the health of high risk residents), there are quite literally dozens of potential management actions that need to be combined in logical combinations in order to make up a management 'plan', and choosing among the fundamentally different management options will involve difficult value-based choices.

This is the classic and most frequent application of SDM – we need to define objectives and measures, identify creative alternatives, estimate their consequences, clarify uncertainties and make collaborative, value-based choices. We are likely to present the alternatives and their consequences in the form of a consequences table to help inform decision makers about key trade-offs. The most common error is probably failing to treat the development of the plan as a decision opportunity – that is, developing a single plan rather than exploring alternatives based on value-focused thinking, or launching a scientific study and assuming it will produce that single best plan or policy, or asking for public input and hoping that a popular – and sensible –choice emerges as a clear winner.

Developing a system or mechanism for making repeated choices

Sometimes repeated decisions that should be relatively routine take up enormous amounts of management time and energy – examples include allocating annual budgets, setting annual harvest rates, and granting routine permit approvals. From a decision-making perspective, what's needed is a system or approach for making efficient, consistent, and defensible decisions and for clearly noting when one or more characteristics of a choice mark it as something other than routine. This implies a clear set of criteria, their relative weights, and possibly a formula-based approach for calculating suggested management actions. And it should be remembered that although these decisions may be routine from the standpoint of the management agency, they can matter a great deal to a variety of stakeholders. Setting annual hunting harvest rates, for example, may

significantly affect the welfare of guide-outfitters, environmental advocates, and local residents. These stakeholders may react to announcements from the management agency, which in turn is likely to be sensitive to media stories and politically-motivated suggestions. For all these reasons, it is important for any rule-based decision-making system to reflect logic that is defensible, both in terms of the underlying analysis (e.g. calculations of sustainable yield) and process (e.g. perceptions of fairness), and is able to incorporate unusual circumstances or unique aspects of the decision context.

Making linked choices

Sometimes we need to make a higher order or strategic level decision before doing a more detailed evaluation of options. Imagine you move to a new city and are choosing where to live. A first order decision is probably whether to rent or buy. Criteria designed to help you make that decision may include the length of time you expect to stay, the availability of suitable rental housing, the expected rate-of-return on residential real estate over the next 5 or 10 years, and – of course – the amount of savings you currently have in the bank. Once you make the higher order rent vs buy decision, you will begin a detailed exploration of alternatives in the rental or purchase category you have chosen. Now you have a new and different set of criteria: square footage, number of bedrooms, distance to schools, etc.

There are many parallels in a resource-management context. For example, you must make a strategic level decision about whether to revise or entirely rewrite a piece of legislation before you decide on the details of the changes. Or you might screen alternatives based on some higher order criteria before undertaking a more detailed analysis. As one example, a decision process for a proposed energy development might first screen alternatives based on cost and greenhouse gas emission reduction. This initial step could eliminate a large number of alternatives, allowing subsequent and more detailed analysis to focus on a smaller number of possible plans. Recognizing that the decision involves linked choices – first this, then that – helps to streamline the decision process, whereas the failure to develop a hierarchy of decisions can lead to wasted effort and time spent on the detailed analysis of alternatives that might have been eliminated early on from further consideration. From a decision-making perspective, this is important because it also can change the role of stakeholders and suggest that some participants either no longer need to be involved in further deliberations (because the focus of the discussions has shifted) or that new participants need to be added.

Another common type of linked or sequenced choice is when a decision needs to be deferred until new information becomes available. A multi-stakeholder process might begin and several meetings could be conducted in order to define the scope of the issue and develop at least an initial understanding of objectives and performance measures. If there are important but easily resolvable uncertainties affecting the development and estimation of performance measures, the process may pause for several weeks or months, with participants reconvening once the required additional information has been collected.

Ranking or prioritizing

Sometimes we need to rank before we can choose. We could be ranking alternatives, hazards, projects, or any other item, usually according to multiple criteria. We still need to define what matters and be clear about how ranking will assist a decision, identify the range of things to be ranked (which could be items, problems, projects, or actions), and decide on the relative importance of the criteria. In such cases, a common practice (and mistake) is to simply identify all the individual items to be ranked, score them, put them in rank order, and then pick off the top items until a budget is reached. However, there are often broader considerations, at what is called the 'portfolio' level. That is, in addition to objectives and evaluation criteria for the individual actions, there are evaluation criteria for the set of actions as a whole. There may for example be an interest in achieving an equitable distribution across geographic areas, departments or interests, a mix of short- and long-term benefits, or a mix of regulatory and market-based actions.

There are two recurring problems with ranking items. First, enormous amounts of resources and time are spent generating the list of ranked items, leaving little or no time and resources to develop thoughtful management plans. Second, sometimes the highest ranking items on the list are those about which you can do little or nothing at all. On the other hand, more moderate risks might be very amenable to easy cost-effective solutions. The net risk reduction from managing these more moderate risks – the low-hanging fruits – might greatly outweigh the risk reduction achievable by targeting the highest risks. This leads us back to the central point of this section: if you are a resource manager, all paths lead to a decision. If you set it up that way from the start, you greatly increase the probability that you will allocate your resources more wisely and focus your analysis more appropriately.

Screening or routing

Sometimes what is needed is not a ranked list of items, but a method to ensure that different items or actions receive appropriately different treatments. The task is to route them into coarse categories, based either on holistic judgments or on some kind of rule-based or formula-based logic. A simple example is the kind of triage managers do regularly to quickly group candidate actions into Must Do (no-brainers, implement immediately), Could Do (needs more analysis), and Don't Do (no further consideration).

Other applications can be more complex, using an explicit set of criteria. Using SDM in these cases involves:

Text box 3.1:   Ranking candidate restoration projects

Federal and provincial resource-management agencies in British Columbia used SDM methods to develop a systematic approach to prioritizing candidate restoration projects within watersheds that have been altered by the development of hydroelectric dams. At the broadest level, the challenge is to narrow in on those restoration program investments that offer the best potential to achieve the joint management objectives of the participating agencies. To begin, we distilled a set of strategic objectives from the core goals and mandates of the lead management agencies:

Conservation   Maintain or improve the status of species or ecosystems of concern.
                  Maintain or improve the integrity and productivity of ecosystems and habitats.
Sustainable use   Maintain or improve opportunities for sustainable use, including harvesting and viewing.
Community   Build and maintain relationships with stakeholders and aboriginal commu-
engagement   nities.

Key questions at this stage of the planning include: what are the priority species to target restoration efforts towards?; which investments in habitat-based or species-based actions are most likely to meet the strategic program objectives? Managers were then asked to assign scores on simple five-point scales to express the degree to which restoration of candidate species was expected to support each of the strategic planning objectives – conservation, sustainable use and community engagement. This initial sketch laid the groundwork for later development of an interactive spreadsheet-based tool which combined and weighted ratings across these three objectives to identify priority species for target restoration investments.

---

1 Figuring out that what you need is an explicit but relatively coarse routing process.
2 Setting criteria for determining the appropriate routes.
3 Identifying the items or actions to characterize.
4 Evaluating them against the criteria.
5 Defining the weights, formulas, or logic to apply to the criteria for routing items to the appropriate categories. It's often done as a scoping tool, to determine which items should go into a more detailed evaluation phase.

3.1.2   Decisions within decisions

Sometimes environmental decision problems involve decisions within decisions. If you're developing an air quality management plan for example, then what the decision is fundamentally about is identifying, evaluating, and selecting air quality management actions. But within that decision there could be several smaller decisions. You may need to decide who will be consulted and how. You'll have a set of objectives and alternatives specific to designing the consultation process. At some point in the process, you may need to identify and evaluate studies to inform decision making. The objectives and alternatives for that will be different than those for the broader decision about management actions. You may have to screen actions that are eligible for consideration, on say cost or feasibility considerations. Each of these sub-problems can be a mini-SDM process, conducted within the context of a larger decision problem. We won't get into the details of that now, but return to it in subsequent chapters. Suffice to say that once a group gets familiar with SDM, they build a vocabulary and skill at dealing with decisions and often can move quickly and efficiently through these mini-decisions.

3.1.3   An introduction to some structuring tools

SDM, not surprisingly, makes extensive use of simple but effective structuring tools from the decision sciences. Some of these tools are used effectively in complex quantitative analyses. However, even qualitative use of the tools leads to substantial benefits. They structure and focus both dialogue and analysis, promote discipline in thinking about objectives and consequences, and focus limited resources on decision-relevant areas of exploration.

We emphasize that they are tools that may be used in a variety of ways throughout a decision process to bring insight to specific sub-problems within the larger decision context, rather than overarching frameworks for structuring the whole decision. Many texts suggest the need to select which tool will be most useful in structuring your decision. While this is not terrible advice, we think it's a bit misleading. As a first step, the most useful organizing construct is the six core steps. What you need to do first, before getting too involved in specific tools, is think. Get a good picture of the decision context, the objectives, the alternatives, the consequences and the likely trade-offs and uncertainties. What are the sticky points of your problem? Is it that the objectives aren't clear? Or is it uncertainty? If it is uncertainty, is it a particular uncertainty associated with one or two of the possible outcomes? Or is uncertainty the dominant characteristic of the problem? Do you need to sequence decisions, in which case decision trees may be useful? Is the real issue going to hinge on differences between science and traditional knowledge? In which case, you may be looking for tools that help you explore competing hypotheses. Once you get

through the sketch, you will have a good idea where some more specific structuring tools will help you think more clearly.

In the upcoming examples, you will see some sketch-level versions of structuring tools such as objectives hierarchies, means-ends networks, and consequence tables. More complete worked examples of these and other structuring tools are found throughout the book. Table 3.2 summarizes nine of the most commonly used tools, and provides a very brief general introduction to what they are and how they typically are used. The short descriptions provided in the table are only intended to highlight these techniques, many of which (we expect) already will be familiar even if the name or key terms are slightly different. Many decisions will use a combination of these tools. In our experience, the most universally useful are means-ends networks, influence diagrams, and objectives hierarchies, followed closely by consequence tables and strategy tables. Relatively fewer decisions will make use of the more quantitative belief networks, decision trees, event trees, and fault trees.

3.2   Some practice with decision sketching

3.2.1   Steps in decision sketching

Once you've confirmed that you really do have a decision on your hands and you've roughly framed out what you need to deliver at the end of the decision process (a single preferred alternative, a system for deciding, etc.), it's time to do a decision sketch. Decision sketching means walking quickly through the first three to five steps of the SDM process at a scoping level. 'Quickly' typically means something between two hours and two days. At minimum, a decision sketch involves defining and framing the decision, identifying preliminary objectives, and identifying a range of possible alternatives. In some cases, it will also involve identifying candidate performance measures, characterizing consequences and uncertainties,

Table 3.2 Summary of structured decision management structuring tools.

Tool	What and why	Relevance/usefulness
Means ends networks	1 Clarify the relationship between means (actions we can take to influence outcomes) and ends (the outcomes we really care about) 2 Demonstrate where and how the concerns of stakeholders are addressed, especially concerns that are not the fundamental objectives 3 Can form a starting point for developing impact models, but is not usually of itself sufficient to define models	Relevant and useful in all management problems – ensures a sound basis for evaluating alternatives
Objectives hierarchies	1 A hierarchy of objectives that will be used to evaluate alternatives. Sub-objectives clarify what is meant by the objectives 2 Objectives represent the things that are fundamentally important to decision makers 3 A measure is selected for each objective or sub-objective, and used to report consequences on the objective	Relevant and useful in all multi-attribute problems – ensures a sound basis for evaluating alternatives
Consequence tables	A matrix that characterizes the consequences of proposed actions with respect to each objective. Generally also contain information about performance measures, used to more precisely define the meaning of objectives	Almost all multi-attribute problems – summarizes the information base to be used in decision making
Strategy tables	1 A tool for creating and organizing alternatives into logical packages or strategies 2 Provides a way to focus assessment on a small number of strategies that represent fundamentally different approaches, rather than dozens of individual ones	1 Helpful for engaging stakeholders 2 Encourages thinking about multiple candidate management actions simultaneously
Influence diagrams	1 Visual presentations of a chain of cause and effect. Can be used for environmental, social and economic systems 2 Influence diagrams, cognitive maps, impact hypothesis diagrams, are all part of a family of conceptual models that illustrate the relationship between causal factors and outcomes of concern. 3 Each of them can start from a simple means–ends diagram and add increasing levels of complexity. In a decision-making context, they are useful for identifying candidate performance measures and designing decision support models	1 All management problems 2 Often useful to construct one for each fundamental objective, as well as one that shows interrelationships between fundamental objectives 3 Useful as a precursor to probabilistic simulation, when variables are causally connected through deterministic functions
Belief networks, Bayes Nets	1 A tool for clarifying, structuring and documenting technical inferences. 2 Often begin with a simpler conceptual model of cause and effect, such as an influence diagram, and subsequently specify/quantify relationships	1 Helpful in problems characterized by high levels of uncertainty and controversy among experts. Useful when estimating the value of new information
Decision trees	1 A tool for illustrating and structuring a sequence of decisions and probabilistic events, with probabilistic consequences at the end nodes 2 Consequences are usually multi-attribute, although analysis can involve calculating either single- or multi-attribute utilities to enter into the tree. 3 Useful when there are discrete events or decisions that stand between the implementation of the initial decision and the eventual consequences	1 More limited application in environmental management 2 Problems likely to benefit include pest invasion problems, endangered species management, natural disasters (e.g. floods, earthquakes, etc.)
Event trees and fault trees	1 Tools for clarifying, structuring and documenting technical inferences 2 Event trees (sometimes called probability trees) begin by identifying an initiating event and trace subsequent events through to a fault or consequence. Probabilities and consequences can be assigned at each step of the tree 3 Fault trees start with the consequence of concern and trace causal events and combinations of events that would have to occur for the consequence to be realized.	1 More limited application in environmental management 2 Problems likely to benefit include some applications in endangered species management, pest invasion management, etc.

and identifying potential trade-offs. Doing this encourages participants to treat the problem as a decision right from the beginning: not a scientific endeavor or an economic valuation exercise, but a multidimensional decision problem seeking the best possible solution(s) to a management or policy challenge.

Undertaking an initial decision sketch can quickly and efficiently shift the focus of a decision, resulting in substantial savings of effort, time, and other resources. In a recent watershed management plan that was setting limits on how much water should be allocated for industrial purposes, over two dozen potential socio-economic impact areas were identified. However, a quick hydrological assessment revealed that the change in water levels in the river would result in a drop of no more than 100 mm across all the options under the worst conditions. This preliminary assessment helped screen all but two of the socio-economic issues and focused all the subsequent planning effort on better understanding the potential impact of water withdrawal on these two areas. This is not to say that the other socio-economic issues were not important, just that – in this decision context – they were not relevant because they were not differentially affected by the range of alternatives under consideration.

The decision sketch is often particularly effective when working collaboratively on a complex multi-objective problem. Different people usually come to the table with different framings of the problem and its possible solution. The decision sketch organizes the issue at hand in a way that helps to build a shared understanding of the key elements of the decision. A key outcome from a decision sketch is often a consequence table – or at least the skeleton of one. A consequence table links objectives, performance measures, and proposed actions and illustrates very visibly that, analytically, the emphasis is on developing information and tools that culminate in an estimation of the consequences of management actions – not baseline studies, or a ranked list of hazards, but the estimated consequences of management alternatives. In addition to providing insight

about the nature of analytical tools that will be required, the consequence table also helps to provide some early insight into some of the potential trade-offs and uncertainties. Identifying these early and thinking about which have the potential to become showstoppers, helps again to understand the relative priority to place on information needs.

There are three key steps to decision sketching:

1 *Frame the decision*. What is the decision and its relationship to other decisions? What deliverable is required from the decision process? Who is the decision maker? These seemingly simple questions can be harder to answer than they first appear (see 'Know Your Decision Maker' and 'Getting the Right Frame', below).

2 *Develop the sketch*. This is the heart of the matter. What is the range of objectives and alternatives under consideration? What information is known, what are the key information gaps, and what trade-offs and uncertainties are likely to be most critical?

3 *Plan the consultation and analysis*. Given what you've learned about the stakes and the nature of the problem, who needs to be consulted and how? What structuring and analysis tools and approaches are likely to be useful? What expertise is required? How will technical analysis and consultation be linked? A key benefit of SDM is that it provides a framework for both consultation and analysis and an understanding of what it will take to produce a defensible decision. Timelines and budgets can be allocated accordingly.

3.2.2 Decision sketching case studies

In this section we provide an overview of some illustrative decision sketches. The first demonstrates the overall process. The second sketch focuses on how sketching can bring clarity about the role and implications of uncertainty for the decision process. The third shows how sketching can be extended to a preliminary exploration of

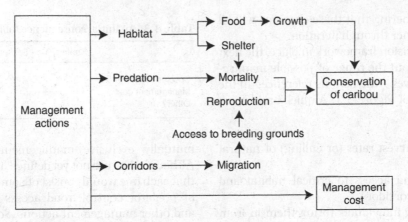

Figure 3.1 Initial influence diagram for recovery planning.

trade-offs. We've based the sketches on real situations, but changed or simplified some of the details, in part to improve the clarity and transferability of key messages, and in part because they really were just that – sketches.

Sketch 1: developing a recovery strategy[3]

Woodland caribou has recently been listed as 'at risk'. The stakeholders include provincial and federal government resource managers, local resource users, consulting scientists, and representatives of interest groups. All parties come to the table with concerns, specific ways of viewing the management issues, and (in some cases) directives from the institutions they represent. The first task is to separate these issues into means and ends, where ends are the outcomes we really care about and means are the things we can do to achieve them (see Chapter 4). To help with this, we first ask each participant to tell us what matters most to them with respect to recovery planning for this species. Through a series of questions and answers, we sketch out a simple influence diagram (see Figure 3.1).

In the course of a short sketching meeting, there is little chance that any initial means–ends discussion or influence diagram will be complete

in terms of all the possible factors affecting caribou. But that's not the goal at this stage. From Figure 3.1, we see that there is a set of management actions (so far not determined, this comes later) that affect three main factors important to species recovery: habitat, predation, and wildlife corridors. These direct effects influence things like food, shelter, mortality, reproduction, etc. The items on the far right of the diagram correspond to objectives, which we can provisionally state as:

1 Conserve caribou (although how to measure this is not yet specified).
2 Minimize management costs (to government, and perhaps also other stakeholders).

This is a good start. There's no need to waste effort fussing with the details at this stage. Whether the objective is 'conserve' caribou or 'minimize risk of extinction', whether (in other situations) an objective is named 'ecological health' or 'ecological integrity', whether we want to 'maximize' or simply 'improve' economic development . . . forget these subtleties for the moment. If the structure is wrong, you're rearranging deck chairs on the *Titanic*. The goal of the decision sketch is to get a framework on the table that's at least close, and then move on to the next steps – remembering that these elements will all be defined further through iteration.

With the decision framework in place, the next step is to flesh out the range of possible management alternatives – the actions referenced in the box at the left of Figure 3.1. A quick brainstorm yields this list:

1 Increase harvest rates (or culling) of natural predators.
2 Restrict road access to critical habitat and migration corridors.
3 Translocate individuals (bring them in from other areas where they are more abundant).
4 Implement captive breeding programs.

As the group brainstorms more alternatives, the influence diagram and the understanding of relationships among ends and means is updated to make sure everyone in the group is following the logic and building a common understanding of the things that influence species survival.

At this point all we have listed are really broad categories of alternatives – there are a variety of ways to alter harvest rates, a variety of road access management strategies, different approaches to captive breeding programs, and so on. Any comprehensive management approach will incorporate a suite of these, not a single one. At this point it's useful to think about provincial strategies for caribou management as opposed to individual actions. For example, one strategy would be to focus all attention and resources on conserving the species in one area where they have the best chance of long-term survival. Another would be to share effort and resources equally across the areas where they are found, reflecting an objective related to distribution, or perhaps an equity consideration. These are critical decision elements, but noting them is good enough for the first run-through. We now have some broad categories of actions, and a sense of how they might be combined into an overall strategy or management plan.

Next we put the objectives and alternatives into the shell of a consequence table. The three strategies shown in Table 3.3 represent three different,

Table 3.3 An initial consequence table.

	Strategy 1	Strategy 2	Strategy 3
Species conservation			
Management costs			
Other?			

mutually exclusive management alternatives. Although they are not yet defined, it's understood that each one would consist of some combination of predator control, road access management, and other management actions. So far, we know that as a minimum, evaluating and choosing among them would require information about their consequences on caribou, and on management cost.

By setting this rudimentary information up in the shell of a consequence table, we fundamentally change the focus of the deliberations and the emphasis of any further technical analysis. Instead of a program of studies and habitat mapping designed solely to explore the biological risk factors to the species, it's evident that we need an information base that could help to estimate the relative benefit of one set of management actions versus another set. This fundamentally changes the allocation of resources.

With the structure of the consequence table agreed to by everyone, an appropriate question is: if this table were filled in, would it summarize all the essential information needed to make the decision? Or are there other things that stakeholders or decision makers might want to ask? Given the set of (admittedly limited) alternatives on the table so far, what else might stakeholders and decision makers want to know before they could support a choice among the alternatives?

Asking these questions reveals that we're probably missing one or more objectives. For example, road closures could have an impact on local businesses. We're not sure how significant it would be, but it's clear that there would be concerns in the community about potential negative impacts. In response, a new objective is added: minimize negative impacts on local businesses (or maximize local economic development – again, wordsmith it later). Remember that participants didn't enter this problem with any notion that they had local business viability as an objective. However, given that one of the most plausible alternatives could affect local businesses (or would widely be perceived to do so), it becomes essential to add it. Hereafter, a key part of the decision process will involve designing road access alternatives in a way that minimizes negative effects on local businesses.

The discussion about provincial strategies leads to a realization that more understanding is needed about the relative importance of distributional concerns. Is it enough to protect a single population in one corner of the province, or is there an objective to maintain the existing distribution of the species? Will 1000 individuals in one location be viewed as more or less valuable, or about the same, as 1000 individuals distributed across several locations? One aspect of this question is technical in nature: distribution and the existence of multiple distinct populations would have implications for the technical assessment of extinction risk. However, this is as much a values questions as a science one – does it matter where caribou are preserved? This leads to the development of a more complete consequence table (Table 3.4), as shown below.

This consequence table shows that there are four major objectives, each with two sub-

objectives. Alternatives will be evaluated based on their effects on caribou populations, predator populations, government cost, and local business. For caribou populations, effects on both species abundance and distribution will be assessed. For predators, effects on both wolf and cougar will be assessed. Government costs will be assessed and for local businesses, both inconvenience effects and permanent closures will be considered. Note that we don't yet have actual measurable metrics in the table – how will we measure and report the effects of the management alternatives on distribution, for example? But we do have a rough outline of the things that matter. This is an important step forward: we now know that we need to figure out how to estimate and report these things and not other things, which helps to focus scarce intellectual, administrative, and financial resources.

The next step in the decision sketch is to think about how to fill in the cells of the table. This aspect of a decision sketch provides a necessary reality check: if critical information is impossible to obtain within the required time frame or is highly uncertain (remember the context is development of a recovery plan for an endangered species, so there is likely to be some urgency), then it may be prudent to consider a different performance measure. From Table 3.5 we see that some of the required consequence estimates can be produced from existing data and modeling, while others require additional modeling or expert judgments in order to make reliable estimates.

Table 3.4 A more complete consequence table.

	Strategy 1	Strategy 2	Strategy 3
Caribou populations			
Abundance			
Distribution			
Predator populations			
Wolf abundance			
Cougar abundance			
Government cost			
Short term			
Long term			
Local business			
Inconvenience			
Business closures			

Table 3.5 Objectives, candidate measures, and potential method of estimation.

Objectives	Candidate measures	Source of informvation
Caribou	Expected abundance in 2020	Population modeling
Local business	No. of business closures	Combination of limited data and expert judgment
Management cost	NPV of total cost over 2010–2020	Combination of limited data and expert judgment

The decision sketch – completed in less than a day – reframes the environmental management problem as a decision. Participants can now work with something close to the full set of objectives. They have a general understanding of the nature and scope of the alternatives and the work needed to develop them further. It is clear where information is available and where new data, models or expert judgments are required.

What are some of the immediate implications? First, we need an economic assessment of the anticipated consequences for businesses of seasonal road closures or full road decommissioning. Would businesses simply adapt by using other nearby viewing areas? Or would this put them out of business? When we ask this question in the decision sketching workshop, we learn that there is very little data on hand about local businesses and how they would be affected by road closures. However, based on past experience, we can expect that there will be enormous public outcry based on the fear of potential economic impacts. Perhaps even more important, we realize we need to explore the range of management alternatives available. Could roads be closed for a particular period of time? Can closures be alternated? Could the area be zoned spatially?

Another important outcome is the realization that the make-up of the current project team is inadequate – it is largely made up of biologists. There are no members of the team with economic expertise and none with expertise in tourism management, yet it's now clear that these will be essential in developing and evaluating potential solutions.

This decision sketch is about developing a recovery strategy for caribou. But it could be about choosing an energy technology to address climate-change concerns, or about deciding whether or not to implement prescribed burns to reduce wildfire risk, or about choosing mitigation options for an oil spill, or selecting a site for a waste treatment plant. Treating the problem like a decision right from the start fundamentally changes what you do, who you do it with, and how you allocate resources.

Sketch 2: working with uncertainty

Some years ago, we were asked to help a group of scientists, managers, and stakeholders on a regulated river in the southwestern United States with a 'target-setting' exercise. Participants realized that they were managing the river for multiple objectives – protecting weak fish stocks, maintaining a viable rainbow trout sport fishery, managing sand and sandbars for the rafting industry, among others. They also realized that many of these objectives were in conflict – that is, changes to the flow regime designed to improve rainbow trout populations might be detrimental to other fish species, and so forth. This complicated the target-setting exercise and the participants were looking for a way forward. This was a complicated management problem, but the first step was realizing that it simply wasn't sensible to try to set multiple targets *a priori*. Instead, we convinced the group to treat the problem as a decision, and took them through a fairly comprehensive decision sketch. Over the course of two days, the group set objectives and performance measures, identified a range of management alternatives (including everything from minor modifications to flow releases from an upstream dam to dam structural changes that would manipulate the river's temperature regime), populated a consequence table using existing data and expert judgment, and examined trade-offs. They selected a short list of preferred alternatives, and spent the last half day talking about difficult and controversial value-based trade-offs.

When the process began, participants identified the main source of conflict as the trade-off between power revenues and outcomes on native fish populations. However as the sketch proceeded, the central conflict turned out to be the individual risk tolerances of the people at the table. At the root of the decision was the choice between one alternative that was a sure thing to enhance native fish stocks, but only modestly, and another that could probably enhance the native fish stocks considerably, but also (if things did not go as expected) carried a small probability of driving them to extinction.

For the purposes of exploring how decision sketching can help to expose and understand the role of uncertainty in a decision, we're going to modify and simplify the problem down to two objectives (maximizing native fish abundance and maximizing power revenues) and two alternatives. Participants clearly have demonstrated that the most probable limiting factor to fish recovery is low water temperatures. The first alternative is therefore a flow change that will create pools of warmer water that can be used by native fish. The second is a structural alteration to the dam that would change the depth at which water is withdrawn from the upstream reservoir, resulting in a complete warming of the downstream temperature regime. The consequences of the first alternative are well understood. The second alternative is another matter altogether; by modifying the depth at which water is released from the upstream reservoir into the downstream channel, the entire temperature regime of the river is expected to change. Experts advise that the most probable outcome is that the abundance of native fish stocks will rise significantly but there is a small (albeit non-zero) chance that warm water exotic species, currently excluded from the river by their intolerance to the cold water, might enter the system and out-compete the more sensitive native species. So the most probable outcome is that the abundance of the native fish stocks will increase threefold, raising it above the listing threshold and out of danger of extinction. But it is also possible that the competition introduced by the newly-entered exotics will drive them to extinction. The estimated probability of extinction doubles under this alternative (Table 3.6).

Presenting the decision this way fundamentally changed the nature of the discussion. Prior to the consideration of uncertainty, the central trade-off was between power revenues and native fish conservation: how much electricity production should we forego to increase native fish abundance? With uncertainty, the central trade-

Table 3.6 Initial consequence table for flow decision.

	Alt 1	Alt 2
Native fish		
Expected abundance	Small increase	Large increase
Probability of extinction	No change	Doubled
Power production revenues	$40 million	$42 million

off became one rooted in risk tolerance: how much risk are we willing to incur in our effort to increase fish abundance? Even among confirmed fish advocates, the preferred alternative was not obvious. Insight gained through the up-front decision sketching exercise resulted in a shift in effort for the remainder of the decision process toward:

1 Estimating consequences probabilistically: in particular, it's not enough to know the range of possible values for fish abundance in this case. Decision makers also need to know differences in the probability of extinction across the management alternatives.
2 Developing strategies for dealing with the worst case or 'surprise' situation – specifically for developing strategies for removing the invasive species if and when it appeared.
3 Exploring the risk tolerances of participants – in this case, reaching agreement will depend on dealing with stakeholders' different risk tolerances.

From a technical perspective, dealing with uncertainty is often one of the most difficult tasks. In this decision sketch, participants went on to express consequences probabilistically. In other cases, participants choose to use a stand-alone objective related to maximizing 'robustness' as a means of dealing with uncertainty. There is no standard procedure or approach. Decision sketching can help to expose which uncertainties are likely to be important, and which aren't, and provide insight into how studies can be focused to bring clarity to the decision with a minimum of resources.

Sketch 3: previewing trade-offs

Most decision sketching focuses on the first three steps – clarifying the context, identifying objectives and alternatives, and getting a rough, mostly qualitative understanding of what kinds of trade-offs and uncertainties are likely to become important. Sometimes, however, it's useful to take the sketch all the way through to making some preliminary trade-offs. There are two main reasons for doing this. The first is to really focus and refine understanding of information requirements – for example, when there are many uncertainties and there's a need to focus limited resources on studies or analysis that support critical uncertainties. As part of a water use planning project, for example, we led participants through the first steps of the process, did a preliminary trade-off analysis to identify alternatives that warranted further attention, identify key uncertainties, and scope out studies. Studies were executed in that same year, and the decision process re-convened. The second good reason is to build support for an SDM approach. If SDM is unfamiliar and the language and ideas of trade-offs are scary to participants, then there's nothing like a quick illustration of how it works to help people see whether and how SDM might help. Below we describe a sketching process that built support for a more in-depth process for Atlantic salmon.

A team of decision analysts from the US Fish and Wildlife Service and the US Geological Survey (USGS), have pioneered the use of decision sketching – or a variation of it that they term rapid prototyping – for a variety of resource planning problems involving natural resource management agencies. A good example is a workshop held in 2007 for Gulf of Maine endangered Atlantic salmon populations[4]. The four-day intensive workshop, which involved participants from the three lead management agencies with authority over Atlantic salmon (USFWS, National Oceanic and Atmospheric Administration [NOAA], and the State of Maine), used SDM techniques to address the short-term goal of avoiding extinction while simultaneously outlining an overall strategy for long-term recovery.

Atlantic salmon in the Gulf of Maine were listed as endangered in 2000, with only about 100 wild adults believed to have returned in 2005. It is, therefore, no surprise that the fundamental objective outlined at the start of the workshop was to maximize the likelihood of persistence of wild Atlantic salmon. Three other objectives also were emphasized: maintain genetic diversity, maintain populations in all currently occupied rivers, and restore ecosystem functions (which was added part-way through the workshop, a good example of iterative thinking). Although it was recognized that these four objectives, broadly representative of the different stakeholders' interests, were preliminary, the discussion helped introduce participants to a SDM decision-organizing framework: the problem was structured in terms of objectives and performance measures, and this foundation was used to identify leading potential recovery actions.

A consequence matrix was used to assess these four actions (fish passage, riparian buffers, adult stocking, and juvenile smelt stocking) against the objectives, as shown in Table 3.7. For purposes of this sketching workshop, the goal was not to gather perfect information for each cell of the consequence table but, rather, to see if these objectives helped to distinguish among the alternatives and to see if any of the actions could quickly be eliminated because they are 'dominated' (outperformed) by other alternatives.

Participants were introduced to simplified methods[5] for assigning weights to objectives and exploring trade-offs. Actions were then ranked, and a sensitivity analysis illustrated how ranks might change if the objectives were weighted differently. Uncertainty was considered explicitly by using probabilistic performance measures for two of the objectives (population persistence and river occupancy).

Table 3.7 Initial consequence matrix for Atlantic salmon.

Objective	Performance measure	Alt 1: passage at Lower Penobscot Dam	Alt 2: riparian buffers on 3 Rivers	Alt 3: Adult stocking on Narraguagus R	Alt 4: smolt stocking on Sandy River
Population persistence	% probability	30	15	5	15
Genetic diversity	Heterogenizity (average)	0.35	0.4	0.3	0.3
Occupation of existing rivers	All rivers occupied (% probability)	80	80	90	100
Ecosystem function	Index	10	30	0	0

Text box 3.2:   Influence diagrams

An influence diagram is a problem structuring or conceptual modeling tool, developed by decision analysts, that graphically shows the key causal relationships between stressors or pressures that are of concern to managers, the actions (means) available to managers and the objectives or outcomes that matter to them (ends), as well as external factors and sources of uncertainty (see Schachter[6] or Howard[7]). As used formally by decision analysts, influence diagrams distinguish among three different kinds of components using different shapes or nodes: decisions (depicted using rectangles), chance events (ovals), and consequences (rectangles with rounded corners), connected by arrows or arcs to depict relationships; the interested reader is referred to Clemen[8]. However, in SDM, we use them at varying levels of complexity and formality. Sometime we use them on-the-fly, just to put some structure into complex discussions as they are taking place – as in Figure 3.1. Often these simple versions are iteratively refined and eventually become the basis for a quantitative model. We use them in a decision sketch to help stimulate and structure thinking about objectives, uncertainties, performance measures, alternatives, and consequences – topics we cover in more detail over the next five chapters.

A structured discussion using an influence diagram typically could highlight and explore:

1   The existence and plausibility of a causal mechanism for an hypothesized effect.
2   The different sources of knowledge that might inform the hypothesis.
3   Key uncertainties, including the existence of competing hypotheses.
4   The relationship between management actions and fundamental objectives.
5   Potential performance measures.
6   Information needs (data collection, modeling, expert judgment) relating to these measures.
7   Candidate management actions (which might include both research and on-ground actions).

Each connecting line in an influence diagram can thus be thought of as a hypothesis about the relationship between an action or other causal variable and an outcome. As a result, they are often called 'impact hypothesis diagrams' because they clearly link the things that can be affected via management actions and the things that people care about.

Some readers will be familiar with a whole range of process flow diagrams or logic diagrams, and the like. What's different about influence diagrams is that they illustrate cause–effect relationships – always. They're easy to use, intuitively appealing to a wide audience, and can help to demystify technically intensive discussions about complex social, environmental, and economic systems. They're among the most common structuring tools we apply in a deliberative setting.

In this case, the exercise served to build confidence in and support for an SDM approach to a more comprehensive decision process. Participants – including biologists from each of the three agencies – concluded that an SDM approach for this type of recovery planning was useful in that it encouraged participation, was pragmatic (i.e. worked within certain constraints) and decision-focused, made good use of the available science, and helped to reveal areas of agreement and disagreement. Progress made at this decision-sketching workshop greatly assisted a later, more extensive SDM process that contributed to development of a recovery plan for Gulf of Maine Atlantic salmon (described in Chapter 7).

3.3   Keys to effective sketching

3.3.1   Scaling the sketch to your problem

We said sketching can take two hours to two days. If you're working on a relatively routine problem, and you're just trying to scope it out, sketching may mean calling a few colleagues in to your office and working together through the key questions – what exactly is the decision? What are the objectives? What range of alternatives are we considering? What's in and what's out? And so on. You don't even need to call it SDM. And you will not do any preparatory work. Start simple, dive in, and try to keep going (i.e. don't let yourself get bogged down by subtleties – remember this is called a sketch).

The same goes for urgent problems. Even if you recognize that it's anything but routine, you still may have only a few hours to make sense of a problem – and a set of possible responses needs to be in front of your manager or cabinet by tomorrow afternoon. How can you organize the problem so that an informed decision will be made? Again, walk through the steps in a couple of hours with a flip chart – the insight gained about objectives, alternatives, and trade-offs will

not be wasted. In either of these cases (routine and non-routine), while it's tempting to rely on your most trusted colleagues or inner circle of managers, consider expanding the set of people involved to include those likely to have a completely different view of the decision. In some cases this will give you a useful new perspective on how to approach the decision problem. At minimum, you'll be armed with greater information about how to communicate to others about what the decision is – and critically, is not – about.

If you have a big decision with big consequences, and a bit more time to think about your approach to it, you will want to do some homework to make sure you get the most out of your decision sketching exercise. Select your participants carefully, identify and interview the decision maker(s) in advance of the session, and do some preliminary thinking about the decision frame, objectives, and alternatives.

3.3.2   Getting the right frame

Veteran decision analyst Ron Howard[9] conceives of decisions as a three-legged stool. The objectives, alternatives, and consequences are the legs of the stool – without these, there is no decision. Equally critical is where you place the stool. This is the decision frame.

We can think of decisions as occurring on a continuum from 'operational' to 'strategic'[10]. Strategic decisions are broad direction-setting decisions, involving exploration of a wide range of objectives and alternatives. Operational decisions are constrained by decisions already made, or by organizational structures and mandates. A key question is how broad or narrow to make the decision context. This will drive the identification of both appropriate objectives and a creative but realistic range of alternatives.

Consider a municipality that is exploring water-management issues. A strategic decision context might consider a wide range of policy interventions to improve the aquatic health of regional rivers – including everything from wastewater treatment options, to stormwater management, to land use practices in upper watersheds, to industrial permitting procedures and programs for neighborhood streamkeepers. The broad objective of 'maximize or improve aquatic health' might be appropriate (with some sub-objectives to clarify what is meant by 'health' – see Chapter 4). However, if the agency has either been mandated by a senior regulatory agency or is otherwise committed to upgrade wastewater treatment facilities, then the range of alternatives under consideration is limited to different treatment options and relevant objectives will be related to this more constrained context (e.g. minimize discharge concentration, maximize reliability, etc.). 'Maximize aquatic health' is a broad fundamental objective, reflecting the broad and strategic nature of the decision-making context. 'Minimize discharge concentration' is more means oriented but is appropriate in this context as it reflects the more operational context of a decision that is constrained by previous policy decisions. Understanding the decision frame is critical to both getting the objectives right, and getting a creative range of alternatives on the table. Decision sketching helps to work through these key components in rapid iterations, before investing greatly in a more detailed analytical or deliberative process.

How do you know if you have the 'right' decision frame? This is complicated, and in some cases the 'right' frame might involve multiple layers or perspectives. Here are three questions that will help (adapted from Clemen[11]):

1 Does the frame respond to the problem? If your problem stems from agricultural run-off, improving centralized waste treatment will not get you very far toward your objective of improving aquatic health.
2 Is it feasible to do the analysis and consultation required for the scope within the resources and timeline available? While we're believers in scaling the analysis to suit the problem and in

using informed back-of-the-envelope calculations to gain insight, you do need a basic level of expertise and resources to tackle the problem. If you don't have it, you're better off rescoping to something you can produce defensible results on.
3 Do you have the authority to make the decision? This means, do you (or whoever you will submit your analysis to) have the authority to approve and/or implement the alternatives under consideration? If you're in the wastewater treatment department, you probably don't have the authority or mandate to examine land use strategies in watersheds. If you or the current 'decision maker' don't have the requisite authority, you need to either change the decision frame or the decision maker.

3.3.3 Know your decision maker

These questions lead to another: who will actually make the decision? This seemingly simple question can be surprisingly difficult to answer, especially in government. There are two aspects to knowing your decision maker. First, who will actually decide. Is it a single individual or a group? Is there an expectation or commitment to shared decision making with external participants outside your organization? Second, when and how will the decision maker(s) be engaged in the process? Ideally, the decision maker will engage on three key questions before you embark on costly and time consuming analyses and consultations:

1 Confirm the objectives and draft measures – are these the key considerations that decision makers will want to see in order to evaluate and select alternatives? Do they address all relevant legal or policy mandates?
2 Confirm the range of alternatives – is the range appropriate to the decision context? Is it both broad enough to be creative and narrow enough to be realistic?

3 Confirm the consultation needs – what is the decision maker's expectations with respect to consultation? If controversial trade-offs are anticipated, what does the decision maker need to know about how stakeholders perceptions and preferences?

3.3.4 Who should participate in the sketch?

The answer varies with the management context and the mandate given to decision makers. In some cases, a decision sketch might involve only the primary decision maker and perhaps one staff person. In other cases, an outside analyst might also be involved. In either case, a quick run-through of the problem will help the decision maker to define the type of decision problem she faces (see Section 3.2), who should be involved, the resources needed for the effort, the nature of other linked decisions, and what might be possible within the specified period of time. In other cases, the decision sketch might be conducted with a small group of five to seven key individuals. And in still others, it could include a larger working group with perhaps 10–20 individuals representing a variety of stakeholder or agency perspectives. In these latter situations the objective will be to create a shared understanding among all participants of the key decision elements: which concerns and alternatives are within the scope of consideration and which are not, which of these might prove easy or difficult to define, where the big information gaps are likely to appear, what types of analysis and consultation might be appropriate, and so on. All of this informs the decision maker, who makes the call on how to proceed.

3.4 Beyond the sketch

In this section we touch briefly on some of the issues related to the design of the decision process that will follow the sketch. Some of these same issues are addressed in more detail in Chapter 11.

Here we're trying to give just enough context to answer design questions that may already have arisen.

3.4.1 Who should participate in the decision?

The answer is found in the sketch. Once you've clarified the range of alternatives under consideration and the endpoints that could be affected you can ask – who will care about these outcomes? This will identify the range of people that need to be involved, but not how they need to be involved. For example, who needs to be fully involved in the decision process? Who needs to be consulted? Who merely needs to be informed?

Some SDM processes will require the involvement of a full multi-stakeholder group at every step of the process – for example, instances where there is a commitment to shared or collaborative decision making. Others will be conducted largely as internal processes. For example, a group of managers and scientists within a particular government or company department may work through the SDM steps internally, going out for well-targeted consultation with either the public or technical experts on particular aspects of the decision. Good discussions on how to select participants in a multi-stakeholder process are found in Forester[12] or English[13].

3.4.2 Breadth, depth, and democracy

SDM is primarily intended for groups of 5–25 people, working intensively on a complex problem over a series of several meetings to develop well-informed preferences over the estimated consequences of different management alternatives. Yet we recognize there is always a balance to be faced between breadth and depth of consultation and analysis[14,15]. Depending on the objectives of decision makers and the regulatory context (in cases where clear legislation or regulations exist), it may be preferred to have: (a) 200000 people express their somewhat informed preferences through a referendum (broad and shallow), (b) 2000 people express their moderately informed preferences through a survey, or (c) 20 people express their very well-informed preferences through an SDM process (narrow and deep).

We are often asked what we do to ensure that these 5–25 people sitting on an SDM committee constitute a representative cross-section of society. The answer is, beyond some basic due diligence, we don't. Because participation in an SDM process requires time and effort, the participants tend to include primarily those with an 'interest' in the outcome – definitely not a neutral and representative set of participants. We think this is OK – great solutions come from people who are truly motivated to find them and, for many environmental management problems, one-shot meetings simply don't provide the opportunities for mutual learning necessary for a full exploration of alternatives and trade-offs. We offer the following caveats:

1 Efforts should be made to ensure that all the major groups who will be affected by the decision are represented.

2 Participants should be asked to make a commitment to learn about the interests of others, and to make recommendations in the interest of society. Efforts should be made throughout the process to think about the interests of potentially affected parties that are not at the table.

3 At key milestones in the progress of the SDM group, it may be important to hold public meetings or open houses can be held for information sharing and feedback – on for example, the set of objectives, the leading alternatives, and then the trade-offs among short-listed alternatives. Surveys can be used in conjunction with an SDM process for even broader input.

4 The need for and role of consensus in developing sound policy should be examined, and care taken not to overweight its importance. The lack of democratic representation at the table should be considered by decision makers when interpreting the outcome of the process and making final determinations.

3.4.3 *To agree or not to agree*

Gaining consensus is of course a great thing. Consensus among managers, agencies, or stakeholders provides decision makers with a very clear message about preferences and can help to ensure timely implementation of actions that might otherwise be controversial. However, pursuing consensus too vigorously, as an end in itself, can be problematic. And failing to achieve it doesn't necessarily detract from the value of the consultation process or the defensibility of the ultimate decision.

That said, there are instances where consensus is not just desirable, but essential – decisions involving two or more sovereign nations (such as transboundary water agreements, airshed management plans, or species recovery plans) or where there is a legal or policy imperative for shared decision making. Several of the examples discussed later in this book involve resource-management agencies from at least two countries, in which case attention needs to be given to the relevant laws of the individual nations as well as to shared legislation or agreements[16].

Just to complicate things, it is often the case that some of the participants in an SDM process have legal authority to approve or reject a particular plan, while others do not. This does not change their role in the SDM process. In Canada for example, the Federal fisheries regulator has ultimate veto power over proposed projects or plans that affect fish habitat. But when they sit at an SDM table, they participate in the same capacity as other participants at the table – they participate in setting objectives, identifying alternatives and evaluating consequences and trade-offs. Upon conclusion of the process, they are in a much more informed position to determine whether what is ultimately proposed for implementation should be approved – not only do they know what alternatives are available and how they affect their interests, they know how these alternatives affect the interests of others and to what extent they're widely supported. Similarly, in many jurisdictions, governments may have a legally mandated duty to consult indigenous communities that exceeds obligations to general stakeholders. There may even be explicit shared decision-making arrangements with these communities. Thus indigenous communities may serve as active participants in a multi-stakeholder SDM process and then use what they learn there to inform subsequent or parallel bilateral consultations that are conducted on a government to government basis.

3.4.4 Identifying analysis needs

Once the decision sketch is done, you're in a position to identify analysis needs – at least on a preliminary basis. Remember that SDM is most useful when it's iterative. You want to start off with a basic level of analysis that gives you reasonable information about implications for all the objectives. Once you know which trade-offs and uncertainties are key, you may target additional resources at clarifying those consequences and uncertainties. The sketch is your first cut at this. Common outcomes from a sketch include:

1 Reallocating study funds to one or more areas of potentially reducible or resolvable uncertainties that will affect the estimates of predicted consequences, rather than spending large amounts of money to 'study' sources of uncertainty that for this problem, in this time frame, are effectively irreducible (we return to this topic in Chapter 10, when describing Value of Information studies).
2 Reallocating study funds to cover a range of environmental, economic, and social outcomes, rather than just one or two. We typically find that the quality of information about local economic effects and social effects is surprisingly poor. A decision sketch serves to reduce the mismatch in information qual-

ity and helps to target analyses to things that will be most relevant to the choice at hand[17].
3 Reallocating study funds and planning time to identification and evaluation of policy instruments or management actions. Too often, this critical task is left to the final weeks of a multiyear decision process. The benefits of a very sophisticated understanding of underlying environmental resources are not realized if there is not a similar understanding of the efficacy of policy instruments.

3.4.5 Integrating analysis and consultation

It's important to think through how technical information will be linked to the consultation process. Will there be technical people as well as non-technical people in the SDM group? Or will there be a separate technical working group? How will technical experts interact with the SDM group? A common pitfall in environmental decision processes is the pitting of one expert against another, each providing compelling but contradictory views of likely consequences, and leaving non-technical participants in a quandary. More productive ways of working with experts are possible with SDM, and are discussed further in Chapters 8 and 10. Given the technically intensive nature of many environmental management decisions, effective mechanisms for making the best available information accessible to non-technical decision makers and stakeholders are essential.

3.4.6 Anticipating change

SDM is an iterative process. The sketch provides a good way of getting people to quickly see the overall nature of the decision, but it's a first cut. Subsequent deliberations may change the decision frame, the objectives, or the alternatives. Don't be discouraged – this is not poor decision making. It's a demonstration that the decision is being taken seriously and that learning is occurring.

Text box 3.3: Drafting a decision charter

A good way to formalize the outcomes of a decision sketch is to write up the results in the form of a decision charter. This isn't needed for every decision. But for major decisions – where deliberations are expected to span several months and involve a diverse group of interests – it's a good idea. In a typical case, a decision charter will summarize:

1 The pending decision, its relationship to other decisions, and confirmed scope constraints (what's in and what's out).
2 The scope of alternatives under consideration (not their details, just the general scope).
3 Preliminary objectives and performance measures.
4 Uncertainties and trade-offs that are expected to be central to the decision.
5 The expected approach to analysis and consultation.
6 Roles and responsibilities including who is/are the decision maker(s).
7 Milestones for decision maker input.
8 An implementation plan, budget and time-line.

3.5 Key messages

Whether the task is to evaluate risks to water quality, rank risks to biodiversity, allocate budgets, rewrite a piece of legislation, or conduct any other policy/management task, a first step is to recognize that the primary task is to make or support a decision. Significant progress in developing a better structured decision can be made quickly through use of a decision sketch, usually a two-hour or half-day quick run – with the participation of key stakeholders – through all the major elements that characterize the decision: initial objectives, candidate performance

measures, key uncertainties, likely trade-offs. Whatever is the original driver of the decision process – legislation about species at risk, a citizen complaint about air quality, a lawsuit concerning forest health, illness due to a waste disposal problem – the solutions will affect multiple stakeholders and multiple objectives, including a wide range of ecological, economic and social outcomes. Recognizing that your problem is a multidimensional decision, and treating it as such from the very beginning, will profoundly change what you do.

3.6 Suggested reading

Edwards, W. & von Winterfeldt, D. (1987) Public values in risk debates. *Risk Analysis*, **7**, 141–158. This article shows how decision-aiding tools can be used to help clarify public values in risk debates, using three examples: structuring energy objectives in Germany, diagnosing public value conflicts in oil drilling off-shore California, and identifying concerns associated with proposed changes to water allocations in Arizona.

Howard, R.A. (1988) Decision analysis: practice and promise. *Management Science*, **34**, 679–695. Introduces influence diagrams as a type of knowledge map that helps to capture, organize, and assess the diverse information held by individuals or members of a group.

Keeney, R.L. (2004) Framing public policy decisions. *International Journal of Technology, Policy and Management*, **4**, 95–115. Clearly outlines how to implement a structured process for eliciting values, identifying alternatives, and understanding consequences in the context of public policy decisions.

3.7 References and notes

1 Von Winterfeldt, D. & Edwards, W. (2007) Defining a decision analytic structure. In: *Advances in Decision Analysis: From Foundations to Applications* (eds W. Edwards, R. Miles & D. von Winterfeldt). pp. 81–103. Cambridge University Press, New York.
2 As every manager knows, a non-negligible percentage of 'one-off' choices will later be reviewed and revised, which means they are not as 'one-off' as first depicted.

3   A few years ago we were asked to help develop and evaluate management alternatives to include in a recovery plan for woodland caribou. This sketch draws on that experience, but adapts and simplifies various details.

4   Led by Mike Runge (USFW) and Jean Cochrane (then USFWS), using methods developed by Tony Starfield and colleagues at USGS.

5   Specifically, the SMART method (which stands for *Simple Multi-Attribute Rating Technique*; see Edwards, W. [1977] How to use multiattribute utility measurement for social decision making. *IEEE Transactions on Systems, Man, and Cybernetics, SMC-7*, 326–340).

6   Schacter, R. (1986) Evaluating influence diagrams. *Operations Research*, **34**, 871–882.

7   Howard, R.A. (1988) Decision analysis: practice and promise. *Management Science*, **34**, 679–695.

8   Clemen, R.T. (2004) *Making Hard Decisions: An Introduction to Decision Analysis*, 4th edn. Duxbury, Belmont.

9   Howard, R.A. (1988). Decision analysis: practice and promise. *Management Science*, **34**, 679–695.

10 Keeney, R.L. (1992) *Value-Focused Thinking: A Path to Creative Decisionmaking*. Harvard University, Cambridge.

11 Clemen, R.T. (2001) Naturalistic decision making and decision analysis. *Journal of Behavioral Decision Making*, **14**, 359–361.

12 Forester, J. (1999) *The Deliberative Practitioner: Encouraging Participatory Planning Processes*. MIT Press, Cambridge.

13 English, M. (2004) Environmental risk and justice. In: *Risk Analysis and Society* (eds T. McDaniels & M. Small). pp. 119–159. Cambridge University Press, New York.

14 National Research Council (NRC). Committee on Risk Characterization. (1996) *Understanding Risk: Informing Decisions in a Democratic Society*. National Academy Press, Washington, DC.

15 Gregory, R. & McDaniels, T. (2005) Improving environmental decision processes. In: *Decision Making for the Environment: Social and Behavioral Science Research Priorities*. (eds G. D. Brewer & P. S. Stern). pp. 175–199. National Academies Press, Washington, DC.

16 For example, SDM approaches were used as the basis for developing a recovery plan for endangered white sturgeon populations on the Upper Columbia River. The geographic area in question included two countries, the United States and Canada, and involved possible changes in the operation of hydroelectric dams operated by Bonneville Power Authority (based in the United States) and British Columbia Hydro Authority (based in Canada). The deliberations therefore required input from resource management agencies from the two countries and were informed by additional discussions with related resource management agencies.

17 Failing to identify key objectives early on in the decision process can result in a disproportionate amount of resources being invested in the evaluation of impacts on a single or a limited set of objectives (often, as in this case, the primary species of interest) while leaving other objectives (local economic effects of proposed management alternatives) virtually unexamined – this disregard of the multiple dimensions of the problem provides an obvious and compelling explanation for why many scientifically defensible management alternatives fail to be implemented successfully.

4
Understanding Objectives

Everyone seems to have something to say about the importance of objectives in guiding our choices. When we suggest to a new group that they need to start by setting objectives, often we're told 'we've already got those'. Usually, what people are describing is a terms of reference, a set of principles, or a list of things to do. Sometimes it's a cold hard regulatory target. Sometimes it's a short narrative from a 'visioning' process. But none of these constitute a set of objectives in the sense we are looking for. Because the decision sciences define 'objective' in a specific way, we need to take a moment to clarify what it is we mean.

The deceptively simple answer is that, in any resource planning or evaluation situation, objectives are the things that matter. We have yet to meet a resource-management context that didn't involve more than one objective; usually there will be four, six or eight things that matter to participants and to decision makers. The clear definition of these objectives is essential for good management decisions. If the objectives are vague, incomplete, or flat out wrong, then everyone will be working to resolve the wrong problem: as Yogi Berra is reported to have said, 'If you don't know where you're going, chances are you will end up somewhere else'.

If your decision context is choosing a restaurant to take your mother to for dinner on Friday night, for example, your objectives might pertain to the quality of the food, the convenience of getting to the restaurant, the price, and how long it may take to get served. You could compare the likely performance of different restaurants on these objectives by looking at their quality rating in a local restaurant guide, the time to travel there by car, the average price of a meal, and the ratio of staff to diners. In an environmental management context such as cleaning up pollution in a coastal estuary, your objectives might include improving water quality, keeping the financial cost of actions low, and enhancing recreational opportunities. The performance of alternative approaches against these objectives might be characterized by corresponding performance measures for each objective, such as the annual reduction in units of pollution entering the estuary, the discounted dollar cost of clean-up actions, and the number of recreational user-days for boating and bird-watching.

Brain surgery this is not. However, clarifying objectives is not always this simple. Many of the things that matter most when making a choice are qualitative in nature and are closely tied to an individual's perceptions or a community's cultural values. When you're choosing a restaurant, for example, the ambience might be as important as the price. But what are the elements

Structured Decision Making: A Practical Guide to Environmental Management Choices, First Edition. R. Gregory, L. Failing, M. Harstone, G. Long, T. McDaniels, and D. Ohlson.
© 2012 R. Gregory, L. Failing, M. Harstone, G. Long, T. McDaniels, and D. Ohlson. Published 2012 by Blackwell Publishing Ltd.

of ambience that you value? Is your definition of ambience the same as that of other people? In environmental management, decisions often hinge on these more qualitative considerations. What is the effect of a proposed development on a community's sense of place? How should a plan assess the effects of a development on the spiritual value of a burial ground for a local aboriginal community? So strong is our usual reliance on the scientific paradigm and our attachment to 'objective' science-based decision making, that these more qualitative values are often either overlooked or, at times, explicitly excluded. Whenever that happens, something that is important to people – something real, that matters to individuals or to their communities – is left out and, as a result, subsequent analyses of the anticipated benefits, costs, and risks of proposed actions are incomplete and the quality and defensibility of the decision is compromised.

SDM does not deliver value-neutral decisions. It does make explicit the subjective basis for difficult choices. Rather than sidestepping values that are difficult to measure or articulate, an effective decision-making process provides both analytical techniques and an overall framework that allows them to be put on the table alongside more conventional objectives (see Gregory, Failing & Harstone[1]). Scientific and economic considerations retain the importance they've always had while less easily quantifiable social and cultural concerns are incorporated in an equally rigorous fashion. This serves to level the playing field and to provide a visible entry point for community members, aboriginal resource users, and others who have often felt excluded from environmental planning processes.

Before going too far down this road, let's back up and cover some basic concepts about the use of objectives in SDM applications. First we'll introduce some core ideas about objectives and sort out some terminology. In the second half of the chapter, we discuss some of the trickier aspects of applying these ideas in the complicated and disorderly world of environmental management.

4.1 The basics

4.1.1 What are 'objectives' in structured decision making?

Objectives are concise statements of 'what matters'

In SDM, objectives are concise statements of the fundamental interests that could be affected by a decision – the 'things that matter' to people. Together with performance measures, they become the basis for creating and for evaluating management alternatives. For use in decision making, the statement of objectives can be kept pretty simple. Essentially, they consist of the thing that matters and (usually) a verb that indicates the desired direction of change:

1. Increase revenues to regional government.
2. Reduce the probability of extinction of wild salmon.
3. Minimize emissions of greenhouse gases.
4. Maximize year-round employment.
5. Minimize adverse effects on subsistence food gathering.

If the verb causes controversy, you can omit it provided the directionality of the objective is clear – that is, whether more or less is preferred, all else being equal. The verb really can become a point of controversy. The difference between 'improve' and 'maintain' embeds a value judgment about the outcomes of the process, and it's way too early to be having that conversation. People may object to 'maximize' and 'minimize' arguing that it's not possible to maximize or minimize and make trade-offs at the same time. So, the practical advice is, as long as the preferred direction is clear, you can move on.

Decision-relevant objectives are context-specific

This means they are defined for the decision at hand, not for universal usage. If we did not say that objectives and measures were context-specific, then it would be difficult – in many cases, impossible – to get agreement on them. By carefully structuring objectives and measures, we can take important but ambiguously defined things that matter – like biodiversity, visual quality, river quality and spiritual quality – and define them, *for the purposes of this decision*, with objectives, sub-objectives and performance measures. This frees us from endless philosophical and technical debates about the proper, universal narrative definition of 'river quality', or 'ecosystem health'. It takes us to a very practical place where the exact name we give the 'thing that matters' is less important than the clear definition given to it by the sub-objectives and performance measures we will use to assess it for the purposes of *this* decision.

Objectives are not targets

In the decision sciences a target is a desired level of performance towards an objective. 'Create 1000 jobs' is a target – a specific quantitative level of performance we want to achieve. 'Increase employment' is an objective (employment is the thing we want, more is better than less). Choosing a particular target usually has implications for multiple objectives. In SDM, we treat targets as alternatives and we compare the implications of one target versus another on the set of objectives. We include more on why we avoid setting targets at the objective-setting stage later in this chapter, and more on treating them as alternatives in Chapter 7.

4.1.2 What are good objectives?

Although there are no 'right' objectives, there are – at minimum – some that are more useful than others. A good set of objectives focuses decision makers on what matters in terms of outcomes of the decision and helps to identify and evaluate alternatives. In addition, a good set of objectives provides a sort of score card to assess outcomes of initiatives. Here we list five properties of a good set of objectives.

1 *Complete*. This means that no essential objectives are missing. A good set of objectives will capture all of the things that matter in evaluating proposed alternatives. In most cases, this will include the key environmental, social, economic, health, and cultural outcomes that may be affected by the decision. In some cases, it may also be important to include process considerations relating to how the decision is made or how actions are implemented.

To know whether your objectives are complete, you need to consider the range of alternatives under consideration. When you start developing a recovery plan for grizzly bear, for example, you may discover that a primary means of creating habitat is prescribed burning (i.e. planned burning of forested areas in order to create better shrub/berry foraging). This may lead you to have an objective related to the impacts of such activities on neighboring communities (loss of tourism income or adverse health effects, for example, resulting from poor air quality), something you might not initially have considered.

2 *Concise*. This means that nothing is unnecessary or ambiguous. Similar objectives are grouped together and there is no double accounting. The harsh reality is that people can keep as many as 6–10 objectives in their minds, but any further additions just muddy the waters. Thus a good set of objectives should ensure that all the important consequences can be described with the fewest possible objectives and measures. Similar objectives are grouped together in hierarchies by creating sub-objectives that define the components of the general objectives.

3 *Sensitive*. This means that the objectives are influenced by the alternatives under consideration. If an objective shows the same level of achievement for all of the alternatives under consideration, then it is not useful in distinguishing among the alternatives. It's just more mud in the water.

4 *Understandable*. The objectives should be stated in a way that they are immediately

understandable to everyone and speak directly to the things that matter. Use commonly understood terms rather than scientific jargon. Avoid ambiguity but bear in mind that many objectives are only made understandable through the use of sub-objectives and eventually specific performance measures. For example, 'conserve biodiversity' is clearly stated, but may be interpreted in dramatically different ways unless it is more clearly defined by lower level sub-objectives and specific performance measures that define how biodiversity is to be understood in this decision context. As a result, conserve biodiversity might have a sub-objective to 'maximize the diversity of native plants' as measured by a 'species richness' index.

5 *Independent.* It is good practice to check that the objectives are also independent – or more formally, 'preferentially independent'[2]. This means that they contribute independently to the overall performance of an alternative, and you don't need to know what's happening on one objective in order to know how important another objective is.

If that sounds complicated, think of it in terms of green vegetables. Imagine you're serving both beans and broccoli for dinner. How important is it to you that your daughter eats a lot of broccoli? Well, it depends on how many green beans she eats, because what you really care about is not that she eat broccoli but rather that she eats some green vegetables, or even more fundamentally, that she gets appropriate nutrition. Thus you want an objective that speaks to this higher-level concern, and the trade-offs you're interested in when preparing or consuming meals are (likely) between nutrition and cost and perhaps time available for leisure activities (if we assume that a good meal takes longer to prepare), and not between consumption of broccoli and consumption of beans. Because preferential independence is almost always implicitly assumed by people using objectives in decision making, it's best to make sure that the assumption is valid to avoid

errors of logic when evaluating alternatives. It is possible to deal with objectives that are not independent, but more complex analysis is required. The good news is that carefully structuring your objectives – especially separating means from ends and grouping similar objectives – will almost always solve preferential dependence problems.

If that all still sounds complicated, don't worry about it. Make sure you have 'fundamental' objectives as opposed to 'means' objectives as described in the next section and you'll satisfy this condition well enough.

Noticeably absent from this short list of properties of good objectives is 'measurable'. Our experience has convinced us that objectives should state the thing that matters, whether we know how to measure it or not. At some point, performance measures will be attached, and yes, these more specific measures should be amenable to clear and consistent assessment. But when eliciting objectives in a decision problem that involves multiple environmental and social outcomes, the last thing you want to do is to brand a concern as illegitimate because it might be hard to measure.

4.2 Developing a good set of objectives

In this section we outline an approach to generating objectives that we can safely identify as a recognized 'best practice'. Based on Ralph Keeney's value-focused thinking approach, it has become the most widely cited and well-used approach to generating objectives in the decision sciences. As a decision maker or participant in a decision process, defining your objectives is one of the most important things you can do to improve decision quality. It's definitely an art, but it's an art that's easy to get good at by systematically practising five basic steps:

1 Brainstorm what matters.
2 Separate means from ends.

3 Separate 'process' and 'strategic' objectives from 'fundamental' objectives.
4 Build a hierarchy of objectives.
5 Test to make sure they are useful.

4.2.1 Brainstorm what matters

A good way to kick off a brainstorming session is to simply ask: what are we trying to achieve or what concerns are we trying to address? To encourage original thinking, it's usually best to ask people to write down their own ideas independently before starting to refine them as a group. By initially freeing each participant to search his or her mind without being limited by (or anchoring on) others' thoughts, the final result will be a more comprehensive set of objectives that more accurately reflects everyone's concerns[3].

If you're having trouble getting what seems like a comprehensive list, you may need to prompt people with more questions. Ask them what would make them really happy? Or ask them to role play and imagine themselves as other stakeholders. What would they be concerned about? If people have anchored on a particular solution, ask them to list what's so good about it; this list may contain great ideas for objectives. Conversely, if participants are in strong opposition to a proposed alternative, ask them why. What would they most want to avoid? What are the hidden agendas or political 'realities' that could thwart things despite a great analysis? The answers to these questions will yield information about objectives that haven't been stated yet.

If the list of issues becomes very long, participants can be asked to start to think about categories of concerns that should be considered and begin to organize sub-points under them.

In the early stages of a decision problem, you're likely to generate dozens (if not hundreds) of 'issues' – a complex mix of actions and objectives, hopes and fears, advice, and accusations. Without structuring, discussions tend to take on the form of what we've come to call the 'conversational swirl' (with a nod to colleague Basil Stumborg). Discussions quickly either bog down in details or blow out in a shotgun of thinly-related concerns, many of which eventually get forgotten or ignored. Taking the time to structure this swirl of issues into a set of 'objectives' and a set of 'alternatives' is essential.

4.2.2 Separate means from ends

The first step in structuring is to start separating these issues into objectives (what matters) and alternatives (what actions can be undertaken). On the objectives front, 'fundamental' or 'ends' objectives are the basic things that matter, the outcomes you really care about regardless of how they are achieved. Means objectives refer more to specific methods of meeting the fundamental objective.

Clarifying means and ends is one of the most important things you will do in developing a useful set of objectives[2,4]. To do this, we get a long way by asking only two questions: why is that important and how could we achieve that?

Consider a case where stakeholders are considering improvements to a park near an urban center. One of the suggestions, to fertilize a small lake, is immediately opposed by other stakeholders who don't want to introduce artificial chemicals into the area. So why might lake fertilization be important? One likely answer: because it increases nutrients in the lake. Why does that matter? At this point, it's often helpful to sketch out a means–ends network, while continuing this same line of questioning. As shown in Figure 4.1, increasing nutrients is important because it increases lake productivity, which in turn increases the number of fish, and so on toward the right hand side of the figure.

You know you've hit bedrock when the answer to the question of 'why is that important' is a perplexed shrug: 'it just is!' You've arrived at an objective that appears to be important, in its own right, regardless of how it was achieved. It's easy to see from this that the distinction between means and ends isn't black and white; it's a continuum. In this case, maximizing the fishing experience is a fundamental or 'ends objective', while increasing lake productivity is a 'means objective'.

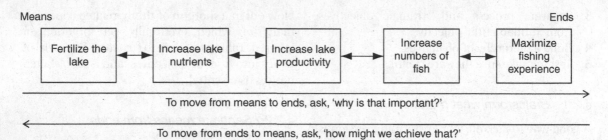

Means   Ends

```
┌──────────┐   ┌──────────┐   ┌──────────┐   ┌──────────┐   ┌──────────┐
│ Fertilize the │◄─►│ Increase lake │◄─►│ Increase lake │◄─►│  Increase   │◄─►│ Maximize  │
│   lake    │   │  nutrients │   │ productivity │   │ numbers of │   │  fishing   │
│          │   │          │   │          │   │   fish    │   │ experience │
└──────────┘   └──────────┘   └──────────┘   └──────────┘   └──────────┘
```

To move from means to ends, ask, 'why is that important?' ────►

◄──── To move from ends to means, ask, 'how might we achieve that?'

Figure 4.1 A simple means–ends diagram.

In Chapter 3 we talked about the tricky business of figuring out the right framing for your decision, which is directly linked to the choice of fundamental objectives. The right fundamental objective is the broadest objective that will be directly influenced by the alternatives and within the control of the decision maker.

Once we learn that the fundamental objective is to maximize the quality of the fishing experience in the park, then we can ask 'how else could we achieve a better fishing experience?' This would result in new pathways being drawn back toward the left side of the diagram in Figure 4.1, leading to the definition of other alternatives, positioned in parallel with 'fertilize the lake', such as stocking with hatchery raised fish, making lake access more convenient for people with boats, banning jet skiers, or perhaps improving access to another nearby lake. Through this process the person who first proposed fertilizing the lake will come to see that fertilization is an alternative, not a fundamental objective, and that it will need to be evaluated alongside other alternatives.

This all might sound a bit elementary. But remember that, when making decisions in a group setting, we need to make sure that everyone sees their concerns captured, can talk about them using a common language, and can understand at what point they will be addressed in more detail. When we talk about objectives, we're referring to the things that later will be used to generate and to evaluate management alternatives. By clarifying the distinction between means

and ends, particularly with the use of means–ends networks, people see that their concerns and interests are not being swept aside; those that are identified as alternatives are important, and will receive more detailed treatment later. But it's not just about process: distinguishing between means and ends is absolutely essential for thinking clearly about the evaluation of alternatives.

4.2.3 Separate fundamental and means objectives from process and strategic objectives

Does this two-way distinction between fundamental and means objectives cover all the objectives that regularly come up as part of environmental management choices? The short answer is no: two further types of objectives frequently are raised by participants. The first is process objectives, which refer to how a decision is to be made. Examples of process objectives include: maximize public participation in the decision, encourage implementation partnerships, complete the analysis within schedule and budget constraints, incorporate uncertainty.

Sometimes, however, things that look like process considerations will directly influence the design of a management alternative; in these cases they are fundamental objectives. Consider the case where the community that is likely to host the incinerator has little trust in the management authority. As a result, one objective of the facility design and siting process may be to 'build trust'. This concern could affect the design

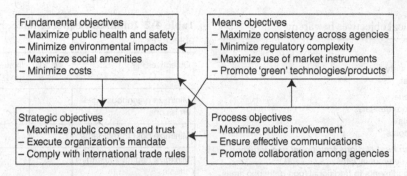

Figure 4.2 Examples of different types of objectives (adapted from Keeney[5]).

of the decision process but it could also affect the selection of alternatives, for example by favouring alternatives that involve local citizens as part of management teams over alternatives that are implemented unilaterally by the municipal government.

Strategic objectives, the second special type of objective, relate primarily to an individual's or an organization's own strategic priorities or direction. They are objectives that are relevant for all the decisions made over time by that organization or individual. Typically, strategic objectives relate only indirectly to the outcomes of the decision at hand. For example, corporations and some governments might have a strategic objective of 'increasing competitiveness' or 'maximizing public consent to operate' (which refers to building public trust and support for the organization). Such objectives are not likely to become shared objectives in a multi-stakeholder process. However, if the decision is an internal one, this objective could play an important role. It might even be useful (or politically important) to use the exact words as reflected in the organization's strategic documents in the statement of objectives.

A useful way to approach process and strategic objectives is to map them to the fundamental and means objectives of the decision at hand. As shown in Figure 4.2, fundamental objectives can influence the achievement of strategic objectives. Means objectives can influence the achievement of both fundamental and strategic objectives and process objectives can influence all of the other three.

4.2.4 Build a hierarchy of objectives

Once you've actually developed some fundamental objectives you may be dismayed to discover that they are rather vague. Now is a good time to ask the next set of key questions: what do you mean by that?; can we identify subcomponents of that?; or sometimes, what are the critical means of achieving that? For example, if we've identified a fundamental objective to maximize fish abundance, the sub-objectives might be related to various species or locations (maximize salmon, maximize whitefish, etc.). Alternatively, the sub-objectives might be related to critical 'means' (as in the water use planning example in Chapter 1, where the sub-objectives were maximize spawning habitat and minimize stranding). We need to disaggregate enough to be clear and to characterize the decision properly, without introducing unnecessary detail.

What typically emerges from this is known as an objectives hierarchy. A hierarchy for a fairly complex decision problem involving development and operation of an open pit mining operation is shown in Table 4.1. As often occurs, objectives are included for environmental, economic, health, and social/cultural considerations. In this

Table 4.1 A typical objectives hierarchy.

Maximize safety
Maximize operator safety
Maximize public safety

Maximize net revenue
Maximize revenue from ore sales
Minimize capital costs
Minimize operating costs

Minimize impacts on First Nations traditional use activities
Minimize impacts to ceremonial sites
Minimize access impediments to traditional food gathering areas

Minimize adverse environmental impacts
Minimize soil contamination from tailings
Minimize material waste
Minimize air emissions
Minimize impacts on groundwater

Minimize disturbances to recreational activities
Minimize disturbance to viewscapes
Minimize trail access blockages
Minimize noise from construction /operations

Table 4.2 Initial consequence table shell, showing objectives.

Objectives	Wood waste	Natural gas	Sewer heat recovery
Minimize greenhouse gas emissions			
Minimize cost to residents			
Minimize negative health effects			
Maximize security of supply			
Minimize noise			
Minimize loss of visual quality			

case all the objectives have several sub-objectives but this is not always the case.

The list of fundamental objectives needs to be both concise and complete, capturing everything important in as few objectives as possible. New objectives may well arise at later stages of the process, and these should be welcomed.

Importantly, this one set of objectives represents the objectives of all stakeholders. Sometimes we start out with one list of 'industry objectives' and another of 'community objectives' and so on. But ultimately, for collaborative decision making, a group needs to agree on a common set of objectives that all warrant consideration when evaluating alternatives. Different participants will weight them differently of course, but it's important to agree that all are relevant.

A good objectives hierarchy is a critical milestone in a decision process. If you're participating in a decision process that is beginning to look in detail at alternatives, and you do not yet have an objectives hierarchy that looks something like this – STOP! Even if you never assess consequences in a formal way, a good set of objectives will help to ensure that you can identify good value-focused alternatives.

4.2.5   Test to make sure the objectives are understood and useful

The litmus test for a good set of objectives is whether it is useful and sufficient for evaluating management alternatives. The best way to test them is to lay them out in the format that's needed to help make a decision – usually a consequence table. You don't need to have identified a detailed set of alternatives to do this; you just need to know the range of things under consideration. Suppose you're examining local energy supply alternatives for an urban community (see Table 4.2). An initial consequence table would clearly set out the range of alternatives under consideration (in this case, neighborhood-scale energy systems fueled by wood waste, natural gas and sewer heat recovery) along with your proposed objectives: minimize greenhouse gas emissions, minimize costs to residents, minimize negative health effects from local air emissions, maximize security of fuel supply, minimize loss of visual quality, and minimize noise (from trucking in waste wood).

Looking at a sketch of a consequence table, imagine it filled with data, text, or plusses and minuses in each cell. Ask yourself (and participants in the process):

1 'In choosing among these alternatives, are these really the things that matter?'

2 'If information is collected about the impacts of alternatives on these objectives, will you then have all the information you need to make an informed choice?'

4.2.6 Don't weight objectives (at least, not yet!)

Strictly speaking, this isn't a step of course. But trying to weight objectives at this point is such a common and serious mistake that we need to address it. First of all, as we discuss more fully in Chapter 9, you can't meaningfully weight objectives without information about the extent to which they're affected by the alternatives. (If the volume of emissions produced by different alternatives can vary by orders of magnitude for example, it will likely be weighted much higher than if it changes very little across alternatives.) But this is information (about predicted consequences) that you don't yet have. Second, by agreeing on a set of objectives, you've just built a shared understanding of what matters. Most, if not all of the people in the room will agree that all of the objectives are important things to consider. At this point in the decision process, you are better off building on this common ground of shared values than worrying about differences in how they are weighted. And third, there is simply nothing to gain in terms of insight about the decision by weighting objectives now – and perhaps ever. In some decision processes (as we'll discuss in Chapter 9) weighting objectives is essential to understanding key trade-offs and developing alternatives. In other situations a well supported solution can be found without ever explicitly assigning weights to objectives.

4.3 Working with objectives in environmental management processes

In this section, we cover some of the trickier territory related to identifying and working with objectives. Many of these insights will be familiar to seasoned practitioners of SDM. However, we've been bitten enough times by these issues to have learned that they merit special emphasis in the world of environmental management.

4.3.1 Do decision makers know what they want?

Once the intent is explained, we rarely encounter any opposition to the suggestion that a good decision process should involve developing a clear set of objectives. It might be expected that career decision makers would find this an easy task; after all, who should be better at knowing what they want than people whose job is to make choices? Yet when it comes to objective-setting, our experience is that decision makers are often in need of help. And we're not the only ones to say so. In three empirical studies of how well individuals are able to generate self-relevant objectives, Bond, Carlson, and Keeney[6] noted that participants consistently omitted nearly half of the objectives that they later acknowledged to be vital for the evaluation of alternatives. This finding held even when participants were given the opportunity to reflect, and it remained true when the participants were experienced strategic decision makers at a private corporation.

Decision makers also need help in clarifying exactly what they mean by the stated objectives. In a group context – for example, when stakeholders representing the multiple perspectives on an environmental management issue come together to seek a joint solution – participants may agree on the importance of an objective only to learn later that they understood the objective to mean profoundly different things. For example, scientists working on a fisheries restoration problem may quickly reach agreement on an objective of 'conserve native fish', only to learn they really have fundamentally different objectives: one seeks conservation of species diversity, another an increase in the production of harvestable fish, and a third avoidance of the low-probability collapse of a single endangered species.

Table 4.3 Illustrative situation of competing objectives.

Objective	Indicator	Alternative Plan 1	Alternative Plan 2	Alternative Plan 3	Alternative Plan 4
Maximize protection of panda bears	Expected no. of panda bears	2800	3000	3200	3400
Maximize community jobs creation	No. of full-time jobs created	100	50	20	10

The bottom line: decision makers often will not think they need help defining objectives. But they do.

4.3.2 How can it be useful to work with 'competing' objectives?

Suppose we are comparing hypothetical industrial development plans and are trying to balance economic benefits with negative effects on a plant or wildlife species that many people care about passionately, such as old-growth trees or polar bears or pandas. When we suggest to participants in a deliberative group that we're trying to simultaneously 'maximize protection of panda bears' and 'maximize economic development', some people look at us like we're nuts. How can we possibly do both? And, from a group process perspective, how can we expect an industry representative to agree that an objective is to maximize protection of pandas, or a conservationist to accept an objective that 'maximizes' economic development?

Although SDM does not offer a panacea, there are two aspects of a decision structuring approach that we find can greatly help in this situation. First, recall that people are being asked to consider public decisions from a societal perspective. While they are of course there to bring their own perspective to the table, they are being asked to work toward solutions that are in the best interest of society as a whole. This requires considering trade-offs across multiple objectives and finding an appropriate balance. An objective is legitimate and should be included if someone affected by the decision cares about it (and of course if it is sensitive to

the decision, as above). However, as noted earlier, each individual can (and will) assign a different importance (or weight) to each objective. Some people may give some objectives no weight at all if they believe that from a societal perspective, it warrants no consideration in the decision. Again, at the objective-setting stage, focus on identifying relevant objectives, not weighting them.

Second, recall that the ultimate task is to use what is known about objectives (and performance measures and consequence estimates) to develop a broadly preferred alternative. Table 4.3 shows a very simple situation where two objectives (pandas and jobs) are what we care about. All else being equal, we want more of (i.e. to maximize) both. It would be perfect if we could find a way to get lots of panda bears and jobs. A key goal of the structured decision process is just that – to explore how and to what extent different alternatives can help to achieve different levels of competing objectives. We first see if it's possible to iteratively improve the alternatives by weeding out inefficient alternatives and seeking joint gains (alternatives that deliver gains on multiple objectives). The remaining trade-offs are then examined to help groups find the alternative or alternatives that represent the best possible balance across the identified objectives. Or, in many cases, to identify a set of alternatives that all deliver an acceptable balance. Which is the best alternative in Table 4.3 depends on how you feel about the trade-offs involved. The table doesn't make the decision; it facilitates a discussion about the best balance between pandas and jobs by trying to maximize both and seeing what the alternatives tell us is possible.

If this seems obvious to you, then great. But be aware that it sometimes isn't immediately obvious to everyone.

4.3.3 The trouble with targets

Targets refer to desired minimum levels of achievement on key environmental, economic or social considerations, and typically operate as constraints on actions by eliminating those that don't achieve them.

The above discussion on conflicting objectives should provide some insight into why we avoid objectives that embed targets. Targets on an objective constrain trade-off explorations in a way that makes us blind to the consequences: if we can't have both 1000 hectares of shorebird habitat and 1000 hectares of salmon spawning habitat, then setting a target of 1000 hectares of shorebird habitat hard-wires a trade-off. And in terms of group dynamics, because the target embeds a value-based trade-off, you'll probably never get agreement on it at this early stage of a decision process.

This is an important point. The act of setting quantitative targets is itself a decision, a multi-objective value-based decision, and should be based on a good understanding of objectives, alternatives and the consequences associated with different targets. If the goal is to conserve a 'viable population' of a particular species then a key question is what constitutes a viable population. An appropriate course of action is to explore a range of alternatives, which might for example represent different levels of abundance or different probabilities of persistence. For example, Tear et al.[7] examined three alternatives that deliver respectively a 75%, 95%, and 99% probability of persistence of the species to 100 years. Each of these implies a different degree of risk for the species concerned. If the 99% alternative has no incremental costs or other negative consequences relative to the 95% alternative, then anything less than 99% may be unacceptable. If enormous economic hardships or severe consequences for another species are associated with the 99%

option, then 95% may be perfectly acceptable. The bottom line is that a key factor in deciding whether risks are acceptable is the availability of alternatives and an assessment of trade-offs across them.

Sometimes, of course, targets have been established by legal mandates, such as a requirement to achieve a minimum level of returns as part of an endangered species recovery program or to set aside a specific percentage of old growth forest. In such cases, the objective can be expressed in terms of the probability associated with achievement of the mandated quantity, which takes the discrete language of a target (yes/no to meeting a number) and replaces it with the continuous language of an objective (measured via changes in the associated probability).

However, unless the threshold is legally absolute and 100% non-negotiable, even this application serves only to unduly limit thinking. Suppose, for example, that an agency wants to restore at least 100 hectares of riparian habitat each year. The target is met if 101 hectares are restored but not met if only 99 hectares are restored. Depending on the relative costs of the alternatives, it seems likely (especially if there's any uncertainty about the predicted habitat) that the 99 hectare alternative could be preferred if it was substantially less expensive or had other merits, but that possibility wouldn't be explored under a 'targets' approach.

4.3.4 Should we worry about double counting?

The short answer is yes. The long answer is yes, but be careful (some things that look like double counting, aren't). And the practical answer is, yes, but sometimes it's the lesser of two evils. We treat each of these in turn.

Double counting in objective setting happens when we count what is essentially the same interest more than once. It is unquestionably a problem, for two reasons. First, it leads to errors of logic in the evaluation of alternatives – decision makers may overweight a concern that is represented

---

Text box 4.1:   How much is enough?

This is the million dollar question that plagues environmental regulators, managers, and advocates everywhere. It's at the forefront of controversy surrounding threshold-based approaches to managing cumulative effects in many parts of the world. Much effort has and continues to be put into the science of thresholds – the identification of indicators and dose-response relationships in order to understand the influence of development pressures on environmental change. But there has been limited success with implementation.

There are two key problems. First, only rarely are there any bright lights or thresholds that demarcate a line between what is 'enough' and 'not enough'. Instead, what exists is a continuum – more habitat is generally better, but there is rarely an objectively definable point at which one can say with certainty that 'enough' habitat has been protected. As a result, no amount of analysis will change the fact that the regulator charged with approving a development project needs to make value-based judgments about how precautionary to be. While some progress has been made in developing systems of tiered thresholds that recognize uncertainty, very little progress has been made in implementing them.

This is because of the second problem, which is a failure to distinguish between setting thresholds – which provide *information* useful for interpreting the significance of change – and making a *decision* to limit development. Such a decision must deal explicitly with both value based trade-offs (across environmental, social, and economic objectives) as well as fact-based uncertainties. Science panels are regularly established to set 'science-based' performance thresholds: minimum populations of valued species, minimum set-asides of 'old growth' forest, maximum road densities. etc. Such initiatives fail to distinguish between providing information about possible response thresholds and deciding on limits. As a result they end in heated conflict between scientists, industry, and environmental advocates. As Tear *et al.*[7] note 'conservation objective setting often mixes scientific knowledge with political feasibility in such as way that one cannot tell where the science stops and the political pragmatism takes over' (p. 836). At the heart of this problem, more often than not, is the mistreatment of targets and thresholds.

The inconvenient truth is that there is no objectively right answer to the question of how much is enough. Regulatory agencies need processes that are socially robust – that is, rigorous, transparent and defensible from a public and legal perspective. They need to know that the answer to the question 'how much is enough' will be inescapably value-based and will need to reflect trade-offs among multiple objectives.

---

twice. Second, it unnecessarily complicates both analysis and deliberations. In environmental management problems already burdened by multiple objectives and high levels of complexity, this is not something to take lightly.

The following is a simplified example of some work we did recently in Australia, looking at protocols for ranking invasive pests in terms of their consequences on the national interest[8]. Like many risk-ranking protocols, this one assigned points for different types of negative consequences, and then calculated an aggregate 'score' to reflect the significance of a pest in terms of the national interest (presumably for the purposes of efficiently allocating management resources). The protocol assigned 'points' for impacts on

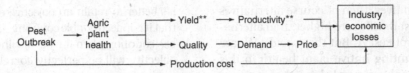

Figure 4.3 Simplified influence diagram linking invasive pest outbreak to economic loss.

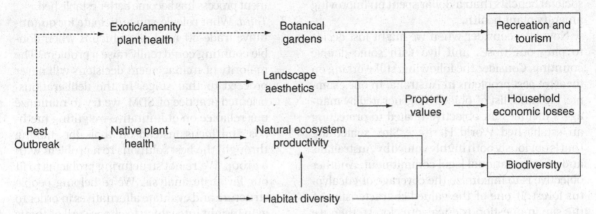

Figure 4.4 Simplified influence diagram linking invasive pest outbreak to multiple objectives.

'agricultural plant health' as well as for impacts on 'economic production'. When we asked 'why is agricultural plant health important', we were told, 'because it results in economic losses'. Figure 4.3 traces this relationship: a pest outbreak causes changes in agricultural plant health, which affects crop yield and quality, which affects productivity, demand, price, and which ultimately affects economic losses for industry. Since in this case, agricultural plant health is important only for its impacts on the economy, then a ranking protocol that counts both agricultural plant health and economic effects would indeed be double counting.

Now consider Figure 4.4. Here we show the (hypothetical but representative) effects of pest outbreaks on Australia's native plants and on amenity plants. Impacts on amenity plants affect recreation and tourism via effects on botanical gardens (amenity plants die, fewer tourists visit the gardens). Impacts on native plants also affect recreation and tourism, but via effects on land-

scape aesthetics (defoliation, etc.). Aesthetic losses also affect property values, resulting in household economic losses. Native plants also affect natural processes and ultimately biodiversity. We sometimes hear the argument that if we consider all three of the fundamental objectives at the right hand side of the diagram (recreation and tourism, household economic losses, and biodiversity) that we are double counting (in this case triple counting) the effects of changes in native plant health, since all of the objectives are significantly affected by native plant health. (we have even seen this argument in textbooks.) However, this is *not* double counting. Suppose that Pest A affects only amenity plants, and results in a loss of 1000 user days in recreation facilities. Pest B on the other hand, affects native plants. Let's assume it incurs the same losses for Recreation and Tourism as Pest A (1000 user days). In addition to these losses, it also causes household economic losses and biodiversity losses. If we are using all three objectives to evaluate alternatives, then of course alternatives that control Pest B will score higher than alternatives that control Pest A. But they *should*! We are not double counting native plant health in this case. We are simply acknowledging that improvements to native plant health have multiple benefits, and a dollar spent there will deliver greater societal benefits than a dollar spent on improving amenity plant health.

Now let's consider when we might just decide to plug our noses and live with some double counting. Consider the following, still working on invasive pest problems in Australia. In one example, an initial list of objectives generated by managers includes an objective related to protecting an established World Heritage Site, something that is obviously both highly valued by Australians and an international legal commitment. Another objective is to maximize the coverage of eucalyptus forest. If one of the valued characteristics of the site in question is eucalyptus forest, then we have some double counting going on. Separating these is possible, but could be both difficult and ultimately quite unnecessary. We suggest three key questions for deciding whether double counting is really a problem:

1 *First and most important.* Does the double counting expose an important value based trade-off that managers are likely going to face when it comes to working with stakeholders in evaluating alternatives? If the answer is yes, then you may be better off leaving it than trying to eliminate it. In the above example, managers might make decisions to allocate control resources to the World Heritage site and accept losses to the broader eucalyptus forest beyond the park borders, or they might accept impacts to the park in order to preserve larger tracts of non-park forest. Further, the actions taken within the park might be quite different than actions taken in the broader land base, in which case leaving the offending objectives in place will help to facilitate thinking about alternatives.

2 *Second.* How bad would it be to leave the offending objectives in for a while? Sometimes it's better to retain an objective until people see that it is indeed double counting – if it's truly double counting then in all likelihood retain an objective will be perfectly correlated (respond to alternatives in identical ways), and people will agree to remove one (or merge them together) later on, once an understanding of the assessment process has become better established.

3 *Third.* What role do you anticipate for quantitative trade-off analysis, which is when double counting could really raise a problem? The majority of management decisions will never proceed to that stage. In the deliberations-oriented practice of SDM, we try to minimize our reliance on quantitative weighting methods and focus instead on tools for 'talking through' the best solutions to a problem with a group. We're not structuring problems to fit our favourite analysis. We're helping people structure and evaluate alternatives in order to gain insight into which to choose. All of this is to say that, although we work hard to avoid egregious examples of double counting, we rarely lie awake at night worrying about it.

4.3.5 Does every objective need to have a preferred direction of change?

Yes. For every objective, we need to know (a) the subject of concern, and (b) whether we want more or less of it: Decrease deaths. Make more money. Protect more pandas. If you're having trouble getting agreement on whether people want more of something or less of it, then you probably are either trying to combine two objectives into one or you're confusing means and ends.

Often we'll hear participants say that they don't want to necessarily increase or decrease something, they want to optimize it. Optimizing and balancing are wonderful things, but they are what we do *across* objectives – indeed optimizing or balancing across objectives is the goal of the overall decision process. We cannot optimize a single fundamental objective. To understand why 'optimizing' an objective is problematic, suppose we say we want to 'optimize forest stand density'.

And suppose we have two alternatives. Alternative A has a density of 2800 plants per hectare, while Alternative B has 3600.

Judging which alternative is preferred, even on the basis of this single 'objective', is a messy mix of fact-based and value-based judgments. Forest stand density is clearly not the fundamental objective – it is important because it influences other things we care about. So judging which alternative is preferred involves both (a) identifying which things really are valued (should we be thinking about the effects of forest stand density on burrowing owls? split toed frogs? carbon sequestration? forest-sector jobs?) and (b) estimating the consequences of forest stand density on those things (which requires specialist knowledge). Setting up an objective to optimize something will almost certainly mean:

1 That you're focused on something other than the fundamental objective.
2 That specialist knowledge will be required to make judgments about which alternative is best, leaving non-specialist participants with little meaningful role.
3 That those specialists will apply not just their technical expertise but also their values about what matters when making judgments about the best alternative, in ways that are not transparent.
4 That you will have little hope of diagnosing the causes of disagreements about preferred alternatives, because preferences are influenced by hidden value judgments and different access to information.

The bottom line: do a good job of uncovering all the objectives and separating means and ends, and the direction will be obvious. If it is not, go back and ask 'why is that important?'

4.3.7 What if what matters is hard to measure?

Count what counts. If something matters when choosing among alternatives, then it should be on the list of fundamental objectives, whether it is easily measured or not.

Some stakeholders may at first be sceptical when informed that concerns such as traditional ways of life, sense of place, or the degree of local participation in decision making can be incorporated into formal analyses of alternatives alongside considerations such as the tons of emitted pollutants or the dollar values of expected tax revenues and resource rents[9]. This capability to enlarge the set of objectives to include all those things that fundamentally matter and meaningfully assess them can be a powerful tool in encouraging stakeholder participation and deliberation, because it provides evidence that decision makers intend to take these things seriously.

If an objective sounds like it would be difficult to measure simply because it's ambiguous, however, then that's another issue. Good structuring at the objectives stage involves a relentless assault on ambiguity. Sometime terms that are popular and in wide use but don't have a clear and consistent meaning, like 'naturalness' or 'sustainability', are often proposed as objectives. There are obviously legitimate values underlying these concepts, but they're not helpful for decision-making purposes in their current form. Why? One reason is that they're really metaconcepts – there are several ideas bundled together that need to be unpacked if we are to have a chance of understanding what the speaker is actually meaning. Interestingly, this unpacking often reveals that people disagree on which of the underlying concepts are most important, leading to some very insightful discussions.

Let's take 'naturalness' as an example. When we ask participants to tell us exactly what they mean by this term, the list we get usually involves issues such as:

1 *Likeness to a previous historical state.* If 200 years ago the natural state for an area was a dense rainforest and now it's a grassland, is this by definition a bad thing? What if 'historical' conditions already had been modified by aboriginal groups: would this matter?

What about natural changes in land and climate characteristics over time?

2 *Ecological performance*. For some people, naturalness is important because it is understood to be a measure of environmental productivity or diversity, based on the assumption that undisturbed areas generally exhibit more of both qualities than disturbed areas. In this case, 'ecological productivity or biodiversity' might be preferable since the meaning is less ambiguous. An interesting and important trade-off occurs when the natural state of a forest has lower biodiversity than the current state.

3 *Visual aesthetics*. If participants believe that a 'natural' environment just looks better, then efforts to improve the aesthetics of a site might be welcomed. This will not be captured if the stated objective is 'naturalness'. Again, an interesting trade-off occurs when naturalness and aesthetic preferences are at odds.

4 *Distance from the equilibrium state*. Arguably a more sophisticated aspect of naturalness is a characterization of how far alternative managed states are from the unmanaged state. In investigating this question at Yellowstone National Park[13], for example, one of the proposed measures is the energy required to keep any given managed solution away from the unmanaged equilibrium state. For parents of kids with messy bedrooms, or for engineers familiar with the notion of entropy, it's easy to appreciate this dimension of naturalness.

Our point in this discussion is not to dismiss the validity of the use of the word 'naturalness' if a participant raises it in decision context (the same goes for 'sustainability' and others of that genre) – far from it. Rather, the discussion needs to continue so that the real value behind this concern, in this decision context, can be identified. If we cannot do this, we get back to the trap of requiring someone to tell us, in an arbitrary way, which alternative is (for example) 'most natural' on a case by case basis. More generally, the message is that ambiguous terms such as biodiversity, sustainability, equity, naturalness, and so on, need to be unpacked and made as unambiguous as possible, either by defining different objectives, or by specific sub-objectives that clarify what is meant. Once again, the question 'why is that important?' is the most important question in your toolkit.

4.3.8   What if what matters isn't what was stated?

This is a broad topic, filled with nuances about people and what they want and how they seek to present themselves to their peers, or to decision makers, in a deliberative group setting. However, at least one common example is fairly straightforward: stakeholders deliberately put forth a false objective in place of what really matters because they feel that their true concerns will not be considered 'legitimate'. In one example, townspeople participating on an advisory committee concerned with the siting of a proposed incineration facility were concerned that the incinerator and the visibility of its emissions stack would hurt their rural image. However, this was not thought to be a legitimate concern in the eyes of the decision makers and so instead townspeople spoke about their fears of adverse health effects – respiratory illnesses, childhood cancers, and the like. Despite the presentation of solid scientific evidence that these health-related fears were unfounded, there was little movement and a serious impasse threatened to block all progress on the negotiations. At an informal side-meeting, held in a local café, one of the townspeople divulged that, in fact, the health fears were not a serious concern but, instead, were a cover for the residents' worries concerning loss of their rural image. Once this issue had been clarified, then the formal consultations as part of the advisory committee shifted focus to ways that the visual impact of the incinerator could be mitigated and the rural character and agricultural reputation of the area could be protected or even enhanced.

4.3.9 What if 'what matters' is outside a manager's mandate?

SDM processes are fundamentally about the integration of multiple objectives and multiple perspectives as part of a multi-issue, multi-stakeholder evaluation or planning process. But sometimes managers' perceptions of their mandate create problems when trying to identify a full suite of objectives.

The problem often arises in environmental agencies. For example, government environmental agencies or departments may have a mandate to 'protect' the environment, to 'conserve' species, or to 'promote' environmental sustainability. Projects or goals are thus defined in terms of these single objectives even though everyone acknowledges that, in practice, multiple objectives are affected and need to be addressed if the implementation of plans is to move forward. Yet within many environmental agencies stating this obvious truth is viewed as heresy, and the recognition that trade-offs will need to be made across conflicting objectives is viewed with fear because environmental goals may be compromised. As a result, tension is created in working with other departments, particularly those whose mandates might emphasize the single objectives of employment or tourism or cultural protection or health, and lengthy delays are created in developing broadly acceptable initiatives.

A similar situation can occur in corporate or business settings. We've had clients who accepted that particular decisions had multiple objectives and wanted to participate fully in exploring trade-offs among them, but simply could not participate in a process that set 'objectives' related to environmental performance, as that would imply a level of responsibility for managing environmental outcomes that they did not feel their organization owned. For such projects, we switched the terminology from 'objectives' to 'accounts', dropped the verb, and otherwise continued the SDM process (so the objective 'maximize fish abundance' became the account 'fish abundance').

4.4 Case studies

4.4.1 Objectives for watershed management: Tillamook Bay, Oregon

The Tillamook Bay estuary management plan provides a good example of the benefits of identifying the multiple objectives potentially affected by an environmental management plan and using means-ends diagrams to facilitate dialogue. With support from the US Environmental Protection Agency (EPA) and the National Estuary Program (NEP), we were part of a team that explored the use of structured decision methods for clean-up and protection of the Tillamook Bay watershed in northwestern Oregon[10]. The stated goal of the effort was to develop a science-based, community-supported management plan for the watershed. Although staff scientists had a legal obligation to incorporate input from an advisory committee, composed of local residents and resource users along with representative of regional, state, and federal resource management agencies, the focus had been on development of a biology-driven plan, with little attention given to social, economic, or process considerations (e.g. who was involved, when meetings were held, etc.). Conflicts between tourism, forestry, and local agricultural interests were handcuffing the effectiveness of the committee.

After initial meetings with staff and the advisory committee, it quickly became clear that the pressing need was ' . . . to find a way to meaningfully involve local residents at a detailed, action-specific level' and to ' . . . develop a mechanism by which community members could learn about and contribute to the more important dimensions of the proposed action' [10] (p. 38). The critical missing element was the articulation of a comprehensive set of fundamental and means objectives. Staff scientists had focused on biological health and water quality in Tillamook Bay, but had failed to involve community perspectives on what actions might be taken. As a result, local farmers were worried about impacts on farm management (e.g. riparian protection limits on

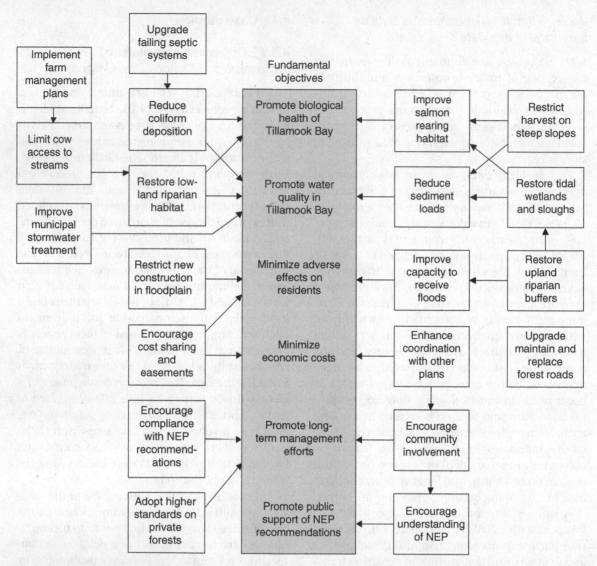

Figure 4.5 Means–ends diagram developed for the Tillamook Bay estuary plan (source: Gregory[10]). NEP, National Estuary Program.

the access of cows to fields), other residents were worried about the cost implications of changes to municipal storm-water and septic systems, and many residents were worried about trade-offs between estuary management and flood protection. In addition, the staff scientists had largely ignored a variety of non-biological objectives important to local residents and to government. After several heated meetings with the advisory committee, NEP staff, and local residents, a new and more comprehensive set of objectives was proposed (shown at the centre of Figure 4.5) that included minimizing adverse effects on local residents, minimizing regional economic costs of the management plan and floods, and encouraging efforts to promote public support of the NEP recommendations.

Figure 4.5 is a means–ends diagram produced as part of the process. First worked out on a blackboard in the community school with participation from tourism, forestry, and local agricultural stakeholders, development of the diagram helped clarify how alternative actions (means objectives) contributed to a small set of fundamental objectives, most of which were shared among all participants.

The figure helps to illustrate the point that both means and ends are critically important to a decision, but that they need to be treated differently. 'Restore Upland Riparian Buffers' (at the right of the network) is an action but also a 'means objective' because riparian buffers (vegetated areas alongside lakes and streams) are a key means for improving the capacity of communities to 'receive floods', and thereby minimize or avoid flood damages. This, in turn, will reduce adverse effects on the quality of life of local residents – something of fundamental importance. Riparian buffers are also viewed as a critical means to restore tidal wetlands and sloughs, which will both improve salmon-bearing habitat and reduce sediment loads. These improvements influence, respectively, the biological health and water quality of the estuary – outcomes that also are fundamentally important.

The restoration of upland riparian buffers may thus turn out to be very important to the success of clean-up efforts in Tillamook Bay. But it's important because it's an alternative, or a component of an alternative that could contribute significantly. It may turn out to be a recommended alternative, but it should be compared against other alternatives for achieving the same fundamental objectives. If we use riparian buffers as a fundamental objective, then we will be evaluating alternatives on the basis of whether and to what extent they improve riparian buffers. For example, 'Restricting harvest on steep slopes' is shown as an alternate way to improve salmon-bearing habitat, but we can't evaluate it in terms of impacts on riparian buffers. So if we want to compare the relative merits of restoring riparian buffers with restricting harvest on steep slopes, then we must do it by assessing their relative effects on some more fundamental objective(s). This logic becomes self-evident with the use of a means-ends diagram, and because it's an accessible visual tool it can stimulate both understanding and helpful deliberations.

In this case, the means–ends diagram served as an essential tool for gaining agreement within a group on a small set of truly fundamental objectives, which then shifted attention in future analyses to those things about which participants cared most strongly.

With these new fundamental objectives in place and with several new common sense process-based initiatives (such as changing meeting times and locations to accommodate local residents), the program began to enjoy much higher levels of public trust, support, and participation – which in turn led to new ideas for improving management actions and for ensuring that recommendations could quickly be implemented.

4.4.2 Objectives for land-use planning: Sabah, Malaysia

In highly visible and controversial settings, the failure to define objectives carefully can lead to political turmoil and social unrest. Once derailed, can a single SDM workshop help? This was the challenging case we worked on in Sabah, Malaysia (the northern portion of the island of Borneo) where we were asked to lead a workshop with a stakeholder group about future development options for the Maliau Basin, a pristine rainforest area where large coal reserves recently had been located[11]. Previous discussions about land-use alternatives had generated anger, frustration, political conflict, and little insight. The charge given the workshop by the Secretary of State was to provide insight to decision makers about the choice between a preservation alternative, supported by regional and international environmental interests, and a mining alternative, supported by national economic interests and industry. We began the three-day workshop by asking representatives of these different interests to describe their concerns and to address the pros and cons of these alternatives. We then met with each of five stakeholder groups for a more structured elicitation of objectives.

By the second day of the workshop the following five fundamental objectives (with no ranking) had been identified and agreed to by participants. The first three of these are common to most environmental management issues, although their interpretation is always a little different. In this situation, for example, social impacts included concerns about workers coming in from other countries. The last two objectives are unusual but considered important because of the international implications associated with selection of a recommended alternative:

1   Maximize direct and indirect economic benefits.
2   Minimize adverse social impacts.
3   Minimize adverse environmental impacts.
4   Maximize positive political impacts.
5   Maximize increases in international prestige.

A comprehensive list of means objectives also was developed, with connections shown to one or more of the fundamental objectives. For example, under economic benefits, a means objective was to 'maximize eco-tourism'. With this information in hand (presented for discussion purposes on large blackboards in front of the group), it quickly became obvious that the initial 'preserve or develop' framing of the problem was very limited. With a clearly defined set of objectives in place, it was possible to generate value-focused alternatives: different ways to develop the coal reserves were possible (at somewhat higher cost) and different ways to preserve the natural amenities were possible

(limiting degradation to less significant areas). Participants were able to see that it might be possible for protection of the site to also yield economic benefits (e.g. through eco-tourism and other initiatives) and that limited development of the site might also yield environmental benefits (e.g. through contributions to a fund that would help pay for active protection of key forest resources).

This new basis for deliberation, built on the clear expression of fundamental objectives, had the added benefit of dramatically shifting the emotional context for the discussions. By dropping the 'preserve vs develop' framing of the problem, which divided the group into two angry and conflicting parties, suddenly there was an active exchange with both sides contributing to the development of new alternatives and providing information helpful to the broader range of objectives (e.g. mining engineers agreeing to release information of interest to environmental agencies). For decision makers, one of the most helpful aspects of this brief workshop was therefore the establishment of a greatly improved basis for future communication, dialogue and negotiation among the different stakeholders.

4.4.3   Objectives for cultural and spiritual quality: the voice of the river

As part of the development of an adaptive management plan for the Bridge River hydroelectric system, resource managers identified objectives related to fish, wildlife, and riparian vegetation as part of the experimental flow treatments. In addition, elders and community members from the St'at'imc First Nation noted their concerns about the impact of water flows on the 'spirit' or 'voice' of the river, which was captured in a cultural and spiritual quality objective.

To obtain information to better define this objective, we collected input from both individual and group interviews with St'at'imc community residents and were able to define four key components of cultural and spiritual quality:

Table 4.4 Objectives used to evaluate experimental flow regimes on the Lower Bridge River.

Objective	Measure	Units of measure
Salmon	Total biomass	kg
	Chinook biomass (Reach 3)	kg
Species at risk	Harlequinn duck abundance	Presence/absence
Riparian health	Adult cottonwood growth	mm/year
	Juvenile cottonwood growth	mm/year
	Recruitment	Yes/no
River health	Benthic community abundance	Millions of individuals
	Benthic community diversity	% EPT
Spiritual quality	Scale: voice of the river	1–5
Finances	Value of electricity	$million/year

1 Sound, including: the voice of the water and birdsong.
2 Smell, including the smell of the water itself and the ambient smell at water's edge.
3 Movement, including the movement of water (seasonally appropriate) and the diversity of movement (pools/riffles).
4 Interaction (of people and water), including shore access and ability to wade in or across the river at desired locations.

These four components clearly do not provide a universal definition of cultural or spiritual quality. They define the aspects of cultural and spiritual quality thought to be relevant for the evaluation by St'at'imc of a suite of alternative flow regimes and habitat enhancement activities on the Lower Bridge River, within the flow ranges considered. This objective became part of the objectives hierarchy that was used to evaluate experimental flow treatments (Table 4.4).

Working with St'at'imc community residents, we developed a scale for reporting differences in spiritual quality across different flow regimes, and an assessment protocol was co-developed by the Stl'atl'imc community and the utility and regulatory management staff to minimize biases and inconsistencies in field assessment over time. Key elements of the protocol include:

1 Assessment to be done by a committee of three to eight experts designated by the St'at'imc community with oversight by a multi-party monitoring team.
2 Observations recorded four times per year, under a range of test flows.
3 Observations recorded at two designated sites per reach, over three reaches.
4 Documentation of a visual record at each observation site, using video camera and still photography.
5 Use of a simple, replicable, and transparent scoring system for assigning scores to each component in each reach.
6 Use of a transparent and defensible methodology for weighting and aggregating scores across observers, components, reaches, and seasons.

This 10-year plan is in the early years of implementation and it remains to be seen how consistent and useful it will prove to be as an aid to future resource decision making. But stop for a moment to reflect on the importance to the St'at'imc elders of what was done here. We've said, 'tell us what's important to you' and have incorporated the answer on an equivalent level to more conventional objectives such as cost and environmental considerations. If we'd hesitated because it might be tricky to measure these things, then we might have missed the objective altogether (and potentially disenfranchised the St'at'imc community).

4.5 Key messages

There is a discipline to defining objectives, If you're designing a decision process, here are some key things to look for:

1 The set of objectives should be complete, concise, sensitive, meaningful and independent.
2 Objectives should state the thing that matters, whether we know how to measure it or not.

Text box 4.2:   A personal perspective on using SDM approaches in consultations with aboriginal communities

SDM practitioners have worked on a variety of environmental management issues that involve aboriginal or indigenous populations; examples include work with Native Americans in the United States, First Nations and Inuit and Metis communities in Canada, tribal groups in Asia, rural communities in Africa, and Maori in New Zealand [1,9, 12, 14, 15]. In Canada, several of these efforts have involved authors of this book in collaboration with Cheryl Brooks (Indigenuity Consulting Group), a member of the Sts'Ailes First Nation, who frequently draws on SDM concepts and methods in her work. Cheryl's observations follow:

'As a resource person working with numerous aboriginal groups to facilitate processes of decision making, identifying priorities and developing strategies I have repeatedly heard peoples' expressions of frustration about outcomes not reflecting their inputs. Through coincidence I was introduced to the decision sciences and in particular a form of decision analysis called Structured Decision Making. Though at first view, with a lot of tables, charts and data, it looked fairly complicated to me, I soon learned otherwise, gaining understanding that this is an excellent and systematic way of organizing and presenting volumes of complex information in a way that people can see all inputs including their own, whether quantitative or qualitative. Most importantly the various organized ways of presenting information tend to stimulate thorough discussions and substantive learning in diverse groups. Then I learned how to tangibly include values in the data and the decision making and have concluded that it is probably the best tool I have personally ever used in complicated discussions with the aboriginal groups, with whom most of my work is done.

So, what am I really talking about here? Start with the proposition that the goal of consultations by government is to identify and implement policies, programs or actions that satisfy lawful, statutory and regulatory obligations, and to do so in an efficient manner. First Nations have been clear that they expect, indeed will only participate in, meaningful processes in which their interests and concerns are identified and clearly expressed, and given at least equal weight as other factors including environmental, economic, social, health, and safety goals. One observation is that most First Nations participants tend to take a much longer term view than many current government decisions reflect and decision analysis works well to build long-term options.

In my opinion, a decision-aiding SDM model is the best choice for many complex decision-making processes with aboriginal groups, for several reasons.

1 It builds on the values and objectives of the different technical and community-based parties, so that their key concerns can be identified clearly and, as a result, a plan or strategy can explicitly be responsive to those things that matter most to people.

2 The emphasis on 'values-based' thinking as part of deliberative processes seeks to increase dialogue and trust among participants while developing a more informed basis for making decisions. This frequently requires overcoming prior misinformation or misunderstandings.

3 It recognizes that being clear about what matters can be difficult, particularly if the decision context is complex or emotionally charged. As a result, particular care is taken to help the participants define key objectives and key information needs, including values such as ceremonial and spiritual values, or intergenerational knowledge transfer, that often are left out of consultations with First Nations.

4 It takes enough time to fully explore issues and solutions so that participants do not feel pressured into choosing options before they are at the point where they are confident they are well informed and able to provide meaningful comment.

5   It can be used as a complete process with some sophisticated models and analysis presented as part of the data, or you can use segments of it in processes – for example identifying what is important and performance measures alone can facilitate excellent discussion. Also, if you are working through the entire process it is relatively easy to loop back and incorporate new or additional information at any time.

Overall, an SDM approach offers three significant advantages in terms of addressing some of the most important challenges facing Aboriginal peoples in working with governments to create sound policies. The first is an ability to help participants think through and express their values and construct a clear perspective on what matters most in the context of the decision at hand – which is crucial given that some of the decisions Aboriginal leaders make today can literally affect survival of their Nations. The second advantage is an ability to help participants understand the factual information relevant to their choices, including the diversity of information sources (e.g. scientific, community, First Nation) that merit close consideration and to better understand the uncertainty associated with estimates of anticipated consequences. Thirdly, this process provides well-organized information to explain why certain decisions were reached and how they dealt with the interests of the people – a valuable resource for getting collective consent and support'.

Besides being important for finding good solutions, this helps to level the playing field and keeps diverse stakeholders engaged in the process.

3   It's important to be diligent in distinguishing between means and ends.

4   Each objective needs a clear, universally agreed directionality, which both clarifies thinking and enables the use of some useful trade-off analysis methods later on.

5   Objectives are not targets: avoid setting targets, and the muddling of facts and values that goes with it, at the objective-setting stage.

6   Objective-setting is intended to build an understanding about shared values among diverse participants and is an important way to build common ground early in the process. Weights should *not* be assigned at early stages of the deliberative process, when the focus is on identifying and clarifying concerns (rather than making trade-offs, which comes later on).

The discussion also emphasizes that, without decision-aiding assistance, decision makers tend to be strikingly deficient at generating a comprehensive set of objectives. How can an SDM process help? To create a good set of objectives, we recommend working with stakeholders and decision makers to:

1   Ask what matters in this decision? In choosing among alternatives, what things are important? What will other stakeholders think is important?

2   Separate means and ends by asking 'why?' Why is that important?

3   Eliminate ambiguity by asking 'what do you mean by that'? Are there sub-components of that?'

4   Organize long lists of objectives into a concise hierarchy.

5   Test objectives to make sure they're useful. Put them in a consequence table with some alternatives – do they help you compare the alternatives?

Objectives lay the foundation for all that follows. They are used to both identify and evaluate alternatives. So take whatever time is needed to get a useful set.

4.6   Suggested reading

Chapter 3 of Clemen, R.T. (2004) (*Making Hard Decisions: An Introduction to Decision Analysis*, 4th edn. Duxbury, Belmont) clearly describes objectives and contains some great problems to get the reader thinking more about issues associated with their identification.

Fisher, R., Ury, W. & Patton, B. (1991) (*Getting to Yes*, 2nd edn. Penguin Books, Harmondsworth) neatly distinguishes between interests (i.e. objectives) and positions, with great examples of how to articulate objectives clearly when working with groups.

Keeney, R.L. (1992) *Value-Focused Thinking: A Path to Creative Decisionmaking*. Harvard University, Cambridge. The discussions and case-study applications covered in this book, appropriately titled *Value-Focused Thinking*, provide many of the principles and techniques that underlie SDM methods with respect to identifying objectives.

Keeney, R.L. (1988) Structuring objectives for problems of public interest. *Operations Research*, **36**, 396–405. A short introduction to methods for structuring objectives of public policy problems.

Turner, N.J., Gregory, R., Brooks, C., Failing, L. & Satterfield, T. (2008) From invisibility to transparency: identifying the implications. *Ecology and Society*, **13**, Article 7. Retrieved from http://www.ecologyandsociety.org/vol13/iss2/art7/ This article emphasizes the need for a broader and more inclusive approach to defining objectives in the context of land-use decisions, including cultural and environmental values that are often not recognized by resource managers or government decision makers.

4.7   References and notes

1   Gregory, R., Failing, L. & Harstone, M. (2008) Meaningful resource consultations with First Peoples. *Environment: Science and Policy for Sustainable Development*, **50**, 36–45.

2   Keeney, R.L. (1992) *Value-Focused Thinking: A Path to Creative Decisionmaking*. Harvard University, Cambridge.

3   Gregory, R.S. & Keeney, R.L. (2002) Making smarter environmental management decisions. *Journal of the American Water Resources Association*, **38**, 1601–12.

4   Clemen, R.T. (2004) *Making Hard Decisions: An Introduction to Decision Analysis*, 4th edn. Duxbury, Belmont.

5   Keeney, R.L. (2007) Developing objectives and attributes. In: *Advances in Decision Analysis: From Foundations to Applications* (eds W. Edwards, R.F. Miles, Jr & D. von Winterfeldt). pp. 104–128. Cambridge University Press, New York.

6   Bond, S.D., Carlson, K.A. & Keeney, R.L. (2008) Generating objectives: can decision makers articulate what they want? *Management Science*, **54**, 56–70.

7   Tear, T.H., Kareiva, P., Angermeier, P.L., *et al.* (2005) How much is enough? The recurrent problem of setting measurable objectives in conservation. *Bioscience*, **55**, 835–849.

8   Walshe, T. & Burgman, M. (2010) A framework for assessing and managing risks posed by emerging diseases. *Risk Analysis*, **30**, 236–249.

9   Turner, N.J., Gregory, R., Brooks, C., Failing, L. & Satterfield, T. (2008) From invisibility to transparency: identifying the implications. *Ecology and Society*, **13**, Article 7. Retrieved from http://www.ecologyandsociety.org/vol13/iss2/art7/

10   Gregory, R. (2000) Using stakeholder values to make smarter environmental decisions. *Environment*, **42**, 34–44.

11   Gregory, R. & Keeney, R.L. (1994) Creating policy alternatives using stakeholder values. *Management Science*, **40**, 1035–48.

12   Arvai, J.L. & Post, K. (2011) Risk management in a developing country context: structured decision making for point-of-use water treatment in rural Tanzania. *Risk Analysis*, in press.

13   Cole, D.N. & Yung, L. (eds.). 2010. Beyond Naturalness: Rethinking Park and Wilderness Stewardship in an Era of Rapid Change. Island Press, Washington, D.C.

14   McDaniels, T. & Trousdale, W. (2005) Resource compensation and negotiation support in an aboriginal context: using community-based multi-attribute analysis to evaluate non-market losses. *Ecological Economics*, **55**, 173–186.

15   Satterfield, T, Gregory, R. & Roberts, M. (2010) Dealing with differences: policy decision making in the context of genetically modified organisms. Report submitted to Foundation for Research in Science and Technology, Wellington, New Zealand. February 15, 2010.

5

Identifying Performance Measures

On a Scale of 1 to 10, How Comfortable Would You Feel in This Bathing Suit?

This is one of the first of approximately 11.6 million returns you can expect to find by Googling the phase 'on a scale of 1 to 10'. Interestingly enough, there seems to a bit of a trend towards the use of this phrase on social networks: 'on a scale of 1 to 10' has its own Wikipedia entry and multiple UrbanDictionary definitions. Through the use of this phrase, these sources point out, numbers are expanding their grammatical territory from adjectives to nouns, as in when referring to a performance, 'that's a two' – meaning that it wasn't very good at all, or 'I'd give that salesman a five' – suggesting mediocrity.

Many people seem to have no problem thinking in terms that require assigning quantitative assessments of quality or performance in essentially arbitrary ways. In developing performance measures for use in important decision situations, we aim to do something parallel but in a rigorous and defensible way.

Spontaneous and implicit scales have obvious shortcomings. How are 'one' and 'ten' being defined? Is one person's idea of a 'one' the same as someone else's? On what basis is a 'three' assigned, and what would it take to make it a 'four' or a 'seven'? How consistent are the ratings across people? How consistent, even, is the same

person's ratings before and after lunch? What aspects of performance are being bundled by this overall rating, and what subjective weightings of these components are being embedded in the aggregation? And in many cases, as we'll discuss, a scale isn't always the best means of communicating performance anyway.

The task of developing sound performance measures is critical but underappreciated and, as a result, their definition is often inadequate. To build some groundwork for how to do these tasks well, we'll list some properties of good performance measures and explore, with (mostly) environmental management examples, what we mean by these properties and why they're important. We introduce a generic approach to identifying performance measures that we've used in a wide range of situations.

Next, we describe different types of performance measures. *Natural* measures are those that directly describe outcomes that matter – such as number of jobs or fatalities when what we care about are employment and human health. Natural measures are in general use and have a common interpretation. *Proxy* measures serve as indirect indicators for something that matters but is difficult to measure – for example we might use air emission levels as a proxy for adverse health effects caused by poor air quality.

Structured Decision Making: A Practical Guide to Environmental Management Choices, First Edition. R. Gregory, L. Failing, M. Harstone, G. Long, T. McDaniels, and D. Ohlson.
© 2012 R. Gregory, L. Failing, M. Harstone, G. Long, T. McDaniels, and D. Ohlson. Published 2012 by Blackwell Publishing Ltd.

We'll give special attention to the third type of performance measure, *constructed measures*, because of its usefulness for incorporating important but hard-to-measure impacts as part of environmental management decisions. We'll discuss what each type of performance measure is, why we might use it in one situation over another, and talk about strategies for choosing among the different types. We end by reviewing some practical issues in creating and using performance measures that those charged with developing them will find useful.

5.1 The basics

5.1.1 What is a performance measure?

A specific metric for consistently estimating and reporting consequences

At the simplest level, a performance measure is a specific metric that can be used to consistently estimate and report the anticipated consequences of a management alternative with respect to a particular objective. Whereas objectives may sometimes be quite broad, performance measures need to be specific because they define how an objective is to be interpreted and evaluated for the purposes of *this* decision. They articulate the exact information that will be collected, modeled, elicited from experts, or otherwise developed and presented to decision makers about how each alternative affects each objective. Performance measures provide the ability to distinguish the relative degree of impact across alternatives, either qualitatively or quantitatively, in ways that make sense and will help decision makers consistently and appropriately to compare alternatives.

Subjectively determined, based on both technical and value-based judgments

While it's widely understood that objective-setting is a value-based exercise, scientists and environmental managers tend to view the selection of performance measures as a largely technical exercise. In fact, selecting performance measures is a subjective exercise, with both technical and value judgments coming into play. Wilson and Crouch[1] provide a nice example. They show how worker safety in United States coal mines could be characterized using any of these apparently reasonable metrics:

1 Number of deaths from coal mining in the United States, per year.
2 Number of deaths from accidents per ton of coal mined in the United States, per year.
3 Number of deaths from United States coal mining accidents per employee, per year.

The interesting part is that each of these metrics suggests a different story with respect to worker safety. Looking backwards, over the 1950–1970 period, the first metric remained relatively constant. The second – accidents per ton of coal – decreased, the result of increased mechanization. The third metric – deaths per employee – increased for the same reason. Looking forward, the choice among these could matter in terms of ranking investments in coal mining. So what's the story with respect to worker safety? Did it improve or not? Well, the answer isn't self-evident – and each of these metrics provides only part of the complete picture.

The insight from evaluating different performance measures may force us to think more clearly about how we want to define our objectives. Suppose, for example, you are comparing two salmon management programs. Your initial objectives are to maximize salmon abundance and minimize cost. Now suppose you have two programs – Program A involves watershed restoration activity that provides habitat for spawning salmon. Program B invests in hatcheries to produce hatchery salmon. If you select 'salmon abundance' as your measure, then Program B is preferred: it produces more salmon overall. If you choose 'wild salmon abundance' as your measure, then Program A will be preferred because hatchery investments don't help wild salmon.

Predicted, not monitored

Performance measures are not the same as monitoring indicators. Monitoring indicators are used to keep track of the state of something, such as the condition of a forest, or the water quality in a river. While monitoring indicators are important in the right context, they are not what we mean by 'performance measure'. There are two main differences:

1 In an SDM context, performance measures are used to report on the *predicted* performance of alternatives, for the purposes of making a choice among possible actions. Predictions are made using some combination of data, models and expert judgment, in advance of an action, whereas monitoring indicators are typically actual measures made after the action has been taken. In other words, we are not interested in how you like your current bathing suit – we want to know how you expect to feel about wearing bathing suit A versus B at some time in the future.

2 In SDM, less is often more. A performance measure is only of use if it serves the direct purpose of communicating key differences in performance of one alternative over another on a specific objective. This takes some careful thinking. Remember from behavioural research that if we have too many objectives or performance measures, thinking can become muddled rather than clarified. So there must be relatively few of them and they must synthesize the essential information about outcomes as concisely as possible.

5.1.2 Why do we need performance measures?

Performance measures serve a host of purposes in decision making. First, they tell us what vaguely defined objectives mean – really mean, for this specific decision. They eliminate uncertainties associated with ambiguity in objectives. Because they define what matters when comparing alternatives, they also define what information we will collect and provide a much needed focus for prioritizing and designing technical studies and predictive modeling efforts. They facilitate the accurate and consistent comparison

Text box 5.1: Monitoring indicator or performance measure?

A list of monitoring indicators for forest management usually includes forest age, which is typically reported in age classes such as Early (0–40 years), Young (40–80 years), Mature (80–140 years) and Old (>140 years). It's an important attribute of a forest from the perspective of understanding ecological health (and a variety of other things), but it's not a performance measure. To be useful as a performance measure, we need to ask, 'what is it about forest age class distribution that is most important to decision makers charged with choosing among alternative forest management practices in a given context'? Usually it's possible to identify a particular age class that's of interest, either because it's most critical to the stated objective or because it's most rare or sensitive. For example, if the objective was to 'improve biodiversity', then 'predicted area of Old Forest in 50 years' might be a suitable performance measure given the unique role of old forests in providing niche wildlife habitat. If the objective was to 'minimize wildfire risk to communities', then 'percentage of Mature forest area within a 5 km radius of town' might be a suitable performance measure given the high fuel loads often associated with this forest age class. So the selected performance measure, based on both technical judgments (about what's a good indicator of forest health or wildfire risk) and value-based judgments (about what it is that managers and stakeholders aim to achieve), is something that is selected based on the need to compare alternatives in a specified management context.

of alternatives. Critically, they provide a way of synthesizing large volumes of technical information into a summary format so that everyone in a multi-stakeholder group can understand critical aspects of performance. As a result, they are the key to leveling the playing field across participants with different levels of technical literacy. Performance measures become the focus of group deliberations, as it is differences in performance measures that expose important value-based trade-offs across alternatives. Ultimately they provide a means for communicating the rationale for difficult decisions. Sound important? Absolutely.

Do we always need them? Well, no, actually. There's a whole class of decisions for which the preferred or at least acceptable solutions become obvious once the objectives are clear and a creative range of alternatives has been brainstormed (recall the Sabah case study in Chapter 4). But for a large majority of substantive environmental management and planning decisions, doing a good job of developing performance measures is absolutely critical.

5.1.3 What's a good performance measure?

In Chapter 4 we introduced the topic of how to choose a good set of objectives, with the intent of focusing decision makers on what matters for the decision at hand. Recall that good objectives are complete, concise, sensitive, understandable, and (preferentially) independent. Performance measures provide a parallel mechanism for reporting predicted effects on the objectives, and so it's not surprising that their desirable properties are closely related. Yet there are also important differences: whereas objectives should state what matters to a decision, whether we know how to measure it or not, performance measures need to be operational and practical. They must be specific and assessable, and also effectively highlight differences in the effects of management alternatives on different objectives.

Text box 5.2:   Leveling the playing field

We mentioned in Chapter 2 the contribution that SDM methods can make to leveling the playing field among the diverse participants in an environmental decision process. Performance measures contribute to this goal in two critical ways:

1 Well designed performance measures allow hard-to-quantify objectives – such as maintaining a traditional lifestyle, protecting aesthetic quality or maximizing local participation in decision making – to be put on the table and treated in the same way as more conventional measures like minimizing greenhouse gases and maximizing old growth habitat. They place hard to quantify things on equal footing with conventional things. We provide examples of this in the section on constructed measures and in the case studies at the end of this chapter.

2 Performance measures synthesize a large volume of technical material into a summary format so that everyone in a multi-stakeholder group can understand critical aspects of performance. As a result, every participant, regardless of their technical background, can participate meaningfully in identifying, evaluating and choosing alternatives, even in very technically intensive decision situations.

Although there are arguably no 'right' or 'wrong' performance measures, there are certainly better and worse ones. Important properties of good performance measures (adapted from Keeney & Gregory[2]) are listed below:

1 *Complete and concise*, meaning that they cover the range of relevant consequences under all reasonable alternatives, as concisely as possible, without double-counting or redundancies.

2 *Unambiguous*, meaning that a clear, accurate, and widely recognized relationship exists between measures and the consequences they describe, and that they would be interpreted in the same way by different people. They should report accurately and consistently on the *relative* differences in performance across alternatives, including differences in the degree of uncertainty associated with the performance estimates.

3 *Understandable*, meaning that performance measures, and the consequences and value trade-offs they describe, can be understood and communicated clearly and consistently by different people. This is critical in a multi-stakeholder context.

4 *Direct*, meaning the measures report directly on the consequences of interest and provide enough information that informed value trade-offs can reasonably be made on the basis of them.

5 *Operational*, meaning that the measure can be readily put into practice within the constraints of the planning or decision making process; the required information can be obtained to assess them, whether from data, models, expert judgments, or other sources within relevant budgets and timelines.

Below we provide some practical illustrations of what we mean by each property.

Complete and concise

Table 5.1 shows three different examples. Each is a possible way to define performance measures for a fundamental objective of minimizing negative health effects of alternative air-quality management strategies.

Despite its clarity, Example 1 is not complete. There are two important aspects to the health impacts: mortality and morbidity (i.e. non-fatal impacts), and the measures for Health in Example 1 describe only the mortality effects. Example 2 is not concise: there are too many

Table 5.1 Example performance measures for the fundamental air-quality objective 'minimize negative health effects'.

Performance measures	Alternative 1	Alternative 2
Example 1		
Fatalities per year (*n*)	5	10
Example 2		
Fatalities per year (*n*)	5	10
Reduced activity days in average population (*n*)	200	25
Reduced activity days among sensitive subpopulations (*n*)	100	10
Hospital visits (*n*)	50	5
Respiratory events (*n*)	60	8
Cardiovascular events (*n*)	20	3
Asthma attacks (*n*)	15	2
Example 3		
Fatalities per year (*n*)	5	10
Reduced activity days in average population (*n*)	200	25

measures, many of which are strongly correlated, and key messages are lost or obscured. Example 3 is a good compromise between being complete and being concise. Because the number of reduced activity days is more or less correlated with all the other morbidity effects listed in Example 2, Example 3 concisely captures the range of effects listed in Example 2 without double counting or redundancies.

Unambiguous

Performance measures should be unambiguous in reporting the effects of policy or management alternatives on fundamental objectives, in the sense that anyone would have the same interpretation of the magnitude and direction of an impact. A change in the value of the measure should represent a real (expected) change in the fundamental objective. Relative impacts across alternatives should be accurately reflected. Performance measures should also express relevant aspects of uncertainty – being unambiguous does not mean ignoring uncertainty! It means expressing it explicitly – more on this in Chapter 6.

Table 5.2 Candidate measures for an objective to minimize cost.

	Measure	Alternative 1	Alternative 2
Example 1	Cost scale	High	Low
Example 2	$/year	$12.5M($11–14)	$9.0M($8–10)

Table 5.3 Candidate performance measures for maximizing water quality.

	Alternative 1	Alternative 2	Alternative 3
Example 1			
Concentration of dioxin in g/ml:90th percentile	0.0015	0.005	0.002
Cost	$20M	$16M	$12M
Example 2			
No. of exceedences per year	2	8	3
Rate payer cost ($/hhd/year)	$100	$80	$60

In the example in Table 5.2, the fundamental objective is minimizing the cost of sewage treatment plant upgrades in a midsized town. The measure in Example 1 is ambiguous, because labels such as 'high' or 'low' are unbounded and could be interpreted differently by different people (costs are 'high' or 'low' in relation to what?). We know one is higher than the other, but we know nothing about the magnitude of the difference. There is a tendency on the part of experts assigning consequences to try to reflect uncertainty by using ambiguous terms such as 'high' or 'low' – this is to be avoided. The example in Example 2 is less ambiguous and the data – including the extent of the uncertainty – are expressed more clearly.

Understandable

People should be able to readily understand what a performance measure is actually referring to. At the simplest level, this may mean using common language rather than technical jargon, but it depends on the level of technical sophistication of the group. However, understandability extends beyond technical comprehension to the design of the measure itself. Table 5.3 illustrates sample measures for maximizing water quality. The expression of water quality impacts in Example 1 is unambiguous, but it may not be particularly understandable to decision makers. By translating the raw water quality data into an estimate of the number of times a recognized water quality standard will be exceeded as in Example 2, the impacts become much clearer to non-technical stakeholders and decision makers (although this introduces the problem of threshold effects, discussed in Chapter 4). Similarly, total costs

reported in millions of dollars may be difficult to think about for some people (particularly if it is not made clear who pays). But if that cost can be translated into what it would mean to an individual household in dollars per year, then the cost information may be more meaningful.

One challenge in delivering understandable information to decision makers is reporting at a scale that is intuitive and meaningful. Consider a proposed initiative designed to clean up contaminated soils in a rural community. Objectives might include improving health for residents, minimizing the cost of the clean-up efforts, and avoiding effects on local livestock and forest animals. Resident health might be measured in terms of avoided mortality and morbidity using a quality-adjusted scale, and cost might be measured in terms of the present value of anticipated costs over 20 years (the length of the clean-up period). Expressing cost figures in terms of billions of dollars (rather than thousands) could lead to an underweighting of costs because the numbers would be so small; expressing health benefits in terms of added days of life (rather than years) could lead to them being overweighted. In both cases, the choice of a scale should reflect what is reasonable and familiar in the context of the choice at hand. In some cases, presenting the information in multiple units is useful and promotes the most robust understanding.

Do performance measures have to be quantitative to be understandable? Not necessarily: the essential requirement is that measures be designed to facilitate consistent comparisons across alternatives. This means that measures need to be constructed to fit the needs of a specific decision context – and they need to encompass all of the properties of good performance measures that we discuss in this section. In some situations, for example, photographs of states of nature (e.g. pictures of how a forest might look after different thinning treatments, the appearance of different reservoir elevations, or the visual impact of a smoke stack) might be a good qualitative means for communicating anticipated impacts and be far more understandable to participants than would comparisons involving numbers (e.g. number of trees of a certain size per acre). Yet qualitative performance measures have possible shortcomings. For example, they require some additional treatment before they can be used in the quantitative trade-off analysis techniques described later, but depending on the decision context and the participating stakeholders, they can be helpful and even essential for adequately characterizing what matters about the alternatives in ways that lend insight to decision makers.

Text box 5.3:   Three rules of thumb for understandable scales

We suggest three rules of thumb that are helpful when thinking about what units to use when reporting a performance measure:

1 Treat all performance measures consistently.
2 Consider both the absolute impact and the incremental change.
3 Provide multiple framings.

In a multi-stakeholder water allocation process we worked on, participants were evaluating effects of different flow regimes on fisheries and industry costs. Initially, effects on 'midwinter rearing habitat' for fish were reported as a change relative to a 'natural' benchmark – specifically, as 'the % reduction in rearing habitat area relative to natural flow conditions'. In contrast, the capital costs for industry to build increased water storage were reported in absolute terms, as 'cumulative dollars spent over 30 years'. This led to participants comparing potentially very large costs (e.g. $3 billion) in order to achieve what appeared to be small gains (e.g. 2%) in fish habitat. When we recast the cost predictions to report the cost effects in incremental capital expenditures over the next 30 years, this incremental cost was less than 1%, providing a different perspective on relative gains and losses.

On the other hand, consider a community evaluating local energy options. In some areas, an attractive option involves the combustion of biomass from wood waste to produce heat (or heat and power). This requires trucking of wood waste into urban communities. When we report this effect as 'three additional trucks per day', this seems significant to local residents. When we point out that under the base case (current conditions), 55 trucks per day pass through their neighbourhood anyway to serve the local grocery store and that a biomass plant will increase this to 58, they have a different perspective. Neither way is 'right', but consistency is a good rule of thumb (if relative measures are used for some objectives, then it's likely a good idea to use relative measures for all), as is thinking about both absolute and relative impacts, and encouraging participants to talk and think about different framings and scales.

Table 5.4 Candidate measures for an objective to maximize fish abundance.

	Alternative 1	Alternative 2	Alternative 3
Example 1 Abundance (no. of fish)	1000–5000	5000–8000	5000–9000
Example 2 Benthic biomass (kg)	600	700	800

Direct

This property speaks to the difference between means and ends and, correspondingly, between proxy and natural measures. A performance measure is direct if it reports directly on the fundamental objective and provides enough information so that decision makers understand key value trade-offs.

Table 5.4 shows candidate measures for an objective to maximize fish abundance. Example 1 uses abundance (number of fish) as the performance measure. It is direct but uncertain, and so in this case is reported as a range. In Example 2, recognizing that fish abundance can't be estimated with confidence, the change in benthic biomass (an important food source for fish) is recognized as an indirect measure or proxy. However, without direct knowledge of how fish numbers are expected to respond to food increases, this information may be of limited use to decision makers. Does a 50% increase in benthic biomass (bugs living on the bottom of the river) correspond to a 50% increase in fish abundance? If yes, and we know that with confidence, then this might be a useful proxy. But if we anticipate that the response may be something other than 1:1, or even non-linear, then this measure will be less helpful, and perhaps misleading, when interpreting and making trade-offs. Which is 'best' is not straightforward. Not surprisingly, there are very often compromises between directness and operationality when choosing performance measures.

Text box 5.4: A word about uncertainty

Performance measures need to be estimated and presented in a way that allows the communication of relevant aspects of uncertainty. So far we've presented only single values or simple ranges of values. But when uncertainty really matters to a decision, it may be important to characterize uncertain outcomes with a probability distribution. We discuss methods to do this, and to represent probabilistic information in consequence tables, in Chapters 6 and 8.

Operational

The adoption of a performance measure leads directly to the practical need to collect data or build models. If it's impossible to collect or develop the requisite data or models within the existing temporal, financial, or personnel constraints of the decision process, then a different measure may need to be chosen, even if it does not link quite as closely to the fundamental objective.

In Table 5.4, the information desired by fisheries managers (fish abundance) is often either not available or plagued by high uncertainty, with ranges of estimates that are substantially wider than shown in this example. There are three options. The first is to accept the number of fish as a performance measure, and defer critical decisions until such time that additional data gathering can reduce the associated uncertainty. A second option is to select an indirect or proxy measure for which relatively good information is available (e.g. benthic biomass); this is intuitively desirable, but can be problematic (we discuss reasons in a later section) but still useful. The third option is to estimate the preferred measure (abundance) using existing data in combination with structured expert judgment. From an operational perspective, which option is better is a context-specific judgment call; a key consideration is whether the relationship between the indirect proxy measure and the fundamental objective is well understood and broadly accepted.

Compromises among the properties

Inevitably there will be compromises to make across these five properties when selecting performance measures. The most unambiguous measure may not easily be understandable to non-technical decision makers. Or the most direct measure of a fundamental objective is not very operational because it's difficult to estimate or model (e.g. data collected in terms of the objective may be highly uncertain). Compromises between directness and operationality need to be openly discussed. Sometimes a more indirect proxy measure will be selected. Sometimes there will be an opportunity to improve the quality of information by collecting additional information or building predictive models. Regardless of the circumstances, it's important to remember two golden rules:

1 Count what counts. Don't count something as important because you have data for it. And don't ignore something important because you don't.
2 Be honest and explicit about uncertainty. Often it will be better to report directly on a fundamental objective of concern (e.g. fish abundance) with an explicit and wide range of possible outcomes, than to report on a proxy measure (e.g. habitat) with narrow bounds.

5.1.4   Steps to a good set of performance measures

Although the organization of this book may leave the impression that SDM processes are quite linear in terms of working through the steps of the decision-making process, in fact, as we discussed in the 'Decision Sketching' chapter, we usually work through these steps with groups in a highly iterative way. The development of performance measures is one area where this is as true as any.

In a first pass through a decision sketch, we would typically:

1 Discuss different decision frames and choose one to work with.
2 Identify and roughly structure objectives.
3 Identify and create a rough list of alternatives.
4 Sketch out a consequence table, using a simple five-star rating or similar scheme.

If people don't like what they see (that is, the structure isn't capturing the essence of the decision they need to make), then it's back to the drawing board.

If, however, this sketched-out structure is holding together and feeling good, we'd then take a second pass, polishing as we go. The decision context is more firmly defined, the objectives more tightly defined and organized, perhaps hierarchically and so on. At this point, we're ready to begin consideration of performance measures. We suggest four basic steps:

1 Brainstorm candidate measures.
2 Develop influence diagrams (if you haven't already).
3 Identify different sources of information for estimating the measures.
4 Evaluate and select the most useful measures.

Brainstorm candidate measures

Start by asking, for each objective: 'what specific information would you like to see to be able to evaluate the impact of these alternatives'? At this point, write down all the possible responses. It's likely this will reveal that some people have different interpretations of the objectives and, thus, different performance measures in mind. You may well find that you need to refine the objectives. No problem: one of the key reasons for structuring objectives and assigning performance measures to them is to build common understanding and improve communication. At this stage active discussions about the 'what exactly are the objectives' and which would be the most useful measures are to be expected.

Develop influence diagrams to explore cause-effect relationships

As we've discussed in Chapters 3 and 4, influence diagrams are structuring or modeling tools that graphically represent the different concerns and relationships important to understanding a decision. We've found influence diagrams to be especially helpful in identifying and facilitating communication about performance measures: they're already familiar to many participants (biologists and ecologists often refer to them as 'effects networks' or 'impact pathways') and they're flexible – you can build them at a conceptual level 'on-the-fly', to put some structure to complex discussions as they are taking place, or they can be constructed more formally, as the product of a structured and in-depth process. Often we start with an on-the-fly version and use it later to develop a more detailed version that ultimately becomes the basis for a quantitative model (see the case study in section 5.4.1). The desired level of detail or formality depends on the decision context. In our experience, however, simple graphical influence diagrams provide a visual structure that's intuitive and user-friendly and are readily accepted by participants in deliberative groups.

Identify different sources of information

A performance measure has to be estimated in some way. Usually there are multiple ways to do it and multiple sources of information that might be used. At the simplest level, experts might be asked to assess directly the consequences of management alternatives. With the use of a structured elicitation process (Chapter 8), this can often be a relatively quick, inexpensive, and reliable way to get good insight about key differences among alternatives. If you have the time and resources, and especially if you need to evaluate many alternatives, then you may consider the development of models to aid in estimating consequences. At this point, it's useful to simply document the existence of data and models that would be directly relevant to the estimation of consequences for each performance measure. This helps in assessing how practical and operational they are within the context of existing resources and timelines. We talk more about some characteristics of decision-relevant models in Chapter 8.

Evaluate and select the most useful measures

As with objectives, the litmus test for good performance measures is whether they're useful when making choices among alternatives. Recall that performance measures serve two very critical and different functions, one related to facts and the other to values. First, they have to report the consequences of alternatives in a way that's technically accurate and defensible, and honest about uncertainty (accurate and unambiguous in the language of our properties). Secondly, performance measures have to expose the key value-based trade-offs that decision makers need to wrestle with. So it's important to review proposed performance measures with both experts who will be involved in estimating them and decision makers who will use them to make value-based choices. Just as we described for objectives in Chapter 4, it's useful to put the performance measures in the shell of a consequence table, and ask:

1 *For experts.* How well can we estimate this consequence using this performance measure? Is it likely to be very uncertain? If so how could that uncertainty be captured?
2 *For decision makers.* Is this the information people will need to see to make an informed choice? Will it expose the key value based trade-offs? If not, what's missing? Or which measures would be better? How important are the uncertainties likely to be?

Text box 5.5:   Choosing performance measures to support air quality management planning

Below is an influence diagram constructed by managers to help managers choose a performance measure for the Visual Quality objective of an air quality management process in Greater Vancouver.

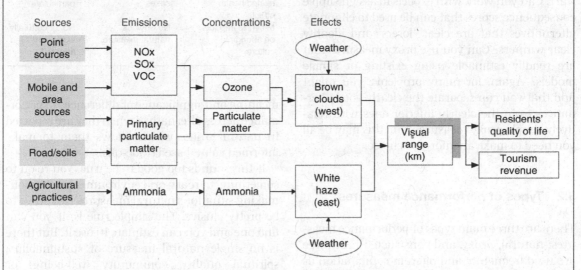

The influence diagram shows the relationship between major pollutant sources (at left) and the ultimate effects or fundamental objectives (at right). To the left of the effects, the major ambient air concentrations that most directly contribute to the effects are shown. To the left of concentrations are the major emissions that cause them, and at the far left, are the broad categories of major sources. Several performance measures are possible. One option is to address tourism revenue and residents' quality of life directly – perhaps by estimating changes in the number of tourist visits and/or associated expenditures for tourism, and developing a constructed scale for quality of life. Another option is to choose concentrations of ozone, particulate matter and/or ammonia as proxies for these more fundamental objectives. These would be less understandable and direct to many people, as the relationship between these concentrations and visual quality has to be inferred. Different people might make different assumptions about how serious the visual quality effects would be from different concentrations. Visual range, reported in kilometres (the distance people can see), is a key contributor to the two more fundamental objectives. It is well understood and easily modeled as a function of concentrations, and is the most significant contributor to the fundamental quality of life and tourism revenue objectives that will be affected by air quality.

Of course, this is not the end of the story; defining a performance measure as 'visual range' is not sufficiently specific. Visual range varies over space and time. A more complete definition is that the performance measure is visual range, measured in kilometres, at a particular site. Since visual range will vary daily, a specific statistic will need to reported (for example, the mean of the worst 20% of days) and it will need to be normalized for weather (the mean of the worst 20% of days under a pre-defined set of weather conditions).

Iterate

Sometimes the best performance measure is hard to estimate or model. Recall that decision makers often find perfectly acceptable solutions long before detailed quantitative modeling. If full-on modeling looks daunting, look for simpler ways to start. Can you work with experts to assign simple consequence scores that can be used to eliminate alternatives that are clear 'losers' and identify clear winners? Can you use proxy measures that are readily estimable using existing or simple models? Again, for many problems, you might find that you can separate the clearly good alternatives from the clearly inferior ones with relatively simple proxy measures, and this may be all you need to make an informed choice.

5.2   Types of performance measures

There are three main types of performance measures: natural, proxy, and constructed[3]. All three are used frequently and often in combination as part of the same project or program evaluation. The lines between the definitions of the different types can be a little grey – the point at which a natural unit becomes a proxy can be subtle. Don't worry about nailing down whether a particular measure is of one type or another; these classifications are offered only as a general guide to their use.

5.2.1   Natural is best

Natural performance measures are quantitative measures that directly report the achievement of an objective. If an objective is to minimize the costs of a proposed facility, then dollars are a natural performance measure (or francs, or rupees). If an objective is to 'minimize fatalities from accidents', then the 'number of people dying from accidents' would be a natural measure. The key advantages of natural measures are that they are direct, unambiguous, and understandable. It's easy for decision makers to intuitively grasp the

Table 5.5 Examples of natural and proxy measures.

Objective	Natural	Proxy
Maximize population of boreal caribou	Abundance (no. of caribou)	Area of habitat
Minimize air-quality related health effects	No. of respiratory cases	Emissions of particulate
Minimize impacts on aboriginal culture	No obvious natural measure	No. of culturally significant sites affected

meaning and significance of differences in performance across alternatives when they are reported in natural units, which allows them to make informed value-based trade-offs.

If this sounds too good to be true, you're on to something. In real world environmental decision-making situations natural measures can prove to be pretty elusive. The simple rule is, if you can find one and you can estimate it, use it. But there is no single natural measure of sustainability, spiritual quality, community well-being, or hunter satisfaction, for example. Even when there is one – say 'fish abundance' for a fisheries objective – it can be difficult to operationalize it due to lack of data, modeling capability, or other reliable means of estimation. Thus a preferred natural measure is not always practical to use.

5.2.2   The promise and pitfalls of proxies

When a natural measure can't be identified or developed in a practical way, we can use proxy measures instead. Like natural measures, proxy measures can usually be measured or modeled, but they report on the fundamental objective only indirectly. An effective proxy measure will correlate well with objectives that are otherwise difficult to measure or estimate (Table 5.5).

Because they are relatively easy to create and operationalize, proxies are certainly an attractive option, and are commonly used in environmental management situations. Habitat area can be used as a proxy for species abundance or welfare. Whereas it is often impossible to estimate with an acceptable degree of accuracy the effects of actions on caribou abundance, it is much easier to estimate the number of square kilometers of undisturbed habitat on which the caribou depend. Often, this is good enough – recall the water use planning example from Chapter 1, where creative win–win alternatives were identified and selected using proxy measures for fish abundance. Proxies can be useful for getting hard-to-quantify things on the table – recall the use of number of days of access to a culturally significant site from the same case study. It didn't capture the entire value of these cultural sites to the First Nations community, but it allowed it to be considered in the decision on equal footing with other objectives.

Unfortunately, the very characteristic that makes proxies easy to use can create problems: proxies report only *indirectly* on the fundamental objective. Decisions makers must make implicit judgments about the actual relationship between the proxy and the fundamental objective. There are four important and interrelated problems that result.

First, proxy measures can hide non-linear relationships between the proxy and the fundamental objective. For example, the availability of forage habitat for migratory birds may serve as a good proxy measure for their abundance, but only as long as availability of forage habitat is the limiting factor; after a point, more forage habitat does not necessarily provide any benefit at all because other limiting factors apply – such as availability of breeding habitat. Alternatives that provide more forage habitat than is useful may result in unnecessary trade-offs (protecting land that would have more value in other land uses).

Second, proxy measures can mask uncertainty in the relationship between proxy and fundamental objectives (whether it's linear or not), even though the proxy itself can be estimated with confidence. For example, although we might be quite certain of being able to produce 25% more habitat to aid migratory birds, we may have no idea whether this is what is limiting the production. If what we really care about is birds rather than habitat, then it's important that decision makers know that there is great uncertainty about whether an action that produces more habitat will in turn produce more birds. By choosing the proxy as the performance measure, subsequent analytical effort may focus on developing accurate estimates of the proxy and leave gaping holes in understanding about how the fundamental objective (birds) will respond.

Third, proxy measures can obscure important value judgments. This occurs whenever one proxy measure serves multiple objectives and the relationship with one objective is different than with another. In such cases, important trade-offs between the objectives may not be exposed[4]. Reservoir managers typically need to consider multiple objectives – dust control, visual quality, and protection of wildlife are examples – in managing a reservoir drawdown zone, and it might seem sensible and intuitive to use the 'area of vegetation' as a proxy for them all. However, the type of vegetation that maximizes dust control (e.g. large areas sparsely vegetated with simple non-native grass communities) is not the same as the type of vegetation that maximizes wildlife values (e.g. relatively smaller areas vegetated with complex shrub and cottonwood communities and a diversity of native grasses). Thus an overemphasis on the proxy can obscure dialogue about what really matters and ultimately lead to management actions that don't address the trade-offs across fundamental objectives.

Finally, the use of proxy measures may have adverse process and deliberative implications. In particular, over reliance on proxies can lead to a shifting of power in a deliberative setting. In environmental risk decisions, the ability to assess the relationship between a proxy and its endpoint or objective usually requires detailed technical knowledge. As a result, non-technical stakeholders will be forced to rely on technical specialists to provide insight into this relationship. It can become difficult to separate the technical judgment about the relationship between the measure and the objective from the value judgment about how much weight (importance) to give the objective. This tends to increase the power or perceived power of technical stakeholders, who should have no more legitimacy (and sometimes less) than non-technical participants for making the value judgments that underlie trade-offs among objectives.

5.2.3   Constructed measures: if you can't find them, make them

Constructed measures are used when no suitable natural measures exist or when the relevance of a proxy measure is tenuous. Constructed measures usually report an impact directly, but use a scale that is constructed for the decision at hand rather than already in wide usage.

A key advantage of well-designed constructed scales is that they facilitate the inclusion of hard-to-quantify things as part of the evaluation framework[5]. Well-designed measures allow consistency in assessing these hard to quantify things. As a result, they have important process advantages. They are a powerful tool for opening up a dialogue with community residents, resource users, or members of an aboriginal community and provide visible proof that the decision focus of SDM is something new and different – the goal is not a number or overall ratio but rather a clear depiction of how the things that matter – all the things that matter – are expected to be affected by potential management actions.

Constructed performance measures are actually quite common in everyday life, even if their generic name is not. Well-known examples include the Richter scale for earthquakes, the Apgar scale for newborn babies (which combines five separate features to indicate the overall health of a newborn), and the Michelin rating system for restaurants[6]. Over time these have become so widely used and commonly interpreted that they function almost like natural measures; most people generally understand that an earthquake of 3.0 is less severe than one of 7.0 (though many fail to realize this scale is logarithmic), and

physicians know to respond immediately when a newborn's Apgar falls below 3. Constructed scales are used in medicine to measure all sorts of physical and emotional responses; the 7-point Stanford Sleepiness Scale, for example, ranges from 'feeling active, vital, alert, wide awake' to 'almost in a reverie, hard to stay awake' and is used to report levels of sleepiness by patients with sleep disorders. As these cases demonstrate, thoughtfully designed constructed indices can greatly facilitate decisions by defining precisely the focus of attention, and facilitating consistency in evaluation.

Constructed measures can range from simple scales to sophisticated and highly specific impact descriptors. In this section we describe three types of constructed measures: simple rating scales, defined impact scales, and weighted indices.

Simple rating scales

Simple rating scales are (typically) three, five, seven, or ten point scales that ask for relative scores on an outcome of interest. They are commonly used in two ways. The first is in public surveys – e.g. 'on a scale of 1–7, how would you rate the quality of your natural environment'? (Just like the bathing suit.) In this context, simple rating scales do quite well, in part because when a large number of people respond to questions using such a scale, any differences in how they interpret it tend to average out. This is not true for the second common environmental management application, when an expert is asked to assign a score on a scale to represent the expected impact of an alternative. Figure 5.1 shows a generic seven point scale for assessing the effects of a management action.

Although usually simple to design and administer, rating scales have potential problems if you're trying to use them to communicate the consequences of alternatives. Suppose the scale is used to assign and report predicted consequences of a management action on mountain goat habitat. Unless the points of the scale are defined unambiguously, there can be a lack of clarity and

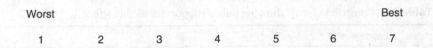

Figure 5.1 A typical seven-point scale.

Text box 5.6: The perils of risk rating scales

A troubling variation on the simple rating scale is a risk rating scale. Experts are asked to rate the risks to an outcome of interest, often using simple labels of low, medium, and high risk. Unfortunately, the language of risk is both ambiguous and value laden. Suppose you are told that 'the risk to wild sheep is low'. What could this mean? It could mean there is little or no probability of an impact (whatever 'impact' means!). Or it could mean the risk to wild sheep is low relative to the benefits for overall biodiversity, and therefore 'worth it' in the eyes of the assessor. This type of assessment happens all the time: assessors implicitly make trade-offs and value judgments in assigning the level of risk. If you hear 'the risk to wild sheep is high', again there are multiple interpretations – it could mean extirpation is imminent (an obviously pressing policy concern) or it could mean abundance may drop 5% rather than rise (the importance of which may be open to debate). If it is costless to reduce risk, this ambiguity doesn't matter. But if there are trade-offs to make, which is nearly always the case, then decision makers need to know more than what they usually learn through simple risk rankings.

---

consistency about exactly what is meant by the different scores: if one alternative scores five and another scores two, how much better is the second alternative relative to the first?; will two people have the same interpretation? The best and worst bounds of the scale are particularly critical, but rarely defined. Is the 'worst' case a total loss

of habitat or a loss of 10% of the habitat? This matters, because at some point a decision maker will have to trade-off this difference against some other measure, such as dollars, or impacts to visual quality. Thus it's critical that the decision maker has some sense of how significant the impacts are in an absolute sense, and simple rating scales don't provide that.

In general, for most decision making purposes, simple rating scales are simply not adequate. That said, they do have a role to play, particularly during the initial, decision sketching phases of a decision-making process.

Defined impact scales

A better approach is to develop a constructed measure known as a 'defined impact' scale, which combines a numerical scale with a narrative description, resulting in the definition of several discrete and well-specified levels of impact. The narrative description should refer to a small set of factors that are used consistently to describe different levels of impact.

For example, to compare the degree of public support or collaboration for different facility siting options, Keeney and Sicherman[7] determined that two factors could be used to define levels of public support: the number of groups who oppose or support the facility and the strength of their support/opposition. They defined the constructed measure in Table 5.6 and used it to evaluate potential power plant sites (from Gregory & Keeney[8]). With the use of such a scale, the level of support can consistently be assessed across communities, and decision makers can reflect on whether it's worth it to spend $XX to ensure that a site is located in a community with a '1' rather than a '−2' in terms of public support.

Table 5.6 Constructed scale showing public support for facility siting.

Level	Description
1	Support – no groups are opposed to the facility and at least one group has organized support for the facility
0	Neutrality – all groups are indifferent or uninterested
−1	Controversy – one or more groups have organized opposition, although no groups have organized action-oriented opposition. Other groups may be either neutral or support the facility
−2	Action-oriented opposition – exactly one group has action-oriented opposition. Other groups have organized support, indifference or organized opposition
−3	Strong action-oriented opposition – Two or more groups have action-oriented opposition

Table 5.7 Calculating a linear weighted index to reflect area and quality of habitat.

Habitat type	Weight (W)	Option 1		Option 2	
		Habitat area affected	Weighted habitat area	Habitat area affected	Weighted habitat area
Habitat A	1.0	200	200	200	200
Habitat B	2.0	300	600	100	200
Habitat C	0.5	400	200	600	300
Total		900	1000	900	700

There are some important methodological issues to address when developing useful impact scales. These are not technically difficult and can be done by any manager, but they do require attention. We cover these in section 5.3.1.

Weighted indices

Recall the description of linear value models from Chapter 2. A weighted index is like a value model for a single objective. It breaks a single objective into its component parts, and then weights and aggregates the parts. This means that weighted indices combine technical and value judgments: after a performance or impact score is estimated for each component under each alternative (a technical judgment), an importance weight is assigned to each component (which involves a mix of value and technical judgments), then a weighted score is calculated for each alternative, and that score is used to compare alternatives in terms of their performance on this constructed measure.

For example, suppose there are two competing site layouts for a mining development, each of which affects three different types of wildlife habi-

tat. Suppose further that Habitat B is twice as ecologically valuable as Habitat A – that is, it takes two hectares of Habitat A to offset a loss of one hectare of Habitat B. Habitat C is less valuable than Habitat A – it takes only a half a hectare of Habitat A to offset a loss of one hectare of Habitat C. Now suppose you want to compare the two options in terms of their impact on wildlife habitat. Table 5.7 shows a calculation of weighted habitat area for each option. Both options affect 900 hectares of habitat. But Option 1 affects 200 hectares of Habitat A, 300 hectares of Habitat B and 400 hectares of Habitat C, while Option 2 affects 200 hectares of Habitat A, 100 hectares of Habitat B and 600 hectares of Habitat C. From Table 5.7, we see that the weighted habitat score demonstrates that Option 2 has less impact on wildlife habitat.

This is a simple example, because it involves weighting the importance of different types of habitat area, all of which are reported in hectares. Some weighted indices combine qualitatively different impacts – impacts that are recorded in different units. The same process applies for weighting and summing effects, but (as further discussed in Chapter 9) extra care

Text box 5.7:   A defined impact scale performance measure for recreational impacts

Suppose an outbreak of stinging ant is expected to result in loss of valued recreation activities. Specifically, the outbreak is expected to result in a loss in the number and quality of beach picnicking opportunities (which matter a great deal to people in some communities). How might the losses on recreation, and the effects of different management actions on limiting those losses, be estimated?

One way to estimate the loss would be to use any of a variety of economic methods (e.g. contingent valuation methods, travel cost models) to develop a measure of residents' willingness to pay to reduce or eliminate this pest. For many people, however, the required line of questioning (e.g. 'how much would you be willing to pay to reduce or eliminate the level of annoyance') would make little sense, since this is not a normal market transaction (beach access being free) and people don't think of the permanent loss of family picnicking (which aids family bonding, improves health and relaxation, provides a way to share with friends, etc.) as directly compensable by money. Affected stakeholders would most likely be more concerned about the loss in the quality of the recreational experience than about its dollar value, and decision makers would be more concerned about public outrage over the loss.

A different approach, in line with SDMs goals of working with peoples' values and encouraging deliberation, would be to develop a defined impact scale based on how many people would be affected and how serious their loss of amenity values would be. One way to structure the measure is to work with stakeholders to construct a defined impact scale, perhaps something like this

Level	Description	Value
A	No loss for any recreationist	0
B	Minor inconvenience for a small number of people (< 1000)	0.1
C	Significant inconvenience for a small number of people (< 1000)	0.5
D	Minor inconvenience for a large number of people (1000–20 000)	0.6
E	Significant inconvenience for large number of people (1000–20 000)	1

Five levels are defined, each described by characterizing two key factors: the number of people affected and the severity of the loss of quality of the experience. Each of these levels of loss is accompanied by a more complete verbal description, so that 'minor' and 'significant' inconvenience are more clearly defined (in this case, referring to the incidence and severity of stings). The third column of the table shows the value assigned to each level. Note that it is non-linear – that is, the loss in value from level B to level C is larger than the loss in value from C to D. This non-linearity may or may not be something you need to explore further – more on this later in 'dealing with non-linear measures'. But at least it's noted here and can be considered in deliberations.

An alternative approach is to develop a quantity–quality measure. This would involve defining several levels of quality of a recreational experience (perhaps based on the incidence of stings), estimating the number of user days at each quality level, and then calculating a weighted sum – just as in Table 5.7.

Which to choose – a defined impact scale or a quality-quantity scale – depends on the preferences of stakeholders and decision makers and the nature of the decision at hand. The defined impact scale offers fewer degrees of discrimination, but allows explicit assignment of relative values. It's focused on the loss of opportunity in a general sense (does the opportunity exist) rather than on actual use. The quantity–quality measure makes an implicit value judgment by its calculation – that is, it assigns the same value to a 1000 user-days at 50% loss of amenity as it does for 10 000 user-days at 5% loss. Decision makers, in discussions with stakeholders, would need to decide which best represents how people view the real relative impact.

and steps must be taken to ensure appropriate weighting and normalizing (see also Clemen[9]).

A special case of the weighted index is a quantity/quality measure. This type of measure is used when an impact is determined both by quantity and quality (as described in Keeney[3]). 'Quality' can refer to a number of things, depending on the quantity in question. For example, it could refer to concentration or toxicity of emissions, the density of an algal bloom, or the severity of a health effect. For example, in the health sector, the benefits of investments in different medical treatments or equipment are regularly compared using 'quality-adjusted life years' or QALYs, which weight the number of years saved by the quality of life. In an environmental context, tonnes of emissions are often weighted by a toxicity rating, recreational user-days are weighted by the 'quality of the experience', and so on.

5.3 Working with performance measures

Here we cover some practical issues in working with performance measures that will be useful for those charged with developing them.

5.3.1 Designing useful impact scales

In section 5.2.3 we discussed impact scales and noted that these have wide applicability. Here we outline some elements of good practice in developing them. Recall how such a scale will be used. Presumably you've already clearly identified the objective and you've defined bounds of the scale that adequately cover the range of consequences for the alternatives under consideration. Now you're developing a measure to facilitate the process of estimating the performance of these alternatives on that objective. The measure you develop will be used (a) to elicit judgments from experts about the consequences of each alternative; and (b) to report the consequences of the alternatives to decision makers, probably in the form of a consequence table. The following

outlines some steps and considerations in developing a useful impact scale.

1   *Choose between two and ten well-defined levels of impact.* The number depends on the context and the level of discrimination that's desirable and realistic. Experts will be asked to assign one of these impact levels to each alternative (or where uncertainty is present, also an upper and lower level).

2   *Define the bounds of the scale.* You must decide whether you're defining global or local bounds. Global means you are considering the entire universe of possible alternatives and their outcomes. More often, it is local bounds that are more appropriate. Thus, defining local bounds means asking, what's the worst and best plausible outcome we can expect in this decision context, given the alternatives on the table?

3   *Decide on a small set of key factors that will be used to characterize each level of the scale.* Why? Because you want to design it so that independent experts could arrive at similar judgments about how an alternative scores on the constructed measure. To facilitate this, it's important to limit the verbal description to a small number (one to three) of considerations relevant to characterizing the nature and magnitude of the impacts. More than this will result in loss of objectivity (there being too many value judgments to make in deciding whether an impact is level 3 vs 4, etc.), continuity (an alternative may fall between levels unless such value judgments are made) and consistency (because different experts will make different value judgments).

4   *Develop verbal descriptions for each level.* Pay special attention to the definition of the upper and lower bounds. These must be clearly described. No exceptions. Interim levels should ideally also be well-defined, but to be honest, we've often cut corners in the middle levels of a five or seven point scale, and developed fairly limited (or no) verbal description of these impacts – see the example in Chapter 9. It's also

useful to make an effort to distribute the levels roughly evenly along the continuum of plausible impacts – meaning if you have six levels, don't cluster five of them near the lower bound.

5 *Label the levels.* Try to use something innocuous. Letters are good. They can be interpreted as 'grades' and thus have an obvious order of preference without implying a quantitative relative score.

6 *Assign relative values for each level.* The levels of the scale clearly imply ordinal ranks. But bear in mind that this ranking says nothing about the relative differences between levels of impact – a 10 is not likely twice as good (or bad) as a 5, nor is 9 likely to be conveniently three times as good as 3. You can use ordinal ranking effectively to facilitate discussion. In these cases, it's important to discuss non-linearities explicitly. If you have an intention of moving on to more quantitative analysis of trade-offs, then you're going to have to translate these ordinal rankings into more appropriate relative values at some point.

Step 6 is tricky. And the reality is, for many decisions you will not need to – or have time to – do it well. As a practical short cut, we recommend you stop and think before doing step 6. Make sure you discuss obvious non-linearities with your group so that they are not overlooked. Then examine what role this scale plays in the analysis. Are non-linearities likely influencing the quality of value judgments? Will quantitative analysis of trade-offs be used? If the answer to either of these is yes, then you will need to more rigorously explore the relationship between points on the scale and value or utility.

5.3.2 When measures have non-linear value functions

In some cases it's important to translate raw performance as reported by natural, proxy or constructed measures, into an expression of value or desirability (formally a single attribute value function; see Clemen[9] or Keeney[10]).

When the relationship between the performance measures and overall desirability is linear, this is pretty straightforward. If performance increases by 10%, value or desirability also increases by 10%. Fundamental objectives, properly defined, are almost always linear. However, when we choose performance measures to report on those objectives we make compromises, and often we choose indirect or proxy measures.

Consider the three 'value functions' shown in Figure 5.2. A value function depicts the relationship between a measure and its associated value to decision makers. The measure in this case is 'habitat', but that's just illustrative. It could equally be 'emissions' or 'cost' or a constructed measure for 'aesthetic quality'. The curve in Figure 5.2(a) is linear: all increments of habitat are valued equally (e.g. this is typical for improvements in habitat for a healthy species). The curve in Figure 5.2(b) is not linear. It demonstrates decreasing returns – the first units of habitat are highly valued but after that additional units count for less (critical habitat is initially limiting, but as it becomes less scarce, other factors such as food or reproductive success become limiting), until at some point, additional habitat has no value. Figure 5.2(c) is also nonlinear. This time the relationship has a sharp step or discontinuity part way through: increases in habitat up to that point count for little, but higher values are meaningful (as when an endangered species needs a minimum amount of habitat to support a population that's self-sustaining).

In each of these three cases, habitat is a proxy measure for the more fundamental objective of protecting a valued species. In the first case, the relationship between the measure and the objective is linear and the measure is a good representation of value, just as shown. In the second and third cases, it might be important to be aware of and discuss non-linearities to make sure they are considered.

This same discussion also applies to constructed measures. The horizontal axis of the charts in Figure 5.2 could equally be the points on a constructed scale. Table 5.8 provides an example of a constructed scale for reporting the quantity

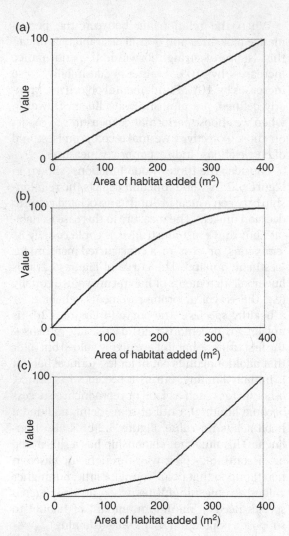

(a)

(b)

(c)

Figure 5.2 (a) Linear value function. (b) Non-linear value function with decreasing returns. (c) Non-linear value function with discontinuity.

of water delivered to rural communities by different water supply options. The first column is the simple ordinal rank label assigned to a performance or impact level. The second describes relative performance in words and numbers. The third shows the value or utility assigned to that level of performance. Here we see that the value of moving from level 1 to level 2 is very high – for example, this may be because an internationally

Table 5.8 A non-linear constructed scale for drinking water supply.

Level	Description	Assigned value
1	Delivers up 10–20 liters per person per day	40
2	Delivers 20–30 liters per person per day	80
3	Delivers 30–40 litres per person per day	100
4	Delivers > 40 liters per person per day	100

Text box 5.8: Psychophysics and the choice of an appropriate scale

For over a half-century, psychologists have been interested in different ways to measure sensory and physical stimuli, which can reflect a subjective magnitude (e.g. the importance of an impact, or the loudness of a sound) or an emotional evaluation (e.g. satisfaction or loss). An extensive and fascinating literature now exists on the pros and cons of different measurement scales. Psychophysics researchers have identified several biases that typically influence how people respond to scales: even if respondents know their own feelings well, strange things can happen because they have to translate this experience into terms (via the choice of scale) set by the analyst or investigator. As one example, Poulton[11] discusses numerous cases of what he terms 'response range' effects; one common effect is participants' tendency to select a value in the middle of a set of responses. Arvai *et al.*[12] observed this effect in the context of wildfire management options: participants asked to rate one of three different fuel management options consistently selected the middle numbers, even though the full associated range of consequences varied widely. These results suggest that whenever scales asking for other than ordinal (ranking) responses are used, it is wise to pay attention to ways in which cues might unintentionally be provided to participants; otherwise, artefacts of the selected method might easily be mistaken for underlying beliefs and preferences.

recognized standard is met for supplying adequate amounts of water to meet basic human needs. The value of moving from level 2 to level 3 is smaller, and from level 3 to level 4 it's zero – this may be because there is no real benefit in supplying more than 40 litres per day, basic needs (in terms of the criteria important to the aid organization) are met and thus there is no justification for spending additional resources for water supplies above 40 litres per person per day.

5.3.3 When does it make sense to monetize impacts?

In the broadest sense, it makes sense to monetize impacts when people naturally think about the consequence in monetary terms, such as profits or financial expenditures. Changes to a royalty that might be paid to the government through mining activities or forest stumpage fees, or changes in the flow of services from environmental resources that typically are sold through well-established markets (e.g. natural gas sales, sawlog production), changes in agricultural production, and local economic benefits from tourism are other examples. Other specific examples of when the use of a performance measure denoted in monetary terms might make sense include:

1 When environmental costs truly are compliance costs (for example, the price of tradeable emission permits).
2 When defensible and widely accepted estimates of the monetary value of damages exist (for example, the repair costs of infrastructure damaged by pest incursion, flooding, or air pollution).
3 When payments or compensation for impacts or damages are under consideration. Although the extent of economic, social and cultural, health and safety, environmental, and other damages – for example, associated with a history of oil or gas exploration and production in an area[13] or associated with long-term exposure to nuclear or chemical contamination – typically extends well beyond

any monetary measure of losses, there clearly exist situations where the legal and administrative context is primarily or exclusively focused on payments of monetary compensation. This does not mean that the estimates of loss are comprehensive, rather that there is a practical constraint and obtaining some compensation for losses is preferred by the individuals or community in question to none. And it is important in such cases for SDM analysts or facilitators to be sure about their own level of comfort in coming up with monetary measures for objectives that are not traded in markets and that can be represented only partially by monetary measures.

5.3.4 What about yes/no issues such as compliance or ethical principles?

Binary measures – where the measure reports a simple yes or no response – can be useful in cases where there are clear thresholds such as the need to comply with a regulatory standard or a legal requirement. If there is broad agreement that there is no value in testing or pushing back on these limits, then such on-off measures can usefully be applied at the screening or decision sketch stage. Even if there is no firm definition of what is acceptable, it may be possible to establish some value for a measure that clearly is not acceptable and to use this to screen out poor alternatives. Care must be taken, however, to ensure that the screening process relates back to the fundamental objectives used to characterize the problem. For example, we often have seen a binary measure such as 'achieving consistency' with a policy directive or with practices in other jurisdictions used as a performance measure (and often as an up-front screen), whereas it is usually much better to identify the underlying fundamental objective that this consistency is meant to address – improving competitiveness perhaps or minimizing cost. This allows for creative alternatives to be developed – see Chapter 7.

Be careful with binary performance measures. In Chapter 4, we discussed at length the 'trouble with targets'. When you set a fixed target, you hardwire in trade-offs in a way that may make it difficult to identify creative alternatives and ultimately to get agreement on a preferred solution. When a measure is proposed that is 'compliance with rule xx', we suggest that it may be useful to explore alternative formats that allow you to report the degree of compliance, such as 'frequency of exceeding rule xx'. If non-compliance with a standard or a policy directive is absolutely not an option at any level, then use this to screen out unacceptable alternatives. By the time you're seriously evaluating alternatives that truly are candidates, there should be no need for this measure.

Binary performance measures can be particularly helpful in capturing moral or ethical values and creating space for discussing them in the deliberative setting. An example comes from a multi-stakeholder water-use planning process at the Campbell River hydroelectric facility on Vancouver Island[5]. The focus was to evaluate alternative flow regimes downstream from a dam, with the goal of balancing power generation, recreation interests, and fish impacts. Beginning with a value-focused thinking approach, participants were encouraged to articulate what mattered to them in terms of water flows and why. It became apparent that the current configuration of the dam facility violated a fundamental principle of the Mowachaht/ Muchalaht First Nation, which is that water flowing naturally in one direction, in this case west, should not be diverted to flow in the opposite direction. A measure was created to report which alternatives under consideration violated this principle, and, in conjunction with measures for power production, salmon habitat, and recreational quality, new alternatives were developed to avoid violating it. It was not appropriate in this case to use this measure as an *a priori* screen, as many participants were not willing to eliminate alternatives that violated this principle. However, they were willing to search for mutually acceptable alternatives, with the consequences for this principle as one legitimate focus of analysis and dialogue.

Clearly, this binary measure was a crude place-holder for a much richer dialogue, but it served the purpose of making the concern transparent and putting the loss created by a violation of principle on an equal footing (i.e. in the same matrix of consequences) with more conventional expressions of impact.

5.4 Case studies in developing performance measures

Here are two case studies where we used some of the approaches in this chapter to develop useful performance measures. The first describes the selection of performance measures for water management on the Athabasca River. The second describes the development of an index for assessing the consequences of alternative strategies for prescribed burning in a provincial park.

5.4.1 Identifying performance measures for traditional use: Athabasca River

As part of a planning exercise to evaluate the multidimensional effects of forecast increases in water withdrawals associated with new production from oil sands, we recently led a multi-stakeholder committee composed of government regulators, representatives of affected aboriginal groups, mining companies and environmental NGOs through a comprehensive SDM planning process in Alberta, Canada. One of the many challenges in the process involved translating concerns over the potential impacts of water withdrawals into objectives and performance measures, and doing this in a way that would encourage defensible analyses and meaningful deliberations and consultations among a diverse group of stakeholders.

In the early stages of the process, we led the multi-stakeholder group through the steps of setting objectives, developing influence diagrams and developing performance measures in three primary interest areas: aquatic ecosystem

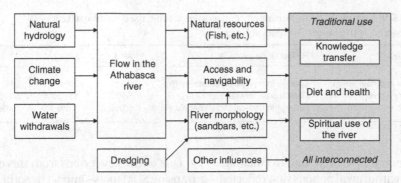

Figure 5.3 Influence diagram showing connections between water withdrawals and traditional use interests on the Lower Athabasca River.

health, sustainable economic development, and traditional use.

For the latter, a comprehensive traditional use study commissioned by the process had provided a series of impact hypotheses linking potential water withdrawals to impacts on the interests of local aboriginal residents. With this basic information in hand, the task was to define a relevant performance measure, which could be used to develop a decision focused model that would estimate the effects of different management alternatives on this multifaceted traditional use objective.

A first step was to synthesize the extensive information available into a concise influence diagram that shows how water withdrawals and other actions (dredging) and stressors (climate change), can affect important traditional use activities in the Lower Athabasca River region (Figure 5.3). Drawing this figure was important because it captured the issues aboriginal communities were concerned about and created a vehicle for discussing the various pathways of effects, or ways that water withdrawals could affect all these issues. After acknowledging the interconnected nature of the fundamental traditional use objectives (knowledge transfer through the generations, diet and health, spiritual use of the river) and the many other influences on these traditional use objectives (e.g. exposure to western cultural influences), the committee soon narrowed in on concerns regarding access and navigability as the primary potential link

between water withdrawals from the river and traditional use objectives.

The effect of water withdrawals on flow in the river depends on the timing, magnitude, frequency, duration, and extent of water withdrawals. For example, a withdrawal of $10 \ m^3/s$ in the fall when flows are naturally low may have a greater influence on navigability than the same withdrawal in the summer when the flows are higher. Likewise, the effect of withdrawal during a dry year may be considerably different than the same withdrawal during a wet year. With these influences in mind, key information for development of the performance measure included:

1 A hydraulic model that could correlate flow in the river with the depth at any point along several test sections of up to six kilometers in length.
2 A depth suitability curve scaled to reflect that depths greater than 1 meter were suitable for all forms of powered water craft used on the river (weight = 1), with suitability declining steadily as depth declined to a threshold of less than 0.3 meters (weight = 0) where shallow sandbars caused significant navigation challenges.
3 Knowledge of the preferred seasons for key traditional use activities (e.g. fall hunting, summer fishing).

By combining this information it was possible to calculate the total suitability-weighted area for

Table 5.9 Sample from the summary consequence table used to inform trade-off discussions on the Lower Athabasca River.

Objective	Performance measure	Alt 1	Alt 2	Alt 3
Access and navigation	Change in navigation suitability (%)	2.7%	2.4%	2%
Industry – cost	Capital cost for water storage	$800M	$1700M	$3100M
Aquatic ecosystem	Change in mid-winterfish habitat (%)	5.9%	4.6%	2.6%

navigation for each test section, and to calculate results for water withdrawal options that reflected differing amounts of permissible water withdrawal during different times of year. After running a series of water withdrawal alternatives through the performance measure calculation, it was determined that the fall period was the most sensitive to water withdrawals. As a result, the final concise performance measure used in the decision process was '% change in navigation suitability from natural flow conditions in reach 4 during the fall period'. These results were then presented with similar performance measure results that were developed for ecosystem health and economic objectives as the basis for informed trade-off discussions (Table 5.9).

Not surprisingly, the entries in Table 5.9 were not immediately meaningful to all participants. To support understanding and dialogue of the performance measure during committee meetings and consultations, we also developed a tool that enabled the visual representation of changing depth along the river test sections during any week of the year and across any flow changes Figure 5.4 shows the navigational suitability for one of several short sections of the Athabasca River, using colours to indicate suitability at any point. As the user turns the top dial to the right, the colours will update to show the navigational suitability at different flow rates. (The second dial presented additional information - not shown here – on the historical statistical likelihood of any given flow rate occurring for a given week in the year). This helped participants visualize the link between water withdrawals and changes in navigability and made the metrics shown in Table 5.9 more meaningful.

One of the key lessons from this example is that by using SDM tools – impact hypothesis statements, influence diagrams, and performance measures – it is possible to apply the same degree of rigor to assess interests as seemingly intangible as traditional use as for other interests such as fish habitat and capital cost. The use of these tools helped to facilitate dialogue and to replace controversy and frustration with a framework for collaboration.

5.4.2   Performance measures for wildfire management: Mount Robson Provincial Park

Mount Robson Provincial Park is a 220 000 hectare protected area on the west slope of the Rocky Mountains in eastern British Columbia and part of the UNESCO Canadian Rocky Mountains World Heritage Site. As a forested area, the park experiences wildfires, which, though part of the natural order of things, constitute a danger to park visitors and nearby communities. Partly due to historical management practices to prevent and suppress fires, and partly, some believe, due to recent climate changes (e.g. reduced precipitation), the recent epidemic Mountain Pine Beetle outbreaks, and consequent threat of large and more intense fires in the region has worrying implications not only for safety and economic reasons, but for environmental reasons too. More than 30% of the park area is rated in a high fire hazard class based on current forest condition. Working with professional foresters and park managers, we were asked to evaluate the effects of proposed wildfire management actions on the fundamental economic, social, and environmental objectives for the Park.

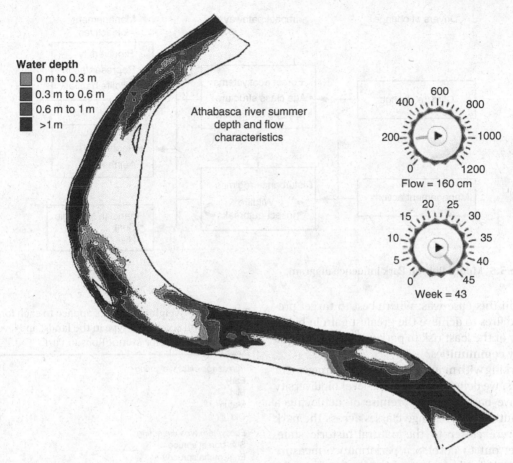

Water depth
- 0 m to 0.3 m
- 0.3 m to 0.6 m
- 0.6 m to 1 m
- >1 m

Athabasca river summer
depth and flow
characteristics

Flow = 160 cm

Week = 43

Figure 5.4 Screenshot of the visual assessment tool that allows spatial representation of depth to be shown for river test sections for user-selected flows (m³/s) and timing (week of year).

Figure 5.5 shows a simplified version of an influence diagram we developed to link the primary drivers of change to fundamental management objectives and to serve as the basis for discussions.

Biodiversity in the park is strongly influenced by forest ecosystem attributes (primarily forest age class structure and species composition) and natural disturbance regimes, such as fire and forest pest outbreaks. Historical management actions, in particular decades of fire suppression, have combined with recent climate changes and resulted in epidemic beetle outbreaks. These outbreaks cause significant tree mortality over large areas resulting in dead wood accumulations that increase the risk of large-scale wildfires. Through this complex relationship, climate change and management actions affect the likelihood and behavior of wildfires, which, in turn affect a range of environmental (in this case, taken to be biodiversity specifically), social and financial and strategic objectives.

Fire management plays a central role in controlling the long-term processes that determine forest ecosystem attributes and hence biodiversity, so not surprisingly conducting prescribed fires is one of the most effective landscape level management tools. The challenging question

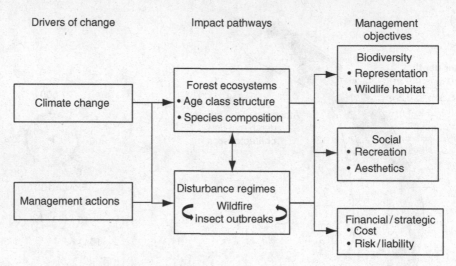

Drivers of change   Impact pathways   Management
                                                      objectives

Figure 5.5 Mount Robson Park influence diagram.

faced in this case was: where best to target pre-scribed fires to achieve the greatest gain in biodiversity at the least risk to park infrastructure and nearby communities?

Working with managers and forest management experts, we defined a landscape-level biodiversity objective based on the premise of achieving a distribution of forest age classes across the park that were closer to the natural/historic state, and set out to develop a performance measure index to help characterize biodiversity under the Current and Projected Post-fire landscapes.

We created a weighted index based on 'forest age class' and 'ecosystem type'. To develop the index, we asked experts to rate the relative value of each forest age class and ecosystem type. First we asked which age class would be of greatest value given the current situation. The answer: a tie between 'Old' and 'Early' – both are less common than they should be, relative to the natural state. We weighted these as '1'. Then we asked the converse – which age class has the least value? The answer – 'Mature'. We assigned a weight of 0 (no value). Finally, we asked for the appropriate weighting that should be associated with the remaining age class. The answer: 0.25 for the 'Mid' class. Performing the same exercise for the ecosystem types yielded a weight of 1 for 'Sub-

Table 5.10 Weighting factors applied to each forest age class and ecosystem type in the landscape-level biodiversity index for Mount Robson Park.

Forest age class weighting	
Early	1
Mid	0.25
Mature	0
Old	1
Ecosystem type weighting	
Sub-boreal spruce	1
Englemann spruce / fir	0.5
Interior cedar hemlock	0.25
Other	0

Boreal Spruce', a weight of 0 for the innocuously labeled 'other' types, a weight of 0.5 for Englemann Spruce/Fir, and a weight of 0.25 for Interior Cedar Hemlock. In other words, one hectare of Old Sub-Boreal Spruce has the same value as two hectares of Old Englemann Spruce/Fir or four hectares of Old Interior Cedar Hemlock (and so on through other combinations of age class and ecosystem types) (Table 5.10).

Prescribed fire options were then developed at strategic locations within the park, taking into consideration the forest age class, ecosystem type, and areas of highest fire and pest outbreak hazard (Figure 5.6). A GIS model simulated a prescribed

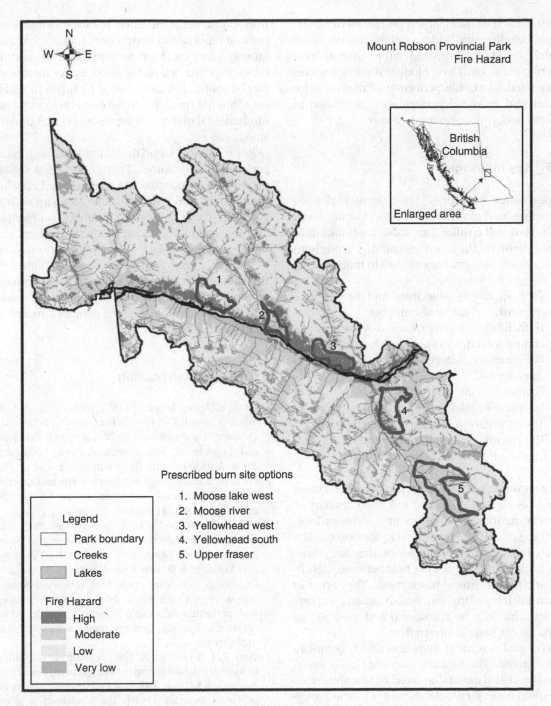

Figure 5.6 Prescribed burn site options in Mount Robson Provincial Park.

burn in each location, and reported out the 'post-burn' biodiversity index weighted-area score which could be compared with current, 'pre-burn' conditions. These biodiversity index scores were used, alongside performance measures for social and financial performance, and used to inform park management decisions.

5.5   Key messages

Performance measures take vaguely defined objectives and make them specific. For any decisions that will require quantitative estimation of consequences, they are essential. Throughout the decision process, they're used in many ways:

1   To help clarify objectives and to generate responsive, creative alternatives.
2   To facilitate discussions about what matters, among stakeholders and decision makers.
3   To compare alternatives accurately and consistently.
4   To prioritize information needs.
5   To expose trade-offs, including trade-offs among different degrees of uncertainty.
6   To communicate the rationale for, and improve the transparency of, decisions.

In summary, performance measures determine what people will think and talk about as part of environmental management deliberations. Nothing could be more critical to decision quality. In technically intensive decisions, they are essential for leveling the playing field between technical and non-technical participants. They are also essential for getting agreement among experts about what will be considered and used as the basis for evaluating alternatives.

The goal is a set of measures that is complete and concise. The best measures are unambiguous, direct and understandable. But compromises are often necessary to get measures that are operational. In general, it's best to use natural measures, but unfortunately they are sometimes hard to estimate, or don't even exist for some objectives. Constructed measures are useful for assessing and reporting consequences that are hard to quantify; they can be important for ensuring that these objectives are not ignored. Proxy measures can be useful, but care should be taken to make sure that the relationship between the proxy and fundamental objective it represents is well understood.

Be prepared to spend time coming up with good performance measures. There is no right set of performance measures, but some are more useful than others, and it takes time to come up with a good set. If you're involved in an environmental decision or looking at decisions that others are making or have made, look at the performance measures closely. Do they accurately represent the things that matter? What value judgments are embedded in them? Who made those value judgments? Would a different choice of measure change the decision?

5.6   Suggested reading

Bazerman, M.H., Messick, D.M., Tenbrunsel, A.E. & Wade-Benzoni, K.A. (1997) *Environment, Ethics, and Behavior: The Psychology of Environmental Valuation and Degradation*. San Francisco: New Lexington Press. A wide-ranging set of conceptual and practical papers discussing how people form and express their opinions and choose to act in relation to environmental and ethical concerns.

Keeney, R.L. (2007) Developing objectives and attributes. In: *Advances in Decision Analysis: From Foundations to Applications* (eds W. Edwards, R.F. Miles, Jr & D. von Winterfeldt). pp. 104–128. Cambridge University Press, New York. A clear and concise primer on methods for developing objectives and performance measures, with examples from corporate, public policy, and environmental applications.

Keeney, R. L. & Gregory, R. (2005) Selecting attributes to measure the achievement of objectives. *Operations Research*, **53**, 1–11. This paper addresses a variety of problems associated with the choice of a good performance measure, illustrates desirable properties with examples, and presents a decision model for selecting among attribute types.

Poulton, E.C. (1977) Quantitative subjective assessments are almost always biased, sometimes completely misleading. *British Journal of Psychology*, **68**, 409–425. The author is knowledgeable and opinionated; in this early book he covers both theory and practice concerning ways to make high quality quantitative judgments and avoid common biases.

5.7 References and notes

1 Wilson, R. & Crouch, E. (1982) *Risk/Benefit Analysis*. Ballinger, Cambridge, MA.

2 Keeney, R.L. & Gregory, R. (2005) Selecting attributes to measure the achievement of objectives. *Operations Research*, **53**, 1–11.

3 Keeney, R.L. (1992) *Value-Focused Thinking: A Path to Creative Decisionmaking*. Harvard University, Cambridge.

4 Fischer, G., Damodaran, N., Laskey, K. & Lincoln, D. (1987) Preferences for proxy attributes. *Management Science*, **33**, 198–214.

5 Turner, N.J., Gregory, R., Brooks, C., Failing, L. & Satterfield, T. (2008) From invisibility to transparency: identifying the implications. *Ecology and Society*, **13**. Retrieved from http://www.ecologyandsociety.org/vol13/iss2/art7/

6 Other common uses of constructed scales include indices of the health or happiness of a country's citizens (extensively used in Europe), rankings of different colleges, or rankings of sports teams (reflecting the relative difficulty of schedules, the magnitude of winning or losing margins, etc.).

7 Keeney, R.L. & Sicherman, A. (1983) Illustrative comparison of one utility's coal and nuclear choices. *Operations Research*, **31**, 50–83.

8 Gregory, R.S. & Keeney, R.L. (2002) Making smarter environmental management decisions. *Journal of the American Water Resources Association*, **38**, 1601–1612.

9 Clemen, R.T. (2004) *Making Hard Decisions: An Introduction to Decision Analysis*, 4th edn. Duxbury, Belmont.

10 Keeney, R. & von Winterfedlt, D. (2007) Practical value models. In: *Advances in Decision Analysis: From Foundations to Applications* (eds W. Edwards, R. Miles & D. von Winterfeldt). pp 232–252. Cambridge University Press, Cambridge.

11 Poulton, E.C. (1977) Quantitative subjective assessments are almost always biased, sometimes completely misleading. *British Journal of Psychology*, **68**, 409–425.

12 Arvai, J.L., Gregory, R., Ohlson, D., Blackwell, B.A. & Gray, R.W. (2006) Letdowns, wake-up calls, and constructed preferences: people's responses to fuel and wildfire risks. *Journal of Forestry*, **104**, 173–181.

13 Gregory, R. & Trousdale, W. (2009) Compensating aboriginal cultural losses: an alternative approach to assessing environmental damages. *Journal of Environmental Management*, **90**, 2469–2479.

6

Incorporating Uncertainty

Uncertainty refers to situations or outcomes for which we lack information that we'd like to have. Of course, this pretty much characterizes the entire scope of human experience beyond trigonometry and sudoku. Given its universality, uncertainty – and skills to handle it – should be second nature to everyone. In casual ways it is, of course. If an appointment is important enough, we'll set a second alarm clock and catch the earlier bus to make sure we get there on time. If we're not sure how much salt to add, we'll add a pinch at a time until things taste about right. We buy insurance, follow rules of thumb, lick a finger and feel the air. The good news is that, despite the all-pervasive presence of uncertainty, most of us are reasonably good at handling it in our everyday lives. The bad news is that, when the stakes get higher and we're dealing with important decisions in our jobs as environmental managers, we often ignore the influence of uncertainty and fail to either anticipate or communicate its impact. As a result, we make mistakes in judging its effects and in developing effective decision strategies.

As you might imagine, there are entire bookshelves dedicated to the analytical treatment of uncertainty. Well-known methods such as sensitivity and scenario analysis, Monte Carlo simulation, decision trees, and Bayesian updating are complemented by methods such as fuzzy sets and information gap theory. Powerful as they often are, each of these methods is relevant for only a subset of the uncertainties affecting environmental decisions. It is a relatively small portion of real-world uncertainties that is amenable to highly quantitative treatment in the best of cases; in practice, given the limitations of time, budget, and technical capabilities, that subset is smaller still. We are not suggesting that complex technical analysis has no role in a deliberative environment – quite the opposite. The small group structure of a typical SDM process allows for a degree of penetration by stakeholders into complex analyses that simply isn't possible with other kinds of collaborative processes. These specialized analyses may be critical to making informed choices in some cases. They're just not the subject of this chapter.

We're aware that some readers may want this level of technical detail. We're also aware that some readers will take one look at graphs of probability distributions and want to skip this chapter entirely. Please don't. Even the most routine decisions in modern resource management require a basic knowledge of the language and implications of uncertainty. Our goal in this chapter is to give resource managers and other decision participants an overview of some useful ways to approach uncertainty using the concepts and

Structured Decision Making: A Practical Guide to Environmental Management Choices, First Edition. R. Gregory, L. Failing, M. Harstone, G. Long, T. McDaniels, and D. Ohlson.
© 2012 R. Gregory, L. Failing, M. Harstone, G. Long, T. McDaniels, and D. Ohlson. Published 2012 by Blackwell Publishing Ltd.

methods of SDM, without using sophisticated quantitative methods.

We begin by briefly introducing different sources of uncertainty and how they arise in decisions. Recognizing that for some readers it may be many years between now and their last course on probability theory (or perhaps they never attended one!), we then provide a brief overview of the basics – just enough to ensure familiarity with principles needed to make smart choices under uncertainty. The rest of the chapter demonstrates the application of these concepts to multi-issue, multi-stakeholder environmental decision processes – the type of decisions where SDM is most often used. The focus is on how to understand and express uncertainty in consequences. How might we structure uncertainty so we can better understand its role in decision making? How can we show that some alternatives are more uncertain than others in a way that will be understood? What is risk tolerance and how does it affect choices? How do we deal with differences or disagreements among experts? What might be done to reduce uncertainty or improve the quality of information? How might alternatives be developed to learn more about, and ultimately reduce, uncertainty over time?

In this chapter, our focus is on uncertainty related to matters of fact. Uncertainty also arises with respect to values. Sometimes managers are not certain of their own values with respect to unfamiliar trade-offs. In others, they may not be certain about stakeholders' values. This question, of uncertainty related to values, was introduced in Chapter 2 (via the concept of constructed preferences) and is discussed again in Chapter 9, in relation to framing difficult trade-offs and testing the sensitivity of decisions to uncertainty in stakeholder values.

6.1   The basics

6.1.1   Sources of uncertainty

There have been numerous attempts on the part of statisticians, philosophers, ecologists, and economists to develop a broadly accepted typology of uncertainty. No single classification scheme is perfect for all management situations. For thinking about uncertainty in the context of environmental decisions, we like the typology proposed by Regan, Colyvan, and Burgman[1], which divides the world into epistemic uncertainties (those resulting from lack of knowledge) and linguistic uncertainties (those resulting from communication failures).

As shown in Table 6.1, epistemic (or knowledge) uncertainties arise in several ways. Measurement error arises from the simple inability to measure things precisely. When this error occurs in systematic ways due to bias in calibration or sampling for example, it can be called (unsurprisingly) systematic error. Natural variation (sometimes called aleatory uncertainty) refers to changes in the natural environment with respect to time, location, or other considerations. While other types of uncertainty can be reduced through better data, measurement, or analysis, natural variation cannot. It can be better understood but not reduced.

Measurement error, systematic error, and natural variation are sources of uncertainty about the value of empirical quantities or parameters[2] – things like temperature, or ambient concentrations or biodegradation rates. In contrast, model (or structural) uncertainty arises when we don't know how the various components or causal factors in a system interact with each other – including physical, chemical, biological, economic, and social systems. For example, we may know the water temperature in a river with some confidence (an empirical quantity) but not know how temperature interacts with processes and properties to influence other biological outcomes, such as algal growth. Debates about dose–response relationships are about model uncertainty: we may know ambient concentrations of particulates in an airshed, but we're uncertain about the resulting human health effects. In decision making, model uncertainty further extends to include not just the interactions of elements in the natural system, but also the interactions of management

Table 6.1 Sources of uncertainty (adapted from Regan[1]).

Sources of epistemic (knowledge) uncertainty	When we are uncertain about facts concerning events or outcomes in the world because …
Natural variation	… outcomes that vary naturally with respect to time, space or other variables and can be difficult to predict
Measurement error	… we cannot measure things precisely
Systematic error	… we have not calibrated our instruments or designed our experiments/sampling properly
Model uncertainty	… we do not know how things interact with each other
Subjective judgment	… we use judgment to interpret data, observations or experience. This results in uncertainty in individual judgments and uncertainty caused by differences across experts
Sources of linguistic uncertainty	**When we are not communicating effectively because …**
Vagueness	… language permits borderline cases. Vagueness can be numeric (how many 'tall' trees?, when does a population of algae become a 'bloom'?) or non-numeric (how to define habitat suitability?)
Ambiguity	… words have more than one meaning and it is not clear which is intended: 'natural' environment, forest 'cover'
Context dependence	… descriptions are not used in context; an oil spill that is 'big' on my driveway would be considered 'small' in the ocean
Underspecificity	… there is unwanted generality; 'it might rain tomorrow' vs 'there is a 70% probability of rain at location tomorrow'
Indeterminacy	… words used at one point in time mean something different at another point in time

actions with those system elements – we may be uncertain, for example, about how water customers will respond to a pricing instrument intended to create incentives to reduce water consumption at peak use times, or how effective industrial codes of practice will be at reducing emissions.

Another important type of epistemic uncertainty – one that is particularly significant in the context of deliberative decision making processes – is related to the subjective nature of judgments. Uncertainty can arise because of an individual expert's uncertainty about the selection of appropriate data or analysis methods, and about how to interpret data or modeling results. It also arises when different experts reach different conclusions based on the same available data and models.

Linguistic uncertainties are distinct from epistemic uncertainties. They flow not from gaps in what we know, but from our failure to communicate precisely what we mean. Vagueness, ambiguity and context-dependence are closely related sources of linguistic uncertainty, and we will use the single term ambiguity from here on as a short-hand for all three. It refers to situations where words have multiple meanings or meanings that haven't been defined. For example, the frequency of algal blooms is uncertain if we haven't clearly defined when a population of algae becomes a 'bloom', the extent of forest cover is uncertain if we have not defined what constitutes either 'forest' or 'cover', and an oil spill that's 'big' on my driveway may not be 'big' in the Gulf of Mexico. Another type of linguistic uncertainty, underspecificity, occurs when we don't adequately define the details of a situation, action, or event. For example, the probability that a prescribed burn[3] will escape and threaten a community can't be estimated unless details of the burn protocol and planned safeguards have been specified. Linguistic uncertainty due to indeterminacy occurs when scientific theories and the meanings of words change over time, such as when a species is reclassified into a new taxon based on new information.

Of course, epistemic and linguistic uncertainties are not entirely unrelated, and it's true that often we're vague in our speech because we don't know or we're not thinking very clearly about what we're talking about. But whether it's influenced by lack of knowledge or not, linguistic uncertainty is worth reducing as a first order of priority. The question, 'will there be a storm tomorrow?' is so (needlessly) linguistically ambiguous as to render any epistemic uncertainties moot.

Why subject you to this academic exercise in definitions? Well, first, the definitions are handy, and we'll use them now and then throughout this book. But the notion that much of the uncertainty we typically deal with as part of environmental management evaluations and deliberations comes from language rather than actual knowledge gaps is important. While the focus of much expertise concerning uncertainty tends to focus on extensive (and often expensive) data collection in the field or complex modeling techniques, we think that linguistic uncertainty is just as important in affecting the quality and outcomes of real-life decision-making situations. And while reducing epistemic uncertainty can often be a real challenge, tackling linguistic uncertainty offers inexpensive, quick, and very real prospects of making a wide range of resource-management decisions more tractable than they'd otherwise be.

6.1.2 How uncertainty arises in environmental decisions

Uncertainty arises at every step of the environmental decision process. In early stages, much of the structuring that is emphasized in SDM is designed to eliminate linguistic uncertainty (affecting both values and facts). Ambiguity at the problem formulation stage may take the form of a lack of clarity about who's the decision maker, what the real problem is that needs to be solved, or what's in or out of scope. It can lead to incorrectly scoped analyses and, if uncorrected, a potentially good answer to the wrong question.

As the decision process continues, ambiguity at the objective setting stage leads to more than linguistic uncertainty. It permits and conceals sloppy thinking. It is addressed in SDM by careful structuring of objectives and well-designed performance measures (Chapters 4 and 5). Means–ends networks, influence diagrams, and objectives hierarchies leave little room for ambiguity, and what's left is looked after with performance measures. Underspecificity in the definition of alternatives is a common cause of problems when it comes to assessing consequences, since it's hard to get reliable consistent estimates of consequences for actions that are not clearly defined. We talk more about this in Chapter 7.

Importantly, substantial uncertainty also arises from a host of socio-political influences on implementation. Technically, this is a kind of model uncertainty – we don't know how human, socio-economic, and political factors interact to support or water down the commitment to recommended actions over time. In decision making we identify and evaluate alternatives. We typically estimate their consequences, including a range of possible outcomes, *based on an assumption that they're implemented*. But what if they're not implemented? What if the promised funding doesn't materialize and they're implemented only partially? What if they're poorly implemented, without regard for key concerns? What if a political champion disappears and support evaporates? Thus there is uncertainty associated with implementation, which encompasses both uncertainty about intent (political support) and uncertainty about the quality of effort as well as uncertainty about the efficacy of outcomes. As a result, when selecting actions to recommend, in some cases it's important to consider the likelihood, timing, quality, and extent of implementation as that can significantly affect both the estimation and valuation of consequences.

In short, uncertainty infuses virtually every aspect of any environmental decision-making process. The challenge faced by resource managers is how to work creatively and respectfully with uncertainty, maintain a healthy humility about what's known and not known, and not be overwhelmed.

6.1.3 What's in a word: risk, uncertainty or ignorance?

By now you've noticed that we're using the term uncertainty to refer to things we don't know that affect our understanding of the outcomes of decisions. Many practitioners distinguish between risk and uncertainty, with risk reserved for situations where the system is well understood, the range of consequences can be defined, and the probabilities associated with different consequences can be quantified. Uncertainty, on the other hand, is used when the system and range of consequences are relatively well understood but the probabilities are unknown[4]. Other practitioners see little utility in making this distinction[5] and we find ourselves in this latter camp. In the context of decision analysis the term uncertainty is more widely used to refer to all the ambiguities and knowledge gaps that prevent a good understanding of the consequences of proposed actions. One advantage it has over the term risk is that it's a neutral term. Unlike risk, which is generally understood to be a negative thing (except perhaps in financial circles), uncertainty does not imply a negative consequence; we could be just as uncertain about beneficial or fun things as about adverse or feared things.

Ignorance refers to situations where not only do we not know the probabilities associated with potential consequences, we don't even understand the nature of potential consequences or the cause and effect mechanisms that could lead to them. This situation is sometimes called 'deep uncertainty'. Assigning probabilities and conducting any meaningful predictive modeling under these conditions is all but impossible. Recall the former United States Secretary of Defense Donald Rumsfeld, in the course of a news briefing on Iraq: 'Reports that say something hasn't happened are always interesting to me, because as we know, there are known knowns, these are things we know we know. We also know there are known unknowns; that is to say, we know there are some things we do not know. But there are also unknown unknowns – the ones we don't know we don't know'. While Rumsfeld was widely ridiculed in the media for this statement, uncertainty professionals point out that he really was on to something – something that has implications for when it might not make sense to seek out probability assignments for uncertain consequences. The very real phenomenon of ignorance is driving a search for creative management approaches in fields such as climate change, nanotechnology, genetics, and environmental hormone contamination.

Regardless of which conventions you adopt, it's worth noting that the inconsistent use of terms can create problems when managers are trying to make decisions or communicate with colleagues and the public[6]. The term 'risk' is particularly problematic. Let's take a minute to look at the many ways we use the word risk. Sometimes we use the word 'risk' to mean 'margin of error', as in 'what's the risk in your estimate?' Sometimes 'risk' is used synonymously with the word 'hazard', as in 'what are the risks to water quality?' Then, despite a widely recognized definition of risk as probability times consequence, in common usage there's also 'risk' as 'probability', as in 'what's the risk of a spill?' and 'risk' as consequences: 'what are the health risks of eating cheeseburgers?' And when we ask, 'is that company a good investment risk?' we are really weighing probabilities along with upside and downside potentials. A further problem – important as part of multi-stakeholder deliberations – is that in many resource-management contexts not all stakeholders will agree that a particular uncertain outcome is a 'risk' at all: a dam failure is not necessarily a 'risk' to people who'd like to see the restoration of a natural hydrograph in the river, and the failure to use Roundup on a public park may create a risk of weeds or wildflowers for some whereas others would argue that these are instead beautiful things, whose existence is to be encouraged.

Further (and perhaps somewhat related), we find that people tend to be sloppy with their thinking when they use the word risk. Experts may rate actions or conditions as 'high risk' or 'low risk' without specifying either what is at risk, what the words 'high' and 'low' mean, or the exact nature of the adverse consequences.

Our advice? Use the term 'uncertainty' as the primary way to refer to knowledge gaps or ambiguities that affect our ability to understand the consequences of decisions. Use the word 'ignorance' as a humble acknowledgment of the possibility of profound surprise – the possible existence of things that we can't even name, let alone quantify. Use the word 'risk' the way people use it in common language – to communicate in general terms about bad things we'd like to avoid. If the term 'risk' is used more specifically in either a technical or a deliberative setting, make very sure everyone is talking about the same thing.

6.2   A probability primer for managers

The communication and use of probability is inescapable in modern resource management. While there are other ways to express and work with uncertainty, probability forms an essential foundation; it is by far the most widely used and accepted way to express uncertainty in the decision, ecological, and economic sciences. For a more thorough review of essential theory of probability and related statistical concepts, readers are referred to Savage[7] or to Wonnacott and Wonnacott[8]. Here we cover only the basics that the less quantitatively inclined reader may need to make sense of the rest of this chapter and book and, more importantly, to support their role as a contributor to defensible environmental choices. If the phrase 'probability density function' leaves you reaching for the exits, this next section is for you – don't run away! If, on the other hand, you eat probabilities for breakfast, feel free to skip ahead to the discussion of Working with Uncertainty in a Deliberative Setting.

6.2.1   Objective and subjective probability

Probability refers to the likelihood that an event or condition occurs or exists, expressed as a number between 0 (*no chance*) and 1 (*sure thing*). As originally developed in the 17th century, probabilities reflected the frequency with which an event would occur over a long sequence of similar trials. This classical or objective view of probability isn't very relevant in environmental management, where precise repetitions of trials are often impossible or impractical. Most management applications involving probability deal with subjective probabilities, which refer to the degree of belief a person has that an event will occur, given all the relevant information known to that person. A common example of probability as degree of belief is the assessment provided by weather forecasters. Based on a combination of observations, models, data, and professional experience, weather forecasters predict the probability or frequency of precipitation, sunshine, and the like on a daily basis. Probabilities can also be assigned to unique events rather than repeating events – the probability that the Brazilian mens' team will win the FIFA World Cup in a particular year, for example.

In this context, probability assignment is informed by subjective judgments. Such judgments should respond to available data, and some are more defensible than others depending on how they are derived and to what extent they correspond to relevant data. Yet there is no such thing as a single 'right' probability. This explains why different people may assign different probabilities to an event: they may have access to different information, or they may interpret the same information in different ways. It also explains why assigned probabilities can change over time. Subjective probabilities incorporate the idea of learning through feedback, which may come through data (e.g. monitoring of management actions) or through discussions with others. Weather forecasters get daily feedback on their predictions, allowing for plenty of opportunities to learn and recalibrate their predictions. Environmental scientists and managers are not so lucky, but the value and role of feedback and learning, where it's available, is still critical.

A subjective probability is thus a formal, quantitative expression of a person's degree of belief that a condition or event exists or will occur, based on the best information they have at hand. Because subjective probabilities reflect a degree of belief, it's generally regarded as more appropriate to talk about 'assigning' rather than 'estimating' probabilities. The remainder of this section (and book) focuses on subjective probabilities.

6.2.2   Discrete and continuous probabilities

Probabilities can be either discrete or continuous. With discrete probabilities (e.g. the outcome of a coin toss), the uncertain quantity can take only a finite or countable number of possible values. With continuous probabilities (e.g. the weight of a newborn), the uncertain quantity can take on any value in a range: there are infinitely many values (or at least a very large number), and correspondingly the probability associated with any specific value is small[9]. In such cases, we don't tend to think about the probability of any one of these outcomes occurring. Instead we talk about the probability that a variable or outcome will fall within a specified 'interval' or range of values (e.g. what is the probability that a carbon tax will be between $5 and $10 per tonne?).

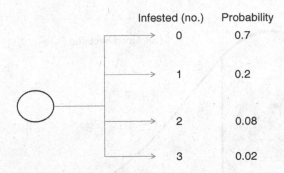

	Infested (no.)	Probability
	0	0.7
	1	0.2
	2	0.08
	3	0.02

Figure 6.1 Probability tree for pest infestation.

In Figure 6.1, biosecurity pest managers used discrete probabilities to report the likelihood that there will be pest infestations in imports of white rice entering the country. To assign these probabilities, the experts first agreed that the universe of possible outcomes can adequately be described by four discrete conditions (annual numbers of infestations between 0 and 3) and they assigned a probability to each. The outcomes listed therefore represent a 'mutually exclusive, collectively exhaustive' set of outcomes and the probabilities sum to 1 – one and only one of these outcomes *can* and *will* occur.

A common way to summarize this probabilistic information is by calculating the 'expected value'. Expected value is a weighted average of consequences, where the weights are simply the probabilities assigned to each outcome. For example, the expected value (or EV) for the set of outcomes in Figure 6.1 is:

$$(0*0.7) + (1*0.2) + (2*0.08) + (3*0.02) = 0.42$$

A simple interpretation of this is that over many years and many pests, we could expect an average of about 0.4 infestations per year, or about one infestation every other year.

While discrete probabilities are useful, a larger proportion of environmental management problems are likely to require continuous probability distributions. If, for example, your concern is that this pest may cross the border and enter a forest,

then you could be interested in estimating the number of hectares that could be deforested as a result. A 'probability density function (PDF)' shows the probability of an outcome falling within a specified range of values (see Figure 6.2). If the total size of a forest exposed to a pest is 10000 hectares then we can be sure that the actual outcome will fall between 0 (no defoliation occurs) and 10000 (the forest is completely defoliated)[10]. The probability that the defoliated area will be between any two values is found by calculating the area under the curve between those two values. Those people who measure such things would find that the total area under the curve sums to 1. In Figure 6.2, the shaded area under the curve represents half the total area; there is a 50% probability that the area defoliated will be between approximately 4000 and 6000 hectares.

However, as you may notice, it is difficult to look at a probability density function and be able to infer probabilities with any precision. Sometimes a more helpful way to look at probabilities is therefore with a cumulative distribution function (CDF). Here is how to read Figure 6.3: there is a 30% probability, for example, that the defoliated area will be less than or equal to approximately 4000 hectares.

Experience suggests that most people, including experts, find it easier to interpret cumulative distributions[11], which also have the advantage of direct relevance to many environmental management questions. For example, we can directly view the probability associated with crossing a threshold we care about from a management perspective: when estimating abundance of a threatened species, we can see the probability that it will fall below a scientifically determined minimum viable population size or below an important regulatory threshold (e.g. that determines whether it will be listed as threatened). If we're managing water resources, we might look for the probability of exceeding a regulatory standard on a particular water quality parameter. If we're allocating budgets, we might ask for the probability that a

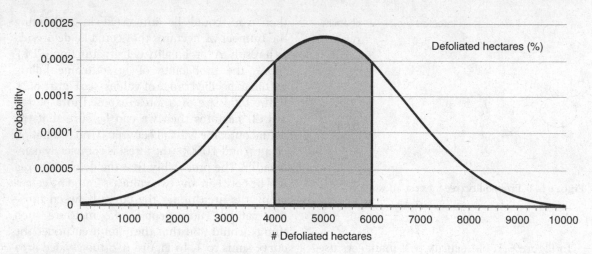

Figure 6.2 Probability density function for pest infestation.

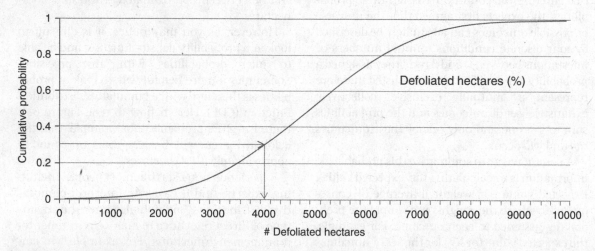

Figure 6.3 Cumulative distribution function for pest infestation.

management action will exceed a departmental budget cap.

The best format for representing probabilistic information will depend on what is being done. Perhaps most salient from a decision-making perspective is that as a manager working with uncertain consequences: (a) you need to understand what a probability distribution is; (b) you should be prepared to ask for and receive information in the form of a probability distribution; and (c) in many decision-making applications you will need

to extract information from the distribution and present that information in different ways to facilitate informed deliberations.

6.3 Working with uncertainty in environmental management decisions

With some probability basics in place, we now turn to discussing various aspects of working with uncertainty in the real world of multi-issue, often multi-stakeholder environmental management. In this section, we identify and describe seven things to keep in mind when dealing with uncertain consequences in environmental decision making.

1 *Organize and structure key uncertainties to lend insight to the decision.* We follow the lead of Morgan and Henrion[5] who (agreeing with Einstein) state that the analysis must be 'as simple as possible and no simpler'. The key is iteration, and the starting point is a decision sketch. Although usually there will be many uncertainties, only a few will matter. A decision sketch will tell you where you need to focus subsequent analyses and point to some useful ways to structure the more important sources of uncertainty.

2 *Be specific when describing consequences and probabilities.* With a little bit of diligence, and with the help of decision-aiding tools outlined in Chapters 4, 5 and 7, linguistic uncertainty about the nature and magnitude of consequences can largely be avoided. There is a widespread tendency to shy away from quantitative expressions of probability – in part, out of fear that numerical expressions will either confuse stakeholders or convey an impression of a higher degree of precision than what we really have. However, being uncertain does not have to mean being vague. We can be – and it's useful to be – very specific about how uncertain we are.

3 *Recognize that assignment of probabilities is not always appropriate.* As much as we champion the clarity that comes with quantitative expressions of uncertainty, we also recognize that there are times when probabilities might not work well as part of multi-party environmental decision-making processes. These situations include when uncertainty is severe, when extremely low probabilities are at issue, or when the assignment of probabilities fails to improve insights for decision makers. Significantly, the involvement of stakeholders

with mixed training or experience is not included on this list; if well presented and clearly explained, we have yet to meet a group that can't usefully discuss management alternatives using probabilities.

4 *Explore the risk profiles of alternatives and the risk tolerance of decision makers.* You won't want to do this in detail for all consequences, as it can make the analysis unnecessarily complex and result in information overload. But where there are important differences in the level of uncertainty associated with the alternatives, these differences are likely to be important to decision makers. Only by seeing the range of possible outcomes can they exercise their risk tolerance. In most cases, it will be useful to express this probabilistically.

5 *Explore and report the nature and extent of disagreement among experts.* The characterization of uncertainty is not complete until the extent of agreement or disagreement among experts is described. The goal is to gain a richer understanding of uncertainty by honestly characterizing and documenting the extent of agreement or disagreement. It may be possible to increase agreement, for example, by examining conditionalizing assumptions, but the primary task is to understand differences in views, investigate how they affect the choice among alternatives, and report them openly.

6 *Look for ways to improve the quality of information.* In some cases, this means reducing uncertainty, and (as discussed in Chapters 8 and 10) SDM offers useful ways to examine which uncertainties are more important to reduce. More often, the goal is to understand uncertainty and its implications for decision making. We can explore and improve our understanding and interpretation of information through improving the quality of data and models, improving the quality of judgments made under uncertainty, and improving the quality of deliberations about uncertain or competing hypotheses. In all cases, a key is providing context to decision makers about the significance of uncertainty for the choices under consideration.

7 *Create alternatives that address uncertainty.* Remember that the goal is a good management solution. Sometimes it's not more analysis of uncertainty that's needed but the development of better alternatives. Alternatives that address uncertainty explicitly include precautionary alternatives (delivering different levels of protection to valued resources in case we're 'wrong' in some assumptions), adaptive alternatives (using monitoring and experimentation to reduce uncertainty over time), and robust alternatives (designing alternatives to perform well under a range of circumstances).

The following sections explore these seven topics in more detail.

6.3.1   Organize and structure key uncertainties

There's no formula for this, but the starting point is to recognize that different tools and approaches for structuring uncertainty will be relevant in different situations. There are three golden rules:

1   Start with a decision sketch.
2   Fit the analysis to the problem.
3   Iterate.

There's no clearer way than starting with a decision sketch to find out how uncertainty affects the decision and what methods might provide the most insight. It will remind you that characterizing uncertainty is only an input to the ultimate decision you have to make and that time and resources will need to be preserved to support the rest of the decision process. We speak from experience when we say that it's very easy to slip down the rabbit hole of examining a particularly intriguing uncertainty to the detriment of other things that might warrant more attention. The sketch will also remind you to focus attention on the sources of uncertainty that are important to the decision at hand, that might help to distinguish among alternative actions in terms of their

effectiveness and likely consequences, and that resource managers can do something about – many sources of potential economic, environmental, legal, or political uncertainty will affect all reasonable alternatives in much the same way and lie outside the bounds or mandate of the decision under consideration. The decision sketch should help to ensure that your characterization of the problem provides useful information to decision makers and orients stakeholders regarding key sources of uncertainty that will require further attention.

Sometimes different approaches are needed to address different uncertainties within the same decision problem. The key is to keep it as simple as possible. You need to fit the analysis to the problem and figure out what might give insight to decision makers. Don't be afraid to try different approaches. And then don't be afraid to admit your approach isn't working. The litmus test for a structuring method is always: is it giving you insight about the problem? If yes, carry on. If not, give it up and try again. Starting simple also means staying open to the possibility of future iteration: if an uncertainty turns out to be critical to the selection of a preferred alternative, you probably will need to do more analysis and ensure that sufficient dialogue occurs among resource managers, other stakeholders, and decision makers so that everyone understands the need for, and goals of, the analysis – even if some of the technical intricacies or math are not as easily accessible.

In some situations, decisions need to be made that consider the implications of soon-to-be resolved uncertainties for subsequent decisions. As complicated as this can get, it is often worth the effort to think through potential decision sequences because the resulting insights can affect the choice of strategy to adopt now, regardless of what might be learned later.

An example comes from work with colleagues at the Australia Centre for Excellence in Risk Analysis (ACERA) and the Commonwealth Scientific and Industrial Research Organisation (CSIRO). We were asked to help support a decision

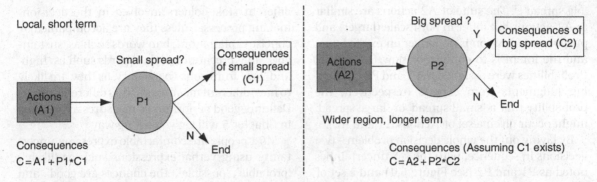

Figure 6.4 Event tree for examining sequential uncertainties.

regarding a recent outbreak of a plant disease called Myrtle Rust, an invasive fungal disease that has the potential to spread to large areas of Australia and could compromise the plant family that includes the iconic eucalyptus. The outbreak was first detected on a commercial property on the central coastal of New South Wales, and at the time of analysis had been found on cultivated plants in several dozen locations. On the face of it, it seemed logical to take immediate steps to eradicate the fungus before it had a chance to spread further. However, there was a key uncertainty about whether any amount of management intervention could have a reasonable expectation of achieving this goal, given the nature of the fungus and the degree of spread at the time the incursion was identified. Furthermore, there was also concern about the very real harm to social, economic, and environmental values that any serious attempt to eradicate would inevitably cause.

Working with a group of 24 senior Australian government biosecurity managers, industry stakeholders, and scientists, we began by sketching the decision. The starting point was to identify fundamental objectives – how would the pros and cons of alternative sets of actions be evaluated? Having worked primarily in a 'risk analysis' context previously, most participants were focused on the consequences of the pest – damage to forests, crops, economies and so on resulting from the pest outbreak. An important insight gained by the group was that there also

are consequences – different consequences – associated with the proposed management actions (e.g. the application of fungicides has potential ecological and human health implications). This realization fundamentally changed how the decision was approached. Another important insight was that the decision could be divided into two, sequenced choices, both subject to uncertainty. The first decision needed to be made more or less immediately, whereas the second was necessary only if earlier actions failed to eradicate or contain the spread of the fungus.

Event trees were used to examine the sequential sources of uncertainty. We first identified the set of actions that should be undertaken locally and in the short term (referred to as 'A1' in Figure 6.4). In this case, the key uncertainty is whether the fungus can successfully be eradicated or contained to a relatively local area, given a set of actions A1. We represent that uncertainty as P1 – what is the probability of a small spread given a set of actions A1? If the fungus is subsequently eradicated or contained, then all is well and the consequences of the management actions (costs, impacts to other environmental species from the application of fungicides and plant removals, etc.) can be estimated. If, however, the fungus spreads beyond a point where eradication efforts are meaningful (a condition defined as a 'small spread'[12]), then a second decision is required.

The second decision concerns which set of actions (A2) should be undertaken to prevent a 'big spread'[13]. The suite of A2 actions are similar to A1 actions but differ in both scale (larger) and purpose: eradication is no longer an explicit aim, and the intent is to contain or slow the spread. Probabilities were assigned to P1 and P2 by eliciting judgments from experts (respectively, the probability that a 'small spread' or 'large spread' might occur given a set of Actions A1 and A2).

By laying out the whole decision problem – two decisions in sequence, with the key uncertainties noted as P1 and P2 (see Figure 6.4) and a set of multi-attribute consequences associated not just with the pest but with the different management alternatives – participants gained an initial understanding of the scope of analysis and discourse needed to deal effectively with the problem.

What does this analysis suggest about ways to generate responsive management alternatives? One alternative might be to throw everything at the fungus to eradicate it in the short term, and to continue to try containing and slowing the fungus in the event of the failure to prevent a small spread. Another might be to try hard to eradicate in the short term, but then to accept the inevitable and conserve resources for adaptation programs should a wider spread occur. A third, more radical alternative might be to forego the opportunity to eradicate in the short term altogether, and to conserve resources for adaptation initiatives later on if required. The 'best' alternative is not self-evident, since it depends on a wide range of factors – how each alternative would affect both the nature of possible consequences and the probability that those consequences might occur (we return to the discussion of specifying alternatives in Chapter 7).

6.3.2 Be specific when expressing probability and consequence

At the highest level, the advice is this: for every objective, you must define what it means using some kind of metric or scale that allows a reasonable and consistent discrimination of the relative effects of different alternatives. And this meaning must enable clear communication among the different stakeholders involved in the decision-making process: Unless they are accompanied by concise explanations, ban words such as 'sustainable' and 'precautionary' or words such as 'high' and 'low' from your vocabulary, as they are likely to have different meanings across different people. Defining good performance measures as outlined in Chapter 5 will get you a long way.

Most people are comfortable expressing uncertainty using verbal expressions such as 'likely', 'probable', 'possible', 'the chances are good', and so forth, sometimes with modifiers such as 'very' or 'highly', or with negations such as 'unlikely' or 'the chances are not good'. There are two problems with using words alone to express uncertainty. The first is ambiguity. A given verbal expression can mean different things to different people, even when they are considering the same event. Granger Morgan, a leading expert on uncertainty in risk analyses, asked members of the EPA Science Advisory Board to assign quantitative descriptors (in this case odds) to verbal expressions of uncertainty associated with the statement that a chemical was likely to cause cancer in humans[14]. Despite the fact that the respondents were all experts, had previous experience working together, and were given a specific problem context, the odds assigned to the word 'likely' ranged from less than 1 in 10 to about 1 in 1000. The range for the odds given to the characterization 'not likely' was even wider – so much so that there was some overlap between the odds assigned to likely and not likely. Scary stuff, when what we are hoping for is clear communication!

A second concern is that verbal expressions have different meanings in different situations (remember 'context dependence'). If I tell you there's a 'high probability' that the water in the lake is safe to drink, then I may mean something like '$p = 0.95$'. But if a hydroelectric engineer working with a utility company says that there is a 'high probability' that a critical component of dam stability is expected to experience a failure within one year, he means something like '$p > 0.05$' or even '$p > 0.01$'. This is because dams are generally designed to a high standard of reliability (given the potentially catastrophic consequences) and 'high' means 'high relative to other dams'.

Of course, the use of quantitative expressions comes with its own set of challenges. What do we know about how well people understand quantitative probabilities? How do interpretations or choices change if likelihood information is presented using frequency or odds rather than probabilities?

Experimental studies clearly demonstrate that people may react differently to different ways of expressing likelihood, even when they are formally equivalent. For example, studies have shown that probability and frequency formats for communicating uncertainty can produce quite different judgments regarding the risks associated with different events. One study, looking at factors that influence people's perceptions of violent risks, found that relative frequencies of 2 out of 10 or 20 out of 100 lead to higher levels of perceived risk than does an assessed probability of 20%[15]. Other researchers suggest that representations of uncertainty based on frequencies (e.g. 1 chance in 10) are 'better' than probabilities, in that they lead to more accurate or more readily understood judgments. For example, Gigerenzer[16] maintains that frequency-based formats are generally easier for people to use when making inferences under uncertainty because the associated mental computations are a better fit with how people naturally think (i.e. our mental algorithms). Not surprisingly, numeracy also has important effects on the interpretation of probability[17]. The bottom line advice from experts in uncertainty analysis:

1 Frequencies, odds, or probabilities can be used effectively, provided they are explained and used consistently.
2 It's helpful to use multiple framings to ensure a robust understanding. If you say: 'there's a 5% probability of a spill', follow it with 'this is equivalent to 5 times out of a 100 or a 1 in 20 chance'.

Sometimes the participants in environmental decision processes will object to the use of very specific probabilities. The idea that someone can assign a '0.33' probability to an event, or say that the chance something will occur over the next 10 years is 1 in 50, seems out of synch with the limited level of understanding that characterizes how we think about many uncertain quantities. A practical middle ground involves mapping verbal expressions to numeric probability ranges. These probability scales are constructed to suit the situation at hand (see Chapter 5 for more on constructed scales). They alleviate both the ambiguity and context dependence problems associated with verbal expressions alone. They also allow experts to assign ranges of probabilities as opposed to a single fixed value. There are advantages and disadvantages to ranges, of course. On one hand, experts are often more comfortable describing probabilities as ranges because they feel that a single figure gives a false impression of precision. On the other hand, ranges are problematic when values fall close to thresholds separating categories.

Table 6.2 provides one example of verbal expressions mapped to numeric probability ranges that has been used for characterizing uncertainty about climate change[14]. Different ranges may be appropriate for different kinds of problems. The important point is that a verbal descriptor is mapped to a specified numeric range. Without this, verbal descriptions are 'highly likely' to be 'nearly' useless, except in 'rare' circumstances.

In sum, to communicate effectively about uncertainty:

1 Use well-designed performance measures as outlined in Chapter 5 to describe consequences unambiguously and consistently.
2 Eliminate the use of verbal descriptions of probability alone.
3 Use multiple framings to build a more robust understanding of uncertainty.
4 Map verbal expressions of uncertainty to numeric probability ranges.

Table 6.2 Verbal descriptions of probability and associated numeric ranges (source: Morgan *et al.*[14]).

Word	Probability range
Virtually certain	> 0.99
Very likely	0.9–0.99
Likely	0.66–0.9
Medium likelihood	0.33–0.66
Unlikely	0.1–0.33
Very unlikely	0.01–0.1
Exceptionally unlikely	< 0.01

6.3.3 Recognize that probability assignment is not always appropriate

We highlight three general cases when probability assignment may not be appropriate.

1 *Problems characterized by overwhelming uncertainty.* In rare cases not just the magnitude of consequences but even the likely direction of response is unknown; a variety of analytical approaches have been developed to help managers make decisions under such conditions of deep uncertainty or ignorance (e.g. see Regan *et al.*[18] or Ben-Haim[19]). A more common situation is that managers are stymied by what feels like an overwhelmingly large number of sources of external uncertainty (i.e. outside their control), with little insight about which of these potential influences on their actions will turn out to be most important. A helpful structuring tool is Scenario Analysis (see Text box 6.3).

2 *Problems involving extremely low probabilities.* Probabilities may be of little help in situations for which consequences are thought to be severe or catastrophic should an event take place but where the probabilities associated with this event are extremely low. An example is a meteor striking the site of a proposed reservoir and releasing water from a dam. The probability of such an event is so low that it will be meaningless to most stakeholders, and even an order-of-magnitude shift in the assessed probability is unlikely to make any difference to perceptions of risk. Under such circumstances

most people tend to ignore the assigned probability (a phenomenon referred to as 'probability neglect') and focus instead on the severity and magnitude of the associated consequences[20]. If – for whatever reason – it is necessary to discuss the highly unlikely event in question, then it usually makes more sense to describe the associated odds as a frequency – 'one chance in about ten trillion' – or something along these

---

Text box 6.3:   Using scenario analysis

A helpful structuring tool for situations when it's difficult to assign probabilities is Scenario Analysis, which involves developing a small number of discrete scenarios in which key external sources of uncertainty are specified[21]. For example, when undertaking long-term energy planning, we can imagine a series of markedly different future worlds: a high climate-change impact world, where the costs of carbon releases have sky-rocketed and large subsidies are given to novel technologies, a medium climate-change impact world, and so forth. The performance of different resource management alternatives can be evaluated for each of these different worlds, with performance tied to explicit objectives and performance measures that have been defined as part of an earlier decision-structuring effort. Through such an analysis, we might discover that one alternative performs quite well across a range of external circumstances, in which case our choice could be made without needing to know more detailed information about the specific probabilities.

Caution should be used in scenario analysis, however, because without reference to explicit probabilities the natural tendency is for people to rely on common heuristics – either (a) they assume that each scenario is equally likely or (b) they begin to believe strongly in a particular scenario that is compellingly presented. Both of these are logical flaws that may be worse than using an uncertainty range that is very wide.

lines rather than as a probability (which would only serve to engage people in counting zeroes).

3 *Problems where probability assignment does not inform a decision.* For the meteor problem above, for example, there is nothing that managers can do to reduce the chance of this occurrence. So there is little reason to spend time or scarce modeling resources on the problem, since no added insight to decision makers will be forthcoming.

A note of caution: guard against the argument that quantitative probabilities should be avoided because we don't know them. Concerns about creating an illusion of false knowledge or precision are relevant, but this is no excuse for willful ambiguity. Even very large uncertainties can be characterized precisely (by asking for confidence intervals, for example), and the information can prove to be invaluable in decision making.

6.3.4 Explore the risk profiles of alternatives and the risk tolerances of decision makers

Risk tolerance, as the name implies, describes an individual's, group's or agency's willingness to accept a specified level of risk. It's an important dimension of how we make choices under uncertainty: some of us are more willing than others to accept relatively higher levels of risk in order to possibly achieve an anticipated level of benefits. Differences in the risk tolerance of management agencies are a frequent cause of disagreement: some people or agencies, for example, might well have a higher risk tolerance for possible adverse effects on a threatened species than others, even if they are within the same organization.

Earlier, we described expected value (EV) as a way of summarizing probability distributions. In some cases, EV is a perfectly acceptable way to handle and communicate uncertainty. However, in many real life situations, EV (alone) is wholly inadequate. Why? If EV were all we needed to know about an uncertain or 'risky' situation, then we should be indifferent to choices that result in identical EVs. In reality we are not, a fact that's comprehensively demonstrated in the economic literature as well as in numerous examples from everyday experience.

Take lottery tickets. When we buy a lottery ticket for $1 with a one in 1 in 5,000,000 chance of winning $1,000,000, we're essentially paying $1 for something with an EV of 20 cents (ignoring taxes). We do this because we don't care all that much about the prospect of losing $1, and the idea of winning $1,000,000 is quite marvelous. Helped by the fact that people really aren't very good at understanding just how fantastically unlikely a 1 in 5,000,000 probability is, many of us exhibit this so-called *risk-seeking* behavior[22]. We're also quite familiar with the opposite phenomenon of *risk-avoiding* behavior. Most of us pay thousands of dollars each year in various forms of insurance, money that all but an unfortunate few will ever receive back. But the prospect (however rare) of losing a house to a fire is so dreadful that it justifies (for many of us) paying hefty monthly fees that would guarantee its replacement in such an event. Many texts suggest that most people are risk averse, but we think it's pretty context-specific: best not to make any assumptions.

To bring this back to a resource-management context, let's look again at the pest infestation example in Figure 6.1, with an EV of 0.42 infestations per year. For the manager looking at investments to manage multiple pests over multiple years, none of which has catastrophic consequences, knowing that s/he can expect about 0.4 infestations per year on average might be exactly the right information for decision making – it allows planning, allocation of resources, and monitoring of progress over time. However, the relevance of expected value diminishes with one-off decisions, and particularly with decisions involving high stakes.

Although there is no way to judge the rightness or wrongness of risk neutrality, decision makers will tend to be more risk neutral – and from a public policy perspective, a good case can be made to justify being risk neutral – when (a) there are many repeated decisions and what people care about is a good average result over all decisions; and/or (b) the consequences of any individual decision will not be catastrophic[23]. As with other kinds of value judgments, it's good practice to ask and expect decision makers to examine and articulate the rationale for their judgments, including an explicit acknowledgment of risk tolerance. As a starting point, decision makers, need to see information about the range of possible outcomes under each alternative.

Consider two alternative management actions designed to enhance recovery of an endangered species. Under a deterministic analysis, experts have indicated that if Alternative A is selected, the best estimate of the species' population after five years is 60 individuals; if Alternative B is selected, the best population estimate is 80. Assuming no other significant differences in performance (relative to other objectives such as costs or employment for example), we can safely expect most managers to select Alternative B.

Asking about uncertainty brings a new dimension to this comparison. Action A (dashed line in Figure 6.5) has a fairly tight distribution of outcomes – the most probable outcome is about 60 individuals and the worst that could happen is about 40. Under Action B (solid line), the most probable outcome is 80 individuals, but there's a much wider distribution of outcomes – and the worst is extinction, as bad as it gets. So this is now a harder choice, dependent not only on the best estimate (or the EV), but also on the distribution of outcomes and the relative importance assigned to the lower and higher portions of the two distributions (i.e. their tails). Depending on their risk tolerance and the legal framework within which decisions are made, two different and equally competent managers could make very different decisions – one may choose the relatively safer Action A, whereas another may feel that it's important to try for the larger gains that are possible under Action B, even if the downside is a small probability of extinction. While these

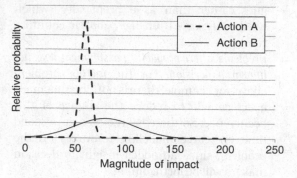

Figure 6.5 Population estimates of alternatives, showing distributions of outcomes.

conditions don't apply to every environmental management decision, neither are they rare exceptions.

It's useful at this point to consider where these distributions come from. There are two primary sources. One is data or more commonly, predictive modeling. The second is expert judgment, which synthesizes available data, predictive modeling results, relevant theoretical arguments, and the practical experience of experts to produce a subjective probability distribution. There are recognized best practices for eliciting probability distributions from experts (which we describe in Chapter 8).

Once you have the distribution, the challenge is how to use it in a deliberative decision making setting. There's no single right way, but here's a useful four-part line of thinking.

Summarize probability distributions

'Box and whisker' plots are proposed by some experts as a 'best practice' way to summarize probability information[14]. There are two primary reasons for using box and whisker plots instead of the full distribution. First, it's useful to have a more concise presentation that facilitates the comparison of multiple distributions. Most often in decision making, you're not looking at one distribution in isolation, but comparing – across experts or across alternatives for example. Second, box-and-whisker plots are easier for non-technical people to read and understand.

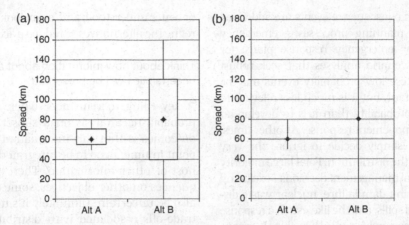

Figure 6.6 Box-and-whisker diagrams showing (a) a five point distribution and (b) a three point distribution.

Figure 6.6 shows two ways that box and whisker plots can be used. The format in (a) shows a five-point summary of the distribution in Figure 6.5. The marker could represent the best guess, median, or mean (or alternatively, multiple markers could be shown). The 'box' shows a designated confidence interval[24] (often 90%, although anything could be chosen) and the 'whiskers' show maximum and minimum values.

The format in (b) is somewhat simpler. The marker typically represents the best guess, mean, or median, and the whiskers represent designated upper and lower bounds. These upper and lower bounds could be the absolute maximum and minimum values. More commonly, we use them to represent upper and lower 'plausible' bounds, corresponding to a particular confidence interval that participants in the decision feel are appropriate.

Figure out which parts of the distribution matter most

The first thing to check is whether the distributions vary across the alternatives. That is, are some very wide while others are narrow? Are some skewed toward upper values while others are skewed toward lower values? If the distributions are relatively similar and narrow, it may be sufficient to compare alternatives using the 'best guess' or a probability weighted expected value. (It is relatively straightforward to calculate an expected value from a continuous distribution; see Goodwin and Wright[25].)

If they differ, then determining which parts of the distribution matter most will depend on the decision context. If individual bad outcomes don't matter very much, the decision maker may be risk neutral and it may be appropriate to use expected value to compare alternatives. If the consequences of being wrong are catastrophic – species extinction, contamination of drinking water, dam failure, invasive species introduction, and so on – decision makers are anything but risk neutral and expected value may hide important information about the relative riskiness of the alternatives.

If the consequences of being wrong matter, it may be most appropriate to focus on the outer bounds. Sometimes people choose the absolute maximum or minimum values. But very often we use what we call the 'worst plausible' case, corresponding to a range of outcomes thought to be relevant to management. It's not uncommon – or unreasonable we think – for managers to state that they're managing for a particular confidence interval – commonly but not exclusively the 90% confidence interval. This means ignoring – for the purposes of the decision at hand – events with less than 5% probability of occurrence. Often such low probability, high consequence events are addressed through other planning processes – emergency preparedness or emergency response plans, for example – or scenario analyses that seek robust alternatives. These low probability events may be critically important, but it is much less relevant to quantify their probability than it is to develop an appropriate management response. At other times, managers may simply decide to ignore the very extreme tails of the distribution. For a large majority of environmental management decisions – excluding decisions involving dam failure, nuclear waste disposal, marine oil spills, and the like – such a response may be quite reasonable and defensible. It means managers are being accountable for uncertainty within limits that are practical for management.

Ask decision makers to think about their preferences, given differences in the risk profiles of alternatives

In the absence of information about other objectives, this is a hypothetical 'if all else were equal' question designed to challenge people to think hard about their preferences under conditions of uncertainty. 'If there were no other differences in performance, which alternative in Figure 6.5 would you prefer?' We find that, depending on the probabilities and consequences of course, this kind of question can be the most difficult kind of question for decision makers to deal with. It's generally a less familiar and sometimes more deeply personal question than the more conventional trade-offs about 'money versus the environment' or 'fish versus wildlife' or 'recreation versus species at risk'. It's useful to expose these trade-offs and encourage people to reflect on them. Sometimes the answer will be dead easy – everyone will agree that the potential benefits don't warrant the risks, or alternatively that the benefits clearly outweigh the risks. But for some decisions – probably most in the case of environmental management – this kind of choice is extremely challenging. Making a decision will rely on: (a) examining the full set of consequences and trade-offs across all the objectives; (b) further analysis to see if uncertainty can be reduced

or better understood; and/or (c) looking for ways to refine the alternatives to reduce uncertainty.

Think about how information about all the uncertainties will be presented

A key thing to remember when working with probabilistic analysis on a subset of uncertain outcomes is that this information will, at some point in time, need to be integrated with a whole host of other information. There will be consequences on other objectives, some of which may also be uncertain. Ultimately it's useful if all key trade-offs associated with distributions of outcomes can consistently be captured in the performance measures and placed in a consequence table. We discuss some ways to capture and present the essential information from probability distributions in consequence tables in Chapter 8.

Every time you extract information from a probability distribution to summarize it for decision makers, you lose information and you make value-based judgments about which parts of the distribution matter. So part of the challenge in presenting uncertainty effectively is technical, but part of it involves understanding decision makers' values and risk tolerances. It's important to challenge participants in a decision process to think clearly about their preferences under uncertainty – in all likelihood, it will be a novel and uncomfortable task for them.

6.3.5 Explore and report the level of agreement among experts

In addition to uncertainty in the estimates of individual experts, best practices now unquestionably involve characterizing the degree of agreement or disagreement among experts. Environmental management issues frequently involve multiple legitimate experts whose views about the anticipated consequences of management actions may be similar or dramatically different. When experts agree, it can be interpreted as an indication (among others) of a well-framed problem and reliable information about the identified uncertain outcome or event. When they don't agree, this is an essential part of the characterization of

Text box 6.4:   Dealing with risk tolerance: a broader perspective

There is an important and interesting ethical dimension to this discussion that arises when we're talking about the distribution of resources for public goods and services. Regulators charged with allocating limited funding to, say, programs to protect the public or an important environmental component from harm from carcinogens or fires should, some would argue, remain focused on expected outcomes of those programs (e.g. perhaps in terms of expected lives/hectares saved). Making an exception for a particular program that has a lower expected value but that, for example, addresses an issue more familiar to, or dreaded by, the public would by definition lower the overall EV of the program. Consequently it could be considered to be 'voluntarily sacrificing' lives or other important outcomes in the name of political or media relations.

Therefore, the argument goes, government decision makers should always be risk-neutral for issues such as lives and money[26]. However, things are rarely so one-dimensional and any such analysis likely balances several objectives, explicitly or not. What matters is that decision makers carefully consider these ethical choices and make them explicitly. The more explicit we can be as analysts in providing information about the range of possible outcomes and their associated probabilities across a range of relevant objectives, the better placed decision makers will be to make difficult choices on the public's behalf. Of course, it is also true that decision makers are then more accountable for those choices. That is arguably a frightening and politically dangerous thing for some. We'll stay out of the ethics of that conundrum, and just remind readers that for analysts, the job is to make sure the decision maker has the information, in a format they can understand and use. How they communicate the trade-off made to a broader public is often outside the analyst's control.

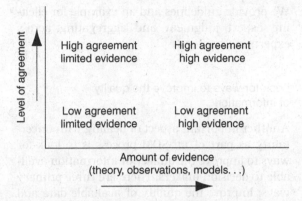

Figure 6.7 Amount of evidence vs level of degree of agreement among experts (adapted from IPCC[27]).

uncertainty and it provides important information to managers and decision makers.

How to characterize differences among experts depends, of course, on the decision context. Sometimes all that's needed is a relatively simple conceptual guide that characterizes the state of knowledge about a particular judgment. One approach comes from the International Panel on Climate Change (IPCC), which characterizes the state of knowledge as a function of the amount of available evidence or data and the degree of agreement among experts (Figure 6.7).

Where a particular uncertainty or judgment falls in this plot has different implications for management. Uncertainties characterized by limited evidence and low agreement among experts are, all else being equal, probably higher priority candidates for limited study funds than uncertainties characterized by high agreement, low evidence. If there is low agreement in spite of high evidence then we might wonder whether some exploration of the underlying basis for the judgments warrants exploration. For example, were some experts giving more credence to some data sets than others? Were they using different conceptual models of how individual elements of a system (human or ecological) work together? What conditionalizing assumptions were they using? What biases might have been influencing experts' judgments?

Sometimes it's possible and useful to compare the judgments of experts more quantitatively.

We provide guidelines and an example for eliciting expert judgment and aggregating across experts in Chapter 8.

Look for ways to improve the quality of information

Another important aspect of dealing with uncertainty as part of an SDM process is to look for ways to improve the quality of information available to decision makers. There are three primary ways: improve the quality of available data and models, improve the quality of (expert) judgments under uncertainty, and improve the quality of the deliberative exchange (where data, models and expert judgments meet).

1 *Improving the quality of data and models.* This is the approach usually associated with improving the quality of information. What SDM contributes is a reminder that the goal associated with any information quality effort is to make better decisions. Improving biological or other information may be valuable for many reasons – creating a comprehensive listing of resident species and their preferred habitats, or establishing an inventory of skilled workers living near a proposed mine site, or understanding social and trading networks within a First Nation community. But SDM approaches have the single-minded focus of helping managers to gain a better understanding of the potential effects of proposed actions on decision makers' expressed objectives, as a means to creating better alternatives. If the potential improvements in information quality won't help managers to make better choices in the context of the specific uncertainties affecting the decision at hand, then the value of the information (at least to *this* decision) is negligible.

2 *Improving the quality of judgments under uncertainty.* Recall from section 6.1.1 that subjective judgment is an important source of uncertainty in consequence estimation. People are subject to a host of systematic and predictable biases (as well as some less predictable ones). Careful research over the past three decades has resulted in best practices with respect to eliciting expert judgment. As Chapter 8 will describe, even some very modest changes in how questions are asked of experts can help to minimize bias and errors of logic, improving the quality of information available to decision makers.

3 *Improving the quality of deliberation.* A third way to improve the quality of information for decision making is to improve the quality of the deliberative environment in which uncertainty is discussed by decision participants. It is here that a mix of expert and non-expert participants will debate the merits of competing hypotheses, examine the available evidence, discuss differences in interpretations and opinions, and compare the risk profiles of alternatives There are many sources of uncertainty in environmental management, and it is a simple reality that precious few of them become the subject of detailed expert judgment elicitation processes. Those that do will only rarely produce a definitive consensus on the questions under consideration. As a result, the quality of the deliberative exchange about consequences in an SDM process is critically important. Key questions include:

What are the competing hypotheses and how uncertain are they?
How much information is enough?
What are the consequences of being wrong?
Are estimated consequences close to a recognized or legal standard?
Are estimated consequences approaching a threshold beyond which they are likely to escalate?

There is a myth that scientific data collected in the field or the results of quantitative, computer-based models of ecological systems are 'objective' or 'real' in contrast to the subjective judgments of individuals, which are therefore demoted to a lower status. The reality, of course, is that any data collection effort requires a number of value-based judgments, and most biological models are simply formalized collections of expert judgments which are supported or opposed by data of varying degrees of quality[28]. Which data sources are considered legitimate, which analyses should be conducted, and what degree of uncertainty is allowed are all examples of judgments where reasonable people may disagree.

6.3.6 Create alternatives that address uncertainty

Sometimes the most important thing to do with uncertainty is not to analyze or reduce it, but to seek creative alternatives for dealing with it. This is true for a host of relatively routine environmental management problems where sophisticated analysis of uncertainty is difficult. It's also true for more controversial environmental management choices, where both managers and stakeholders struggle at an ethical level with making decisions about valued environmental decisions because of the associated uncertainty. It is particularly true in problems characterized by ignorance or deep uncertainty about environmental responses, such as large-scale global climate change, where an urgent need for new decision making approaches that are precautionary, adaptive, and robust is now broadly recognized[29].

Here we briefly outline three broad types of management alternatives that can help (to varying degrees) to provide managers and decision makers with defensible approaches to managing under uncertainty.

Precautionary approaches

The precautionary principle is famously stated in the United Nations' (1992) 'Rio Declaration on Environment and Development': 'Lack of full scientific certainty should not be used as a reason for postponing cost-effective measures to prevent environmental degradations'. As generally interpreted, the precautionary principle assigns the burden of proof to those proposing an action rather than those protecting an endpoint (e.g. conservation of an endangered species). Rather than a community having to prove that forest practices will pose a risk to ecological values, the onus should be on the forest industry to prove the lack of such an impact. More loosely, the principle is sometimes invoked to justify limits on any activity that has the potential to harm a valued environmental component, or to justify allocating large amounts of money to their protection – 'just to be sure'.

The problem with the precautionary principle is that being precautionary with respect to one endpoint often means being profligate with respect to another. On the north coast of British Columbia, several communities and governments have identified two pillars to guide natural resource decision making: ecological integrity and community well-being. If we are concerned about ecological integrity, then being precautionary may mean limiting forest practices with the potential to affect key ecological attributes. But if local communities are relying on the economic benefits of forestry activity, then we may be putting local communities at risk. The precautionary principle assumes that there is one (usually environmental) endpoint that is inherently more important and more at risk than another. Unfortunately, it's typically impossible to be simultaneously precautionary with respect to environmental, social, cultural, *and* economic considerations. In fact, it's often not possible to be precautionary with respect to all of the diverse environmental attributes that are valued (e.g. different species, ecosystems, and unique site-level features). Changing forest practices to benefit one species may adversely affect another. Making choices about where to be precautionary requires explicit *value-based*, not technical, decisions.

So the precautionary principle, like all principles, is useful in a meta-guiding sense – we should be conscious of the need to be precautionary with respect to some objectives and make explicit choices reflecting this. But it provides only limited guidance in actually choosing alternatives. One concrete way the precautionary principle can be usefully applied in decision making is in the *development* of alternatives. When uncertainty in the estimation of outcomes is high, we can develop alternatives with different degrees of precaution and compare the implications for all the objectives. For example, we can develop an alternative designed to deliver an 80% probability of recovery for boreal caribou, and compare it with an alternative designed to deliver a 95% probability of recovery. Then the costs (economic and possibly other social or health and safety implications) can be compared and a value-based judgment made about whether the additional precaution is 'worth it'.

Adaptive approaches

Adaptive strategies encourage learning and are flexible enough to respond to that learning and improve their performance over time. Actions that are taken – including experiments and monitoring – are designed to allow decisions to be changed in the future, as new information becomes available. Thus adaptive strategies include the idea of keeping options open, which typically means viewing management as a set of sequential decisions, and adopting a strategy that, at every decision point, incorporates the best information that is currently available. In the ecological management world, adaptive strategies have been promoted for many years under the label of adaptive management (discussed again in Chapter 10).

Some researchers have suggested that adaptive strategies are most likely to be successful when the decision context is relatively simple, so that the number of interested groups is small, the impacts are not severe, responses are not non-linear, and the risks are low. However, our experience is that adaptive strategies are most needed in just the opposite conditions – complex decisions, multiple stakeholders and jurisdictions, potentially severe impacts, high uncertainty, competing hypotheses and polarized views. When all that can be done to improve information quality has been done, and groups still face tough decisions under high uncertainty, people are left with a challenge: making tough trade-offs is hard enough at the best of times, and when the consequences are uncertain it's even more difficult. If multiple stakeholders hold different perceptions

of and tolerances for risk, the prospects for agreement are slim. Add to that the reality of competing hypotheses put forward by parties with a long history of conflict, where choosing one over another is seen as siding with one party versus another, and reaching agreement becomes an ever more distant possibility. In these cases, adaptive strategies – meaning a real, well-resourced commitment to formal learning over time and a formal, institutionalized commitment to review and revise choices as new information is gained – can provide people with a way to make responsible, defensible decisions.

Robust approaches

Although there's no single definition of 'robustness', robust decision making generally involves moving away from trying to select an optimal plan or action and identifying alternatives that are likely to perform reasonably well over a range of uncertainties and plausible future conditions[30]. The concept follows from the observation that sometimes alternatives that are preferred under most-probable conditions perform particularly badly should less likely conditions occur – their performance is said to be brittle, or extremely sensitive to key assumptions. The idea behind robustness is to focus not just on the most-probable events and conditions, but rather on 'unlikely but plausible' events and conditions and to identify a small set of alternatives that might be reasonable under them all.

Robust and adaptive alternatives will together contribute to more resilient complex systems; that is, systems that are able to absorb shocks or extreme external events while maintaining key functions[31]. For example, many maritime ports are now faced with a situation where the 'best' response to future potential climate-change impacts will depend on precisely how climate change happens to unfold. How strong a wind surge should cranes be designed to withstand? How high should a sea wall protecting wharves be built? A conventional approach might compare one alternative designed for a low sea rise scenario with another designed for a high sea rise scenario. However, it also may be possible to develop an alternative that involves first building at the (cheaper) lower level but designed, at varying costs, to readily be increased in height by one or two meters at some point in the future. Now the suite of alternatives is more robust to the uncertainty about how climate change might manifest itself (if at all), and the debate about which is the best alternative is reduced to more mundane (and much less uncertain) technical and administrative questions about the most cost-effective way to build in the necessary flexibility. Of course, robustness to uncertainty doesn't usually come free, and most often 'robustness' is just one more issue to be considered as part of multidimensional trade-offs: would you prefer an expensive alternative that's robust to a wide range of climate-change impacts, or a cheaper alternative that will be catastrophic if water or wind levels increase in the future?

Similar issues and trade-offs arise in response to a host of natural sources of risk with uncertain frequencies and magnitudes; in the case of west-coast maritime ports in North America, for example, a key question facing decision makers is the selection of appropriate design criteria and decision-making processes for building in responsiveness to future seismic events[32].

6.4   Key messages

Uncertainty about facts influences nearly every environmental management opportunity. We firmly believe that everyone – including people without technical training – can understand the implications of uncertainty for decision making provided uncertain information is clearly presented as part of a structured, open deliberative environment. We also know from both practical experience and research that even technically trained people don't always understand various representations of uncertainty as well as they think they do[34]. What should you do (or expect to see) to ensure that uncertainty is being appropriately treated?:

1   *Explore ways to structure key uncertainties to lend insight to the decision.* Start with a decision sketch. Fit the analysis to the problem. The key to this is iteration: a simple first pass will tell you where you need to focus. Sophisticated quantitative methods and results can be used,

---

Text box 6.5:   Using SDM to develop robust forest-management strategies under climate change

Decisions about adaptation to climate change in large-scale human–natural systems involve selecting alternatives under deep uncertainty. As part of a study conducted in British Columbia, we worked with professional foresters to develop and apply a judgment-based approach to selecting robust forest management strategies, defined as those which are reasonably likely to achieve stated management objectives over a range of uncertain future conditions[33].The management context involved managing forestlands that had been severely affected by an infestation of mountain pine beetle. The four fundamental objectives were:

1   Reduce community fire risk.
2   Promote ecological resilience.
3   Maximize net economic value.
4   Maximize non-timber values.

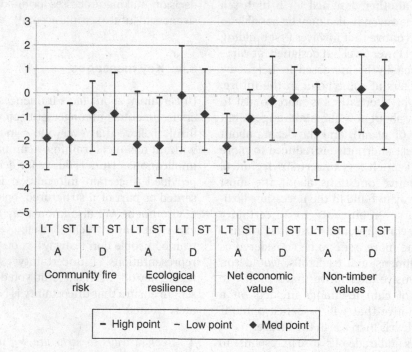

Comparison of averaged judgments for strategies A and D for each objective for the high climate change scenario[33].

We elicited subjective probability distributions from 15 forestry experts regarding the relative performance of a set of management strategies. Judgments were made about the performance for each objective over both short and long term, and across a set of three defined climate change scenarios – *no change, low change* and *high change*. The judgment task involved rating future performance relative to recent (past 20 year) performance on a scale (−5 to +5). The figure shows the averaged ratings of all experts for the 'high' climate change scenario for management strategy A (the status quo) and strategy D under both long term (LT) and short term (ST) timelines. The vertical bar shows the range of average performance ratings from all experts; the diamond represents the median performance. For the high climate change scenario shown, strategy D outperforms strategy A over all four objectives and over both short- and long-term time periods. Similar relative results were found for all three climate change scenarios leading to the conclusion that strategy D was the most robust of all strategies considered.

with care, as part of deliberative processes. But addressing uncertainty as part of many management problems is well-served simply by logical structuring, disciplined dialogue, and the use of expert judgment.

2 *Be specific about probability and consequences.* At a minimum, question linguistic uncertainty in the description of consequences. Express probabilities quantitatively or in combination with verbal expressions.

3 *Recognize that there are times when estimating probabilities is meaningless.* Look for alternate approaches – scenario analysis is often helpful.

4 *Explore the risk profiles of alternatives and the risk tolerance of decision makers.* If some alternatives are more uncertain than others, make sure key differences are exposed for decision makers. Recognize that choices made about how to summarize probabilistic information involve value judgments. Press decision makers to examine their preferences for alternatives with different levels of uncertainty.

5 *Characterize the degree of agreement among experts.* View the diversity among experts as an opportunity to gain a richer understanding of uncertainty and its implications for the choice at hand. The goal is to honestly characterize and document the extent of agreement and disagreement. It may be possible to reduce disagreement by reviewing conditionalizing assumptions, for example, but beware the false convergence that can occur when people are looking for a happy consensus.

6 *Look for ways to improve the quality of information.* This includes improving the quality of data, modeling, and research, improving the quality of judgments under uncertainty, and improving the quality of deliberations. It also includes providing a better understanding of the relevance and significance of uncertainty. For example, are results close to a recognized or legal standard? Are the results near an ecological threshold or severe impact level? What are the consequences of being wrong? Are some alternatives more robust to uncertainty or more precautionary than others? If so, for which objectives is this particularly true?

7 *Remember that the goal is good environmental management decisions.* Sometimes it's not more analysis of uncertainty that is needed but the development of better alternatives. Generate and evaluate alternatives that deal with uncertainty in different ways, including precautionary alternatives (delivering different levels of protection to valued resources), adaptive alternatives (involving monitoring and experimentation) and robust alternatives (that perform well under a range of circumstances).

8 *Use an iterative approach to exploring uncertainty.* Relatively simple methods may provide all the insight you need. Move on to more sophisticated analysis only when you need to.

Above all, when dealing with uncertainty, keep your eye on the ball. The goal is insight, not clever methods. Count on spending time discussing uncertainty, and realize that you may need to experiment with different presentations.

6.5 Suggested reading

Bernstein, P.L. (1996) *Against the Gods: The Remarkable Story of Risk*. Wiley: New York. The author is a financial and economic consultant who begins his exploration of risk and uncertainty with the Greeks and ends with the modern stock market, providing insights about what our relationship to uncertainty shows about our nature.

Conroy, M.J., Barker, R.J., Dillingham, P.W., Fletcher, D., Gormley, A.M. & Westbrooke, I.M. (2008) Application of decision theory to conservation management: recovery of Hector's dolphin. *Wildlife Research*, **35**, 93–102. An excellent case-study example of how structuring the objectives and key sources of uncertainty that underlie a difficult conservation choice can improve natural resource decisions.

Cullen, A. & Small, M. (2004) Uncertain risk: the role and limits of quantitative assessment. In: *Risk Analysis and Society*. (eds T. McDaniels & M. Small). Cambridge University Press, New York. The authors cover the basics of dealing with uncertainty in the context of conducting risk analyses and provide a succinct discussion of reasons to adopt either more or less sophisticated approaches to uncertainty assessments.

Gigerenzer, G. (2002) *Reckoning with Risk: Learning to Live with Uncertainty*. Penguin Books, London. Written in an entertaining fashion, and using a variety of everyday examples, this book notes the omnipresence of judgments about uncertainty and demonstrates the failure of most people, including experts, to understand common presentations of uncertainty.

Gunderson, L.H. & Holling, C.S. (2002) *Panarchy: Understanding Transformations in Systems of Humans and Nature*. Island Press, Washington, DC. These innovative ecologists introduce the compelling ecological and social concept of panarchy, which emphasizes the importance of resilience and adaptation as responses to uncertainty.

Morgan, M.G. & Henrion, M. (1990) *Uncertainty: A Guide to Dealing with Uncertainty in Quantitative Risk and Policy Analysis*. Cambridge University Press, New York. Despite the intimidating title of this book. it's reader-friendly, filled with examples, and demonstrates the theoretical knowledge and practical experience of its authors.

Morgan, M.G., Dowlatabadi, H., Henrion, M., *et al.* (eds) (2009) *Best Practices Approaches for Characterizing, Communicating, and Incorporating Scientific Uncertainty in Climate Decision Making*. US Climate Change Science Program, Washington, DC. Retrieved from http://www.climatescience.gov/Library/sap/sap5-2/final-report/sap5-2-final-report-all.pdf. Much of the most useful recent work on characterizing, communicating, and incorporating scientific uncertainty in decision making is being done in the context of climate change. This report from the US Climate Change Science Program provides a succinct summary of best practices.

Tversky, A. & Kahneman, D. (1974) Judgment under uncertainty: heuristics and biases. *Science*, **185**, 1124–1131. This short paper, published in *Science*, was one of the first to demonstrate how judgments under uncertainty are affected by a series of systematic heuristics and biases; the compelling logic and general niftiness of these experiments initiated entire new fields of study in psychology, economics, and law.

6.6   References and notes

1   Regan, H.M., Colyvan, M. & Burgman, M.A. (2002) A taxonomy and treatment of uncertainty for ecology and conservation biology. *Ecological Applications*, **12**, 618–628.

2   Following Morgan *et al.*[14], by empirical we mean properties of the real world that can, at least in principle, be measured.

3   Deliberate managed burning of an area to reduce fuel loading and avoid the possibility of a future catastrophic fire.

4   Luce, R.D. & Raiffa, H. (1957) *Games and Decisions*. Wiley, New York.

5   Morgan, M.G. & Henrion, M. (1990) *Uncertainty: A Guide to Dealing with Uncertainty in Quantitative Risk and Policy Analysis*. Cambridge University Press, New York.

6   Slovic, P. (1999) Trust, emotion, sex, politics, and science: surveying the risk-assessment battlefield. *Risk Analysis*, **19**, 689–701.

7   Savage, L. J. (1972) *The Foundations of Statistics*. Dover, New York. (Original work published 1954.)

8   Wonnacott, T.H. & Wonnacott, R. J. (1990) *Introductory Statistics for Business And Economics*, 4th edn. Wiley, Hoboken.

9   Some texts will reserve the term 'continuous' for outcomes that truly can assume an infinite number of values, such as the area of a forest fire. However, continuous probabilities are commonly used for variables that have a finite but very large number of possible values, such as dollars, numbers of wolves, and so on.

10   The simplest form of a probability density function is thus a flat line or rectangular distribution, which implies that every value within the possible range is just as likely as any other value.

11   Burgman, M. (2005) *Risks and Decisions for Conservation and Environmental Management*. Cambridge University Press, UK.

12   A 'small spread' was defined as occurring if, in the next five years, myrtle rust is detected in other locations and the infestations cannot be explained by tracing.

13   A 'big spread' was defined as occurring if the myrtle rust is detected in at least four states or territories over the next five years.

14   Morgan, M.G., Dowlatabadi, H., Henrion, *et al.* (eds) (2009) *Best Practices Approaches for Characterizing, Communicating, and Incorporating Scientific Uncertainty in Climate Decision Making*. US Climate Change Science Program Washington, DC. Retrieved from http://www.climatescience.gov/Library/sap/sap5-2/final-report/sap5-2-final-report-all.pdf.

15   Slovic, P., Monahan, J. & MacGregor, D.G. (2000) Violence risk assessment and risk communication: the effects of using actual cases, providing instructions, and employing probability vs fre-

quency formats. *Law and Human Behavior*, **24**, 271–296.

16 Gigerenzer, G. (2002) *Reckoning with Risk: Learning to Live with Uncertainty*. Penguin Books, London.

17 Peters, E., Slovic, P., Västfjäll, D. & Mertz, C.K. (2008) Intuitive numbers guide decisions. *Judgment and Decision Making*, **3**, 619–635.

18 Regan, H., Ben-Haim, Y., Langford, B., Wilson, W., Lundberg, P., Andelman, S. & Burgman, M. (2005) Robust decision making under severe uncertainty for conservation management. *Ecological Applications*, **15**, 1471–1477.

19 Ben-Haim, Y. (2001) *Information Gap Decision Theory: Decisions under Severe Uncertainty*. Academic Press, London.

20 Levi, A. & Pryor, J. (1987) Use of the availability heuristic in probability estimates of future events: the effects of imagining outcomes versus imagining reasons. *Organizational Behavior and Human Decision Processes*, **40**, 219–234.

21 Schoemaker, P.J.H. (1995) Scenario planning: a tool for strategic thinking. *Sloan Management Review*, **37**, 25–40.

22 Kahneman, D. & Tversky, A. (1979) Prospect theory: an analysis of decision under risk. *Econometrica*, **47**, 263–291.

23 One way to think of it is that when these conditions hold, then probabilities are a good proxy for importance weights, and EV becomes a good predictor of the real value of the alternative.

24 A confidence interval designates a range of values in which the outcome is expected to occur x% of the time; for example, 9 times out of 10 for a 90% confidence interval.

25 Goodwin, P. & Wright, G. (2004) *Decision Analysis for Management Judgment*, 3rd edn. Wiley, Chichester.

26 Lichtenstein, S., Gregory, R., Slovic, P. & Wagenaar, W.A. (1990) When lives are in your hands: dilemmas of the societal decision maker. In: *Insights in Decision Making: A Tribute to Hillel J. Einhorn* (R.M. Hogarth, ed.). pp. 91–106. University of Chicago, Illinois.

27 Intergovernmental Panel on Climate Change (IPCC) (1995, July) *Guidance Notes for Lead Authors of the IPCC Fourth Assessment Report on Addressing Uncertainties*. Retrieved from http://www.ipcc.ch/meetings/ar4-workshops-express-meetings/uncertainty-guidance-note.pdf

28 Gregory, R., Failing, L., Ohlson, D. & McDaniels, T. (2006) Some pitfalls of an overemphasis on science in environmental risk management decisions. *Journal of Risk Research*, **9**, 717–735.

29 Morgan, M.G., Kandlikar, M., Risbey, J. & Dowlatabadi, H. (1999) Why conventional tools for policy analysis are often inadequate for problems of global change. *Climate Change*, **41**, 271–281.

30 Lempert, R.J., Groves, D.G., Popper, S.W. & Bankes, S.C. (2006) A general, analytic method for generating robust strategies and narrative scenarios. *Management Science*, **52**, 514–528.

31 Gunderson, L.H. & Holling, C.S. (2002) *Panarchy: Understanding Transformations In Systems of Humans and Nature*. Island Press, Washington, DC.

32 Gregory, R., Harstone, M., Rix, G. & Bostrom, A. (in press) Seismic risk mitigation decisions at ports: multiple challenges, multiple perspectives. *Natural Hazards Review*.

33 The study team included Tim McDaniels, Tamsin Mills, Robin Gregory, Dan Ohlson, and Bruce Blackwell.

34 Dieckmann, N.F., Slovic, P. & Peters, E. (2009) The use of narrative evidence and explicit likelihood by decision makers varying in numeracy. *Risk Analysis*, **29**, 1473–1488.

7

Creating Alternatives

An emphasis on the active construction of alternatives is one thing that sets SDM apart, philosophically and practically, from many other assessment methods. Rather than coming up with one reasonably good alternative and then figuring out whether its impacts are 'acceptable' to people, or conducting sensitivity analyses to explore minor variations – seeing how acceptability changes if we create 10% more restored habitat, add 40 fewer jobs, or increase expenditures on emission reductions by a million dollars – SDM uses what has been learned about the nature of the problem, the context-specific objectives, and explicit performance measures to develop and to compare a suite of alternatives. It's a bottom-up rather than top-down way to build alternatives for a resource-management problem.

In most textbook examples of decision making, alternatives are 'just out there'. They appear as simple, clearly defined lists of items to choose from, like cars or flights. In environmental management, that's occasionally the case: sometimes a list of projects has been proposed for funding, or sites suggested for an activity, or technologies proposed for investment. But these initial alternatives typically are limited: they fail to address the full range of multiple objectives, respond to uncertainty, or stand a reasonable chance of being implemented on time and on budget. Not surprisingly, they often are received with strong opposition from some stakeholders. As a result, the usual task facing a resource manager or decision maker is not simply to compare an already defined set of alternatives but to generate and examine new and creative ones.

Are there always alternatives? We think there are. When environmental managers undertake a routine or familiar action, there is often the assumption that there is no decision to be made and thus no reason to even think about alternatives. An example is licensing and permitting decisions, which are often sufficiently routine that one situation is so much like another (or the previous 100) that it's clear what to do, a pattern is established, and the next choice looks the same as all the others. Yet even if the outcome is set, by habit or by legislation (as with some licensing or listing decisions), there still will be alternatives regarding how actions are undertaken – choices about timing, documentation, which personnel are involved, what level of enforcement is maintained, and the like.

As in previous chapters, we begin with some basics: what we mean by an alternative, what constitutes a good alternative, and how to generate and then structure a good set of alternatives. We emphasize the contribution that can be made by strategy tables, a tool for developing and comparing alternatives. In the last part of the chapter we

Structured Decision Making: A Practical Guide to Environmental Management Choices, First Edition. R. Gregory, L. Failing, M. Harstone, G. Long, T. McDaniels, and D. Ohlson.
© 2012 R. Gregory, L. Failing, M. Harstone, G. Long, T. McDaniels, and D. Ohlson. Published 2012 by Blackwell Publishing Ltd.

cover a few practical tips for working effectively with alternatives in a deliberative environment.

7.1   The basics

7.1.1   Why do we need alternatives?

First and most obviously, we need good environmental management alternatives because the action that ultimately is selected will only be as good as the best alternative that is considered. When an alternative is generated and examined, the process involves a close investigation of how stakeholders think it will affect their fundamental objectives – perhaps through quantitative modeling, perhaps through disciplined thinking, perhaps through structured dialogue. We learn by generating and exploring alternatives: we're able to propose new and potentially better actions, and we ultimately arrive at solutions that do a better job of achieving the fundamental objectives.

A second reason for creating good alternatives is less obvious. We need alternatives so that value judgments will be made explicitly by those with legitimacy to make them. Too often, in government as well as in the private sector, resource managers end up providing recommendations to political leaders. In so doing, they are making important and potentially controversial judgments based on their own values. And to be fair, decision makers are typically asking for recommendations. Yet what decision makers – senior managers, elected and appointed officials, or their designates – really need is a small, carefully thought out and documented set of alternatives, along with clear information about the objectives that might be affected, their anticipated consequences, key value trade-offs, and the likely responses of different stakeholders. Then their task is to make the final choice.

This leads to a third reason why we need alternatives. It is the presence of alternatives that provides context for people to evaluate choices. Research in behavioral decision making shows that knowledge of characteristics across a range of alternatives actually changes our expressed preferences (see Text box 7.1). In simple terms, we often don't know how to value what we have until we can compare it to another alternative; we only know what we have, or what we're missing, once we see an inferior or more complete alternative[1]. Ultimately, judgments about what is acceptable are made only with reference to the set of alternatives under consideration[2].

Finally, involvement in the process of identifying and evaluating alternatives changes people's willingness to make tough choices. Participants in a decision process often resist (quite reasonably) making a difficult trade-off until they're sure that all reasonable alternatives for avoiding the trade-off have been explored and exhausted. This shouldn't be too surprising. It is, after all, how we deal with tough or unexpected trade-offs in our personal lives. Suppose you plan a trip to Hawaii, and an initial search turns up two flights – one takes twice as long as you planned, the other costs twice as much. You don't immediately sit down to think about which compromise you will make. Instead, you look for more alternatives. Only after you've called a travel agent or talked with friends or thoroughly searched Expedia, and determined that there really aren't any other flight options, will you be willing to address the trade-off and feel good in the long term about your choice.

7.1.2   What's an alternative?

This sounds like a trivial question, but it's not. Alternatives in a decision process are not just any proposed solution to the problem. Alternatives are *complete* solutions to a given problem that can be directly compared by decision makers. What is complete depends on the context. If a department is going to fund a single project, then the individual candidate projects are alternatives. If the department is going to fund multiple projects up to a budget cap, then the alternatives are the different sets of projects that could be funded – and individual projects are elements of the alternatives.

Text box 7.1: Evaluating alternatives

Behavioral researcher Chris Hsee has studied how people judge alternatives, focusing on the question of whether alternatives are evaluated individually or through a comparison among two or more alternatives[3]. In one group, people were asked to assign a value to Dictionary A. They were told that Dictionary A had 20000 entries, and that it was new but with a torn cover. A second group of people were shown Dictionary B, which they were told had 10000 entries and was new and in good condition. Based on these descriptions, participants assigned a value of $20 to Dictionary A and a value of $27 to Dictionary B, implying that Dictionary B was valued more highly than Dictionary A. However, when a third group of people was shown both dictionaries the assigned values remarkably reversed, with Dictionary A now preferred (valued at $24) over Dictionary B (valued at $20). Why? Because the presence of alternatives provided a benchmark that was useful in determining relative value. When group B assigned a value of $27 to Dictionary B, they were not aware that an alternative existed with twice as many entries. Group A was apparently put off by the torn cover, but when faced with a choice between this dictionary and an alternative with half the entries, the torn cover looked like a reasonable trade-off to make. What's the moral of this experiment? People may make a different decision when they are able to compare alternatives across their relevant dimensions; this argues for an assessment process that facilitates choices across transparent and easily understandable (or, to use Hsee's term, easily 'evaluable') dimensions of value.

Sometimes the elements are similar things, such as research programs. In other cases the elements could be qualitatively different kinds of things. If a decision involves developing a recovery plan for a threatened wildlife species, for example, the elements of a complete alternative will likely include some combination of harvest regulations, habitat enhancements, and monitoring or enforcement programs.

It follows that complete alternatives are necessarily *mutually exclusive* – one and only one of the proposed alternatives can be selected – and that the alternatives under consideration in any given decision problem must be *directly comparable*. We talk more about these and other properties of useful alternatives in the next section.

7.1.3 What's a good alternative?

From an outcomes perspective, a good alternative is one that has a good chance of providing a meaningful solution to the problem at hand. In other words, it addresses participants' fundamental objectives. From a process perspective, a good alternative is one that is amenable to analysis and helps decision makers and stakeholders to learn, and ultimately gain clarity about, what would be a good solution. Alternatives with a good chance of delivering these share several common properties. They are complete and directly comparable, value-focused, fully specified, internally coherent, and distinct.

Complete and comparable

Combinations of actions do not make true alternatives in a decision process unless they are complete and directly comparable. To be complete and directly comparable, alternatives must address all the key aspects of the same problem, over the same time period, using the same underlying assumptions about events or conditions beyond the influence of the decision at hand. Suppose your decision context involves developing a municipal solid waste management plan. You can't compare a waste treatment facility that has a 10-year life with one that has a 20-year life, or one that handles 50% of the waste stream with another that handles 100%. To be complete and comparable in this case, you need to design two

20 year plans that each fully addresses the projected waste stream. One may involve a single investment that lasts 20 years, the other may involve two or more sequential investments as facilities with shorter lifetimes are retired. One may involve a single landfill with the capacity to handle the entire waste stream, the other may involve several elements – for example filling the existing landfill, building a new incinerator, and expanding recycling facilities.

Value-focused

A focus on values means that alternatives are explicitly designed to address the fundamental values or ends of the decision – the things that matter, as defined by the objectives and the performance measures. Of course, different alternatives will vary in terms of how well they address different objectives. Some value-focused alternatives will seek win–wins (gains on multiple objectives) or a balance across objectives; others will emphasize one objective more than others. There's no saying which is right. The key is that the alternative is designed with the ends (the fundamental objectives) in mind.

Value-focused thinking can guard against the common mistake, which we often observe, of defining alternatives too narrowly. Defining a choice as concerned only with purchasing a new car, for example, encourages an immediate focus on the pros and cons of three different makes, but misses the fact that the car is only a means for accomplishing something of value (getting the individual from A to B safely, at a reasonable cost and with a minimum level of comfort, etc.). As a result, other possible alternatives such as moving closer to where one works, joining a car-pool, or taking public transit (in addition to looking at used cars or paying for yet another round of repairs on the existing vehicle) are likely to be ignored. The same logic applies to resource-management choices: lengthy deliberations by resource managers on alternatives to enhance riparian vegetation along an urban stream or to improve nesting habitat for a threatened bird species only make sense if other means for achieving the respective fundamental objectives have been considered (e.g. improving hiking opportunities or perhaps the health of urban residents, or increasing bird populations or perhaps regional biodiversity).

Fully specified

Fully specified means that alternatives are defined to a sufficient level of detail, using logically consistent assumptions. Ambiguity in the specification of alternatives is an important source of errors and inconsistencies in the subsequent estimation of consequences. An alternative that proposes a 'prescribed burn' of a forested area would need to specify what safeguards would be built into the alternative: what time of year, under what weather conditions, with what provisions for firefighting backup. Without this level of specification, the consequences of the alternative can't be evaluated consistently and different experts are likely to make different assumptions. Further, all of the alternatives should be specified to a similar level of detail. A family looking for a new home can't fairly compare 'a five-year old, 2000-square foot house in the 4500 block of Brown Street' with 'a 'downtown house' because one is much more specific than the other.

Internally coherent

The combination of elements that make up an alternative need to be internally coherent, which means that, when considered together, the individual elements are both feasible and make logical sense. They should work together to address effectively the relevant multiple objectives, and there should be a plausible rationale for why an alternative might be expected to be effective in addressing these objectives. At a very simple level, if you're evaluating personal transportation alternatives, purchasing a good bicycle and an annual bus pass might make sense together. Purchasing a good bicycle and an annual train pass might make sense together. But purchasing an annual train pass and an annual bus pass probably don't make sense together.

Distinct

Distinct means that alternatives are different enough to offer decision makers a real choice. If alternatives are distinct and value-based, then they will succeed in exposing difficult but unavoidable value-based trade-offs. Despite the fact that it's common practice, giving decision makers one favored alternative along with two others that essentially bracket it – showing a little more of this or less of that – is a sham. Alternatives should represent either very different ways to achieve the objectives, or very different levels of performance across at least some of the objectives.

7.2   Developing alternatives

There are three basic steps in developing alternatives: brainstorm a range of potential management responses, organize them into fully specified alternatives, and iteratively refine the alternatives under consideration.

7.2.1   Brainstorm a range of potential responses

With objectives, we were always asking why? With alternatives, we ask how? – how can the stated objectives best be achieved or fulfilled? A value-focused way to generate alternatives works through the list of objectives and identifies actions that satisfy each concern: what project would best restore habitat?; be least expensive?; improve other species?; create the most employment?

Because these 'bookend' alternatives have a single-minded, one objective focus, they're unlikely to provide the basis for a widely supported solution. Yet they play an important role early in the decision process, for two reasons. First, they allow room for creative thinking: by looking at one objective only, unconstrained by worries about what might happen to other objectives and whether people would find those effects acceptable, a wide range of ideas is put forward. Many may ultimately prove unworkable, but others will form a starting point for creative win–win solutions. It's critical at this early stage of alternatives generation to allow and encourage a creative, no-boundaries, value-focused exploration of a wide range of ideas.

A second function of bookend alternatives is to create an atmosphere where it's safe to explore alternatives. The fact that someone puts an alternative forward for evaluation does not mean that they will ultimately support it as a solution. It means they think that something useful may be learned by exploring its consequences. Particularly in a complex decision process where people will meet on multiple occasions to work through alternatives, it's useful to start with a set of alternatives that facilitate learning and allow people to start talking about their values and desired outcomes and the pros and cons of different alternatives, without feeling that they really need to defend any one solution.

Once you have a long list of possible solutions, you may discover either that it includes a variety of items that you really can't consider (given your mandate) or that the sheer number of items or actions is overwhelming (given the resources available). In this situation, it may be useful to screen individual actions or elements on the basis of a subset of objectives and measures. For example, a municipality with firm emission reduction targets for climate change might screen energy options on the basis of cost and greenhouse gas reductions, with only the resulting short list undergoing detailed analysis. Or if there is a suite of possible actions, all of which deliver exactly the same sort of benefit (emission reduction, energy or water supply/savings, etc.), it may be helpful to screen them on the basis of unit cost – this then becomes a relatively simple cost-effectiveness analysis.

Use caution in screening, however. First, screening puts enormous weight on the objective used for screening. Is this what you want? Is it appropriate in your case? Are you the right person to make this value judgment? Be particularly wary of screening on the basis of 'public acceptability.' Often, this is based on someone's premature

---

Text box 7.2:   Watch out for new objectives

When brainstorming alternatives, be on the lookout for new objectives that weren't previously identified. When someone proposes an alternative, ask why – what is it that makes this a good alternative? If their answer doesn't appear in the list of objectives then it could be that a new objective needs to be added. For example, in setting up a plan for cleaning up contaminated wastes, one of the federal regulators said that a particular plan was great because it provided lots of flexibility for changes in the scale of clean-up over time. This was said to be a concern because the volume of wastes in question was largely unknown and any plan might need to be drastically scaled up or down, so flexibility was important. But when this same regulator had listed his agency's objectives, flexibility had not been included.

presumption about what stakeholders will want. Remember that stakeholders' views often change as they learn about the alternatives and the values at stake.

Second, and related, there is a danger in prematurely screening out candidate actions that seem unacceptable at first glance. In fact, once combined in a logical strategy portfolio, these actions might be completely viable. For example, strategies that involve slightly more risk in managing an endangered species might be unacceptable at first glance, but become acceptable if packaged with more intensive monitoring. Similarly, actions involving greater cost might be acceptable if packaged with appropriate cost-recovery mechanisms, for example new regulatory or pricing policies, or including a transition period for industry to respond. We discuss the use of strategies in more detail in the next section.

Screening nonetheless remains a useful approach to keeping your decision problem focused on realistic alternatives. Some common

screens include legal constraints or requirements, technical feasibility, cost, cost-effectiveness and, with caution, minimum performance standards on particular objectives.

7.2.2   Organize actions into complete alternatives

The next challenge is to combine a long list of potential solutions into complete and comparable alternatives. A common approach is to find a quick way to score and rank individual actions or elements and then select the top ranking ones. Only rarely will this be a sensible approach. Most of the time it's useful and necessary to think more strategically about logical combinations of actions that make sense together. We think of this as developing strategies.

Developing strategies

Imagine you own a deli and you're creating your sandwich menu. You have many possible ingredients: brown bread, white bread, bagels, sourdough buns, ham, turkey, roast beef, hummus, avocado, tomato, lettuce, cucumber, mayonnaise, mustard, chipotle sauce. There are a lot of possible permutations, but not all of them make sense. A combination of brown bread, mustard, and mayonnaise doesn't constitute what most of us would think of as a decent sandwich. So you will need to combine the ingredients into logical combinations. To do so, you need to group ingredients into categories. In the bread category are rye bread, bagels, croissants, and so on. Then you might have a category for sauces, one for meat and meat substitutes, and one for vegetables. A sandwich (and by analogy a strategy) would have elements from multiple categories (but not necessarily every single one). Note also that some categories contain elements that are mutually exclusive – you won't want both rye bread and bagels on the same sandwich. However, others are not – you might well choose more than one meat, and multiple vegetables. There are an apparently endless number of possible combinations; you can't possibly put them all on the menu. So you come up with a few themes – you create the 'meat lovers' sandwich, the 'low-fat', the 'Mexican', and so on. These sandwiches are listed on your menu. Customers order them as is, or tweak them to suit their particular tastes.

The strategy building we're talking about follows much the same pattern. In a management or policy context, just like in the sandwich-making business, there are usually several categories of possible management actions. Creating a strategy involves selecting none, one, or more than one action from each category and combining them to create an interesting alternative, normally based around some kind of theme. There's no magic way to develop themes. When developing alternative plans for managing air emissions, for example, one strategy might focus on market instruments, another on regulatory approaches, and another on a hybrid strategy. Alternatively, the 'theme' for creating strategies might also be oriented to specific standards: one strategy may be designed to achieve a 25% reduction in emissions vs another that is designed to achieve a 50% reduction. Some strategies account for actions that are sequenced in time (see next section), whereas others address fixed performance targets (we often call these 'portfolios'). Or the themes guiding strategies might be value-focused: one may be a low-cost option, while another is particularly focused on improving visual quality and a third focuses on minimizing localized health concerns.

A 'strategy table' is a useful structuring or organizing technique for developing strategies. Table 7.1 shows a strategy table with two strategies for a park-management problem. The rows are the individual management actions or elements, grouped into categories. The categories relate to fundamental objectives – in this park-management example, the objectives might be to maximize user satisfaction and to protect wildlife habitat – and usually there is more than one way to express the objectives; as a result, the choice of rows will often elicit substantial discussion. The columns are the strategies, and the checkmarks

Table 7.1 Strategy table for park management.

	Strategy A	Strategy B
Habitat management		
Do nothing		
Build barriers	Y	
Enhance existing habitat	Y	
Create new habitat outside the park		Y
Create management zones		
Access management		
Do nothing		Y
Limit access at all times	Y	
Limit access at biologically sensitive times		
Cost recovery instruments		
None	Y	
From treasury		
From user fees		Y
Monitoring		
None		
Basic	Y	
Enhanced		Y
Enforcement		
None		Y
Basic		
Enhanced	Y	

shown in the table indicate which elements are present in each strategy. In most cases, some categories will contain mutually exclusive elements (e.g. you can choose only one element under access management) while others do not (e.g. you can choose multiple elements under habitat management).

In practical terms, it helps (both for clarity and to encourage learning and discussion) to identify strategies as expressing a theme and, often, to name them. In Table 7.1 Strategy A ('Preserve What We Have') maximizes habitat by preserving or enhancing existing habitat and limiting access. Strategy B ('Build It and They Will Come') creates substitute habitat elsewhere while allowing free access to the park. Each strategy is designed to address both habitat management and access management concerns, albeit with different degrees of emphasis. Each is supported by a plausible rationale about why it might be effective (e.g. it's coherent), but note that no analysis has yet been done. We don't know if either strategy will be effective, or to what degree the fundamental objectives will be achieved. All we've done is put heads together and come up with something that is worth investigating further. At this point, it's like sandwiches on the menu – you can't possibly list all the possible combinations, so you choose a few to stimulate ideas. By examining their consequences and talking with others, you identify themes that warrant further attention without having to identify and evaluate every single possible alternative.

One of the key benefits of strategies is that it's often possible to develop creative alternatives that meet multiple objectives out of initial ideas that, on their own, weren't thought to be viable. When the idea of building substitute habitat was first proposed in this case, there was opposition to it. It was seen as a risky alternative in the sense that it was uncertain whether or not the targeted wildlife would actually use the new habitat. Yet instead of rejecting this proposal, we asked: what could be done to make it more viable? This led to the suggestion that it might be more acceptable if it were accompanied by close monitoring. Monitoring costs money, so this suggested a need for creative cost recovery instruments. So Strategy B was born – and broadly supported as an alternative that warranted further exploration.

Clearly the strategies in Table 7.1 aren't ready for consequences assessment. They are not fully specified: an expert could not assess the performance of a strategy with elements called 'build barriers' without further definition (what kind of barriers, where and when). The strategy table allows people to think critically about combinations of things that make sense together without getting lost in the details; however, the details of specifying what those elements are must still be done (see Text box 7.3).

At the end of this section, we provide two case studies that demonstrate the use of strategy tables as an aid to deliberation when working with small groups charged with developing (a) a recovery strategy for an endangered salmon species and (b) an integrated energy strategy with fixed performance constraints.

Sequenced alternatives

Sometimes strategies have a temporal aspect. This can occur for several reasons, but it often happens when we want to think about the wisdom of doing one thing, learning more, then doing another thing based on what's been learned. By looking ahead at subsequent decisions in a sequence in advance, it's sometimes possible to gain insight into the wisdom of the first decision before making it.

The classic decision analysis example of this is the oil company that needs to choose whether or not to develop an oil platform in a particular area. A first question might be, 'should we drill a test well in the area'? Drilling a test well has a certain probability of resulting in a positive sign, but also has predictable costs in terms of time, resources, and possibly environmental impacts. If the test well were to be drilled and come up positive, then a second question might be, 'is this well promising enough to justify building a platform'? This second question has its own probability of being true.

By thinking through different variables for probabilities of success, and ranges of values for costs and revenues, it's possible to make a judgment on whether or not to sink the test well. Decision makers often use 'expected value' as defined in Chapter 6 to help them decide (where expected value is the sum of probability-weighted consequences across a range of possible outcomes).

As an example of this, recall in Chapter 6 we introduced the example of a sequenced decision problem faced by the Australian Government, which needed to make an immediate choice about whether or not to attempt to eradicate an invasive fungus. In that case, there were several 'outcome' state possibilities – e.g. the fungus might not spread at all, or it might spread in a small way but not in a big way, or it might spread over a large area. For each of these states, we could estimate both the notional outcome consequences (assuming each outcome had come to pass), as well as the

Text box 7.3: Defining sequenced alternatives

In Chapter 6, we introduced a case study that concerned an outbreak of an invasive fungus that threatened to spread the disease myrtle rust across Australia. We introduced how management responses could be divided into two sequential stages: local, short-term actions that may (or may not) be directed at eradicating the fungus, followed by wider scale activities that might be undertaken should the spread continue beyond a certain point. We introduced the idea of developing strategies such as 'eradicate'. To work with these strategies further, we need to more fully specify what is meant by an 'attempt to eradicate'.

The table below shows the list of specific management activities that might be undertaken to a greater or lesser disagree in both the first and second decision points. On the left hand side are the individual elements that can make up a strategy: surveillance (monitoring) activities, fungicide application, infected hosts clearing, and so on (left hand column). Some of these elements refer to activities that would help an attempt to 'eradicate', whereas others might help to slow the spread of the disease. Five pre-spread strategies (V through Z) and three post-spread strategies (1 through 3) were developed. Each strategy is defined by assigning a level of effort (shown in the cells of the table) to each individual element.

(a) Pre-local spread activities for each of the five policy alternatives (A1 Actions). (b) Post-local spread activities for each of the five policy alternatives (A2 Actions).

(a)	'V – Do nothing'	'W – Live with it'	'X – Slow the spread'	'Y – Eradicate lite'	'Z – Eradicate full'
Surveillance	Minimal	Minimal	Large	Moderate	Large
Fungicide application	Minimal	Minimal	Moderate	Moderate	Large
Infected hosts clearing	Minimal	Minimal	Small	Small	Large
Movement controls	Minimal	Minimal	Large	Minimal	Large
Public-use restriction	Minimal	Minimal	Minimal	Minimal	Large
Resistance breeding	Minimal	Large	Minimal	Minimal	Minimal
Ex-site and in-site conservation	Minimal	Large	Minimal	Minimal	Minimal
Hygiene measure implementation	Minimal	Large	Large	Minimal	Minimal
R&D of impact mitigation	Minimal	Moderate	Minimal	Minimal	Minimal

(b)	'1 – Do nothing'	'2 – Live with it'	'3 – Slow the spread'
Surveillance	Minimal	Minimal	Large
Fungicide application	Minimal	Minimal	Moderate
Infected hosts clearing	Minimal	Minimal	Small
Movement controls	Minimal	Minimal	Large
Public-use restriction	Minimal	Minimal	Minimal
Resistance breeding	Minimal	Large	Minimal
Ex-site and in-site conservation	Minimal	Large	Minimal
Hygiene measure implementation	Minimal	Large	Large
R&D of impact mitigation	Minimal	Moderate	Minimal

For example, an alternative called 'V1 – do nothing' proposes a 'minimal' level of effort in all the listed activities in both decision points. In contrast, the 'Z2' alternative is to 'eradicate full' until we learn whether eradication is successful, and then to switch to a 'live with it' strategy if eradication is unsuccessful. In that case, we would see 'large levels' of activity in surveillance, fungicide application, infected hosts clearing, movement controls, public-use restriction in the short term, which would switch to resistance breeding and other activities if eradication fails.

To minimize ambiguity as much as possible under the circumstances, we developed a reference table to describe in more detail what we mean by these different levels of activity. The table below shows the descriptions developed for the 'fungicide application' activity. Similar descriptions were developed for all the activities.

	Minimal effort	Small effort	Moderate effort	Large effort
Fungicide application	Do not apply any fungicide	Apply fungicide only in infested plants and the 'batch' where they located	Apply fungicide in infested plants, the 'batch' where they located, and the surrounding batches	Apply fungicide in all susceptible batches and in the surrounding bush areas

Thanks to these tables, people could refer to alternatives such as 'slow the spread then live with it' in conversation with a much greater chance of being understood than might otherwise be the case.

probabilities of each of the outcomes happening. An overall 'expected value' for each alternative could then be calculated by first multiplying the consequences of each state by the probability of that state occurring and then summing the resulting values.

This invasive species example also tackled the potentially tricky issue of designing sequential alternatives, as discussed in Text box 7.3.

7.2.3 Establish a reference case

As part of the alternatives generation process, it's often helpful to define a reference alternative, which then is used to compare with other alternatives that are generated through the analytic and deliberative process. Often the reference alternative is closely related to the status quo. In keeping with the requirements noted earlier in

this chapter, it has to be specified completely and unambiguously.

If you are defining a reference case, we have three pieces of advice. First, you'll need to think hard about what the 'reference case' should be. Is it a snapshot of current conditions? If so can you define what that is? Or is it a snapshot of some historical condition? If on the other hand, you're projecting into the future, what does it mean to reference the status quo or business as usual? Will all current actions continue to be implemented? What new actions will be done 'anyway', regardless of this process?

The second piece of advice is: don't underestimate how difficult it may be to nail down exactly what constitutes 'what we're doing now' to manage a particular problem and what it's costing. Dialogue about this alone can lead to important insights (see Text box 7.4).

Text box 7.4:   Defining the status quo may not be straightforward

Recently we were engaged to help a working group composed of senior managers and scientists from three United States agencies with overlapping mandates for the protection of Atlantic salmon (USFWS, NOAA, and the State of Maine). Our task was to help the group develop a framework for coordinating their species recovery activities. A starting point included the review of existing ideas for inclusion as part of the draft recovery plan; this comprised a lengthy inventory of projects that the three agencies proposed to undertake. Unfortunately, they conceded that no new funds were likely to be forthcoming, and in fact future budget cuts were probably more likely. This meant that implementation of this long list of projects was unrealistic, and a first step was to reframe the task as a budget allocation exercise. After establishing objectives and performance measures, we therefore helped to create and characterize the likely performance of different strategies or 'portfolios', composed of sets of coordinated activities, that could be undertaken for different levels of funding.

At an early point in the deliberations, all parties agreed that they should compare all proposed strategies against the 'status quo'. This was expected to be a straightforward information-collection task, but it turned out that defining the status quo wasn't easy. Nevertheless, the working group agreed that it was essential to do this because in order to initiate new actions, some existing activities might have to be scaled back or cut. After several weeks of administrative research and analysis, a characterization of combined 'status quo' resource efforts was finally available and presented as a 'reference strategy' (i.e. continue to do what is now being done) to the working group. The primary response of participants was surprise, closely followed by disappointment; the main concern was the high proportion of their combined resources being used for general stock assessment purposes, in contrast to funding activities that might directly lead to population recovery or to addressing specific stock assessment data gaps that were explicitly tied to a specific recovery initiative.

Ultimately, this clarification of current strategies led to the development of a decision tool to help participants create new strategies within various levels of budgetary constraints. Each of these alternatives was evaluated based on its anticipated performance relative to the specified objectives and performance measures.

Finally, watch out for anchoring. One of the most common traps in decision making is to anchor on a single alternative (often the status quo) and tweak it. One way to minimize this bias is to develop a set of creative, value-focused bookend alternatives first, and then come back to develop a base case in detail.

7.2.4   Iteratively improve alternatives

The development of alternatives typically occurs iteratively, and new and better alternatives are the result of the learning gained from estimating their consequences. The process is iterative. It starts with a brainstorming session that leads to the generation of a first round of alternatives – perhaps a reference case, a couple of bookends, and a few more alternatives. These alternatives are then analyzed, and may undergo further refinement by technical specialists to fully specify them before their consequences are estimated (using models or experts' judgments; see Chapter 8).

In many cases, the results are brought back to a working group, which reviews the results and deletes clearly inferior alternatives, proposes refinements to the remaining alternatives or, based on what has been learned, identifies totally new ones. How many rounds there might be depends on the decision context. In a relatively simple system, acceptable alternatives may be

Table 7.2 Wind farm example to illustrate simplifying the decision process.

Objectives	Measures	Preferred direction	Alt 1	Alt 2	Alt 3
Birds	Number of fatalities per year	Lower is better	1200	3200	5600
Savings	$ per household per year	Higher is better	$300	$450	$250
Employment	Number permanent full-time jobs	Higher is better	23	22	12

found after one or two rounds. In more complex situations, four or five rounds might be needed before an acceptable alternative (or set of alternatives) is identified. Whether these refinement rounds are all conducted in a full multiparty deliberative environment, or whether some occur with a smaller subset of experts or managers, also depends on the context. The important point is that the process of identifying and evaluating alternatives is iterative. Each round of analysis produces insights that help to (a) delete inferior alternatives from further consideration, (b) develop new alternatives that are likely to perform better in terms of achieving objectives, (c) build an understanding of what the key trade-offs are and how all parties view those trade-offs, and (d) narrow in on alternatives that have a good chance of meeting multiple objectives and also being supported by a broad range of stakeholders as well as decision makers.

An obvious candidate for omission from further consideration is a dominated alternative – one that is outperformed in every way by another alternative. Simple as this might sound, it turns out to be a very useful concept in the development of acceptable alternatives. Suppose you have a proposal for a wind farm development, as illustrated in Table 7.2. The main concerns are about effects on resident and migratory bird populations. The main benefits are financial savings to local residents (as they will receive subsidized electricity rates) and local employment. Assume that three objectives have been set: minimize bird fatalities; maximize household savings; and maximize employment. Now suppose that in round 1, three alternatives are developed (consisting of different configurations and sitings of the turbines). These alternatives are then analyzed and

their consequences on each of the three objectives are estimated. The objectives, measures and estimated consequences of the three alternatives are shown in Table 7.2.

A review of the consequence table shows that alternative 3 is dominated by alternative 2. That is, for each of birds, savings, and employment, alternative 2 outperforms alternative 3, so we can eliminate it from further consideration. Left only with alternatives 1 and 2, we see that employment doesn't vary meaningfully across the two alternatives – therefore this objective is no longer necessary or useful for discriminating between them, and it also can be removed from further consideration. (Note that we're not saying that employment is not useful or necessary in a general sense, just that it isn't useful in choosing between these two alternatives.) Now participants in the decision can either (a) choose between alternatives 1 and 2 on the basis of the two remaining objectives, in which case they'll have to address the trade-off between bird mortality and household savings, or (b) they can try to develop new alternatives, for example a new site location or a different turbine configuration that aims to improve bird performance while retaining household savings.

Despite the seemingly straightforward nature of this simplification-by-elimination process, it can create a real sense of collaborative decision making within a group. The collaborative removal of dominated alternatives and unhelpful performance measures can greatly simplify – through a defensible, easily documented process – the task facing resource managers and decision makers. Although this is obviously a simplified example, the same idea can be applied to decisions involving dozens of complex alternatives and performance

Text box 7.5:   Be patient and thoughtful when creating alternatives

Paul Nutt[4] reviewed over 400 decisions to evaluate the quality of choices made by managers and concluded that one of the primary blunders was a frequent 'rush to judgment' and a miscoding of decisions as urgent when, in fact, very few were actually crises. He discusses how the responsibility of making a decision, combined with a perception that delay may signal an inability to cope or foot-dragging, encourages managers to favor rapid (and sometimes rash) actions that have the result of 'making it difficult to wait for alternatives to emerge'. Nutt's research suggests that managers often will unduly limit the search for alternatives and, instead, go for the comfort that comes from 'settling on a single idea early in the decision process'. The result is a failure to uncover innovative solutions and, often, creation of exactly the future that was least desired: unanticipated delays and/or costs and a waste of resources spent in defending an inferior alternative.

A willingness to iterate and to review and rethink alternatives leads to better resource-management choices. In some cases iteration will lead to a larger set of alternatives, by introducing possible new initiatives or by adding important objectives. In other cases iteration will result in a smaller set of alternatives, by avoiding dominated actions or by focusing analyses and deliberations of influential objectives. One of the most important roles of the analyst/facilitator is in knowing when to push for more clarity and structure and when to recognize that the basic information needed for a good decision is in hand. Just to make explicit what may be an obvious point: as much as decision aiding is about creating smart and more broadly acceptable and high-performing alternatives, it is also about avoiding stupid, indefensible or poorly performing alternatives.

measures over multiple rounds of iterative improvement. Eventually, you will run out of alternatives that are strictly dominated, and begin to look for and use 'practical dominance' to simplify the set of alternatives. This involves value judgments, and we deal with this further in Chapter 9.

7.3   Practical tips for working with alternatives

This section provides some tips and advice for dealing with some of the tricky situations that arise when dealing with alternatives in a deliberative environment.

7.3.1   Watch for decisions within decisions

In complex decisions, there are often decisions within decisions. Making the sub-decisions requires a kind of 'SDM-lite' that takes place within the main decision process. In case this sounds confusing, here's an example.

We were asked to lead a decision process to reconsider water allocations at a hydroelectric facility; the primary task was to identify, evaluate, and select a range of water-management alternatives that primarily involved different flow release rates in rivers and water levels in the reservoirs. Objectives and measures used to identify and evaluate alternatives were related to fisheries, wildlife, power production, recreational quality, flooding, and First Nation culture values.

The overall decision process was conducted over an 18-month period. At an early stage, a decision was required concerning how to consult with the resident First Nation. Making this choice also involved setting clear objectives, although obviously these were not the same objectives as those used to identify and evaluate water-management alternatives. For this decision-within-a-decision, objectives included concerns such as participation, efficiency, fairness, and effectiveness in having the First Nation's interests heard. After discussion, it was agreed that the set of possible alternatives included: (a) a single consultative table, including First Nation participation; (b) a separate, parallel First Nations consultative process; and (c) direct bilateral consultation between the utility and the First Nation. This decision process–choosing a consultation method for First Nations – occurred over about one month.

Once this issue was settled and the main decision process initiated, attention turned to addressing uncertainties associated with the consequences of the various proposed water-management alternatives. Studies were proposed to address these uncertainties and facilitate consequence estimation – a long list of candidate studies (alternatives) with total costs exceeding the allocated study budget. A decision was required to select a set of studies to fund. Objectives and measures specific to the selection of studies were set, and these were related to the timing of information, the quality of information, and relevance to the decisions that had to be made. This decision process occurred over about a month (subsequently the studies were conducted over a six month period before the overall decision process reconvened).

Finally, after a short list of water management alternatives was on the table, it became clear that a few specific uncertainties were unresolved and profoundly important. These uncertainties affected just one part of the system (one river in a series of three interconnected rivers and reservoirs), so decisions about this part of the system were handled separately. An experimental program was established to reduce the identified uncertainties; it involved a sequence of experimental flow releases, each lasting two years. Different experimental designs were proposed, involving different flow rates, durations, and intensities of monitoring[5]. Objectives were related to a subset of outcomes including salmon biomass, power revenues and monitoring costs.

This example highlights a simple but subtle message. Decisions about the management of environmental resources evolve, and sub-problems arise in the course of deliberations about the main considerations (we'll revisit this as part of Chapter 10's focus on learning).

In many cases, these smaller but nevertheless critical decision problems can effectively be addressed by an abbreviated SDM process that takes place within the context of the broader decision.

7.3.2 Labeling alternatives: what's in a name?

You might be tempted to give your alternatives catchy shorthand names: 'precautionary approach', 'compromise solution', or 'Bill's wish list'. This usually is fine in the early stages of developing alternatives; we've developed lots of 'recreation friendly' alternatives and 'low-cost' alternatives at the bookend stage. But you should shift into more neutral labels once more fully developed alternatives are on the table, for several reasons:

1 People will latch onto labels and fail to notice that the estimated performance doesn't live up to the promise of the label. Remember that when you build strategies and label them – 'fish friendly' for example – your intention may be to create the best alternative for fish. Yet with iterative development of alternatives that incorporates learning, you may ultimately end up with another alternative that performs as well or better for fish but has an innocuous name such as Mary's idea, or hybrid A.

2 People may respond emotionally to terms that resonate with their values, without really checking on the actual estimated performance of the alternative. 'Protect what we have' may resonate with protectionists more than 'build it and they will come', but if the estimated probability (based on more detailed analysis) that management actions will succeed in protecting what you have turns out to be low, then the label becomes misleading.

3 Other people may associate a particular technical meaning with a term that does not

match the understanding of the decision participants. As a result, although the people in the room understand the history of an alternative and its evolution over time, the ultimate decision maker may not. We know of one case where an alternative developed and recommended by participating stakeholders was rejected by the decision maker because it was characterized as having involved 'consultation' with a community group rather than 'engagement' – for this key decision maker, the two words held quite different meanings.

7.3.3   How many alternatives?

Barry Schwartz opens his book, *The Paradox of Choice* (2004)[6], with an anecdote about a shopping trip to buy a pair of jeans, where he quickly was overwhelmed by a friendly shopkeeper who asked question after question about the desired fit, style, and fabric. Schwartz's thesis is that, contrary to conventional thinking, people are not most contented when faced with infinite choices. Rather, people very quickly become overwhelmed when faced with an excess of options and they may end up either becoming distressed by the task (because the burden of finding the 'best' choice is onerous) or disengaging from the decision altogether.

Remember that this example concerns an individual engaged in the familiar act of buying a pair of jeans; imagine the confusion when the context shifts to a multi-stakeholder planning process engaged in coming up with a broadly acceptable environmental management alternative that works well for five different groups across seven objectives and 12 performance measures! The key difference between these two settings – and the reason for optimism rather than despair – is that the choice among jeans is typically unaided whereas structured decision-aiding processes such as SDM seek to help participants clearly define relevant objectives and performance measures and, through shared discussions and

thought, to create an appropriate evaluation framework. Once this decision structure is established, and particularly if the context is such that the assignment of weights and the calculation of scores is readily understood, then individuals and groups can become highly skilled at considering large numbers of alternatives and agreeing on an efficient path to selection of a preferred plan.

So back to our opening question: what number of alternatives is appropriate? That depends first on where you are in the process. At the early, brainstorming stages of planning, dozens of individual actions may be identified as elements of a potential solution. Once you start creating alternatives that will undergo serious consideration, it's important to think about two things: (a) how many can you analyze and (b) how many can you present for careful group deliberation. How many alternatives you can analyze largely depends on what technical resources and analytical tools you have available (e.g. quantitative modeling tools, or whether analysts or technical teams are on call). The number of alternatives you can present for deliberation by a group, on the other hand, is more limited. At any given sitting of a deliberative group, it's hard to work meaningfully through more than six or eight alternatives. Of course that depends on the nature of the alternatives, the complexity of the consequences and trade-offs, and the skills and dynamics of the group, but as a guideline, the preferred number is more than four and less than a dozen. If you're presenting a final set of alternatives to senior decision makers (in either the public or private sector), you'll need to keep the number down to something very small – usually no more than three or four.

7.3.4   Dealing with constraints

Constraints serve two purposes at the alternatives stage. One is useful, even essential: constraints serve as a reminder that alternatives must be realistic. If they require additional personnel or more money or more information than what is available, then they are unlikely to be enacted or to result in any change in outcomes. The other role of constraints is not useful, and is even harmful: constraints serve as an impediment to innovative thinking and to creativity.

The trouble with constraints is that it's hard to know how fixed they are. Are they real fixed constraints that cannot be challenged? Or are they soft constraints, meaning that with an appropriate rationale they could be removed or relaxed? Often, of course, what is presented as a constraint merely reflects the zone of influence of the people in the room, or their range of vision. The hugely successful inventor Henry Ford is reported to have said 'If I'd listened to customers, I'd have given them a faster horse'.

Consider the first meeting of a multi-stakeholder group that has come together to try to define a management alternative for clean-up of a hazardous waste site that all parties can support (albeit, perhaps, with varying degrees of enthusiasm). A first meeting is often filled with reasons why suggested alternatives cannot be undertaken:

1   Industry proposes a level of residual land-use contamination that an environmental organization claims exceeds regulatory standards.
2   An environmental organization proposes alternatives that address water quality concerns downstream of the site, and industry claims this is out-of-scope.
3   Government insists on a 2-year clean-up schedule (time) that industry says is not feasible and cannot possibly be met.
4   The community proposes a new technology that other stakeholders consider impossibly expensive.

It's a dog's breakfast of issues, including some real constraints as well as some other concerns masquerading as constraints. How should you proceed? Start by distinguishing between different kinds of constraints[7]. Legal constraints are obligations or prohibitions defined by legislation or regulatory standards (as in 1, above). For most environmental management decisions, these are pretty 'hard' constraints, as they are unlikely to

be relaxed within a time frame that's relevant to the decision at hand. You might as well meet them or work within them. In many cases, they can be used to screen alternatives.

Other constraints are the result of administrative boundaries and/or policy decisions – we can't do something because it's outside the defined scope of the problem, beyond the jurisdiction of the decision maker, or exceeds the budget, resources, and timeline allocated to the process (as in 2, above). This kind of constraint can be challenged with a good rationale; if you work through your decision sketch and discover that the problem could be solved more effectively or creatively if there were a different scope or a modified mandate, then make the case for the necessary changes – which might require the involvement of more people, larger budgets, or (in some cases) suggested changes to the set of decision makers or even to the guiding legislation or regulations.

Time constraints are a particular kind of policy constraint that deserves special mention. We have mixed feelings about this. In some ways we like the advice that 'done trumps perfect'. There's something to be said for setting timelines, sticking to them, and getting things done. Every week or month that a solution is not implemented, benefits are foregone. On the other hand, recall the findings of Paul Nutt (discussed earlier), who reports that one of the main errors made by managers is to falsely code decisions as urgent[4]. We've seen plenty of alternatives get rejected because they couldn't be implemented in time. Then the whole process slowed, and the original timeline extended well beyond what it would have taken to implement the alternative.

A third kind of constraint is feasibility (point 3, above). Be on guard with this one. Many objections that are framed as feasibility constraints are really either premature (and potentially very inaccurate) predictions about performance, or value judgments masquerading as technical judgments. So let's unpack the feasibility argument. What could the statement: 'that's not feasible' really mean?

1 It could mean it's not technically possible. There are of course plenty of theoretically desirable alternatives that are not physically or technically possible due to physics or local conditions. True technical feasibility constraints can be used to screen alternatives.

2 It could mean it's not economically attractive to the speaker. This is a very common one. In some cases, it might mean it won't pass a specific financial screen – say a company's internal rate of return requirements. In some contexts, this is a valid screen. In other cases, this may just be an individual's personal value judgment about how much something should or shouldn't cost. This should be addressed at the trade-offs stage (Chapter 9).

3 It could mean the speaker doesn't believe it's 'practical' or would get public or political support. If that's the case, then it's time to try a little more creative thinking about what it would take to make it practical or gain support – perhaps it's more money or an implementation partner, or a political champion, and so on. If the objection is lack of public support, push back. It's probably premature to make that judgment, and while it may be a consideration, from the perspective of managing public resources in the public interest, the objections of a few vocal stakeholders shouldn't necessarily be used as an upfront screen.

4 It could mean the speaker doesn't believe it will perform very well. This then is a prediction – a technical judgment about consequences – regarding the success or effectiveness of the alternative. But it's the wrong time for this judgment. Estimates of consequence need to be treated later – in a disciplined way by recognized experts at the consequences stage (Chapter 8).

5 It could mean that the alternative's performance is unacceptable with respect to some other objective – residual impacts on the environment, recreation, culture, or financial outcomes (as in 4, above). Such arguments are really value judgments and should be addressed at the trade-offs stage.

So in some situations technical feasibility (meaning technical possibility) will serve as a constraint; if so, use it to screen alternatives, but recognize that these cases are rare.

7.3.5 What to do with alternatives you know won't work

When stakeholders are invited to participate in an SDM process, often they are joining something that is new to them, at least in terms of having an in-depth technical understanding of the system. Some of the alternatives they propose early in a decision process may be correspondingly naïve – great-sounding ideas that others' already have thought of, analyzed, and found lacking. So the question is what to do when participants in the process propose alternatives that more experienced managers or technical experts know are flawed. The answer depends of course on the context, perhaps primarily on the level of trust.

One obvious response is to explain that it just won't work for reasons 'a, b and c'. Where there is a high level of trust or the problem is relatively simple to understand, then this can be an acceptable approach. However, for other decisions where trust among lay stakeholders and the government or corporation leading the decision is more tenuous, then remember that – towards the end of the decision process – participants will only make tough trade-offs if they believe that the full range of alternatives has been identified and explored. When the stakes are high they're looking for that silver bullet, a win–win solution. SDM works because it promotes mutual learning, and sometimes the best learning comes not from a manager or company representative saying 'we've looked at that and it doesn't work'. Instead, use the iterative generation and evaluation of alternatives to level the playing field in terms of the knowledge about how alternatives affect the objectives. The key is iteration. Use the first round of alternatives to learn, and importantly, to bring everyone up to the same level of understanding. Seeing is believing. Once people see that their alternative doesn't perform as well as they thought, or that it's harmful to an objective they hadn't even thought about, they'll use what they learn to develop more refined alternatives. (This also relies on them having trust in the estimation of consequences, a topic we discuss more in Chapter 8.)

Sensitive work may be required to find the right balance between process and outcome – the process of generating alternatives needs to be responsive to participants in the process, and they need to promote learning. On the other hand, it takes time and money to evaluate alternatives and no one wants to waste time rehashing what they know won't work. As facilitators, we've lived through plenty of uncomfortable foot shuffling on the part of the agency or industry people who've hired us yet can't believe we're committing them to model and bring back results on alternatives they consider duds. Remember that modeling and presenting the results of an alternative does not suggest you support it. When the predicted consequences are presented for discussion, even the person who proposed it may not support it. In fact, with clients who've become familiar with the SDM process, we often see them putting forward alternatives they don't support, because they understand that others need to see how they stand up when evaluated against all the alternatives. The mutual learning that results is worth the investment.

7.3.6   Common pitfalls and traps

As in all the steps of decision making, people's ability to think clearly and creatively about alternatives is hampered by a host of psychological, cognitive, and behavioral traps. Here are some of the most common traps and pitfalls:

1   *Anchoring and tweaking*. Starting from the status quo, or the first proposed alternative, or what's going on in a neighboring jurisdiction, and then making only minor incremental changes.
2   *Accepting constraints*. Accepting constraints as inalienable when, under defined circumstances,

they could be removed or softened. Or perhaps value judgments are masquerading as constraints.
3   *Giving in to affective responses*. Once an alternative is associated by linking its label ('adaptive', or 'participatory' or 'precautionary') to a particular feeling, it may be difficult to adjust or eliminate. It's critical to ensure that participants look at the actual estimated performance of alternatives rather than relying on a name to inform them about what the alternative will achieve.
4   *Relying on sunk costs*. This refers to choosing alternatives that justify past decisions or financial, political or intellectual commitments – past expenditures of money or time, or personal or political commitments to an approach or policy.
5   *Avoiding trade-offs*. Not presenting an alternative because it would be controversial, or glossing over difficult value-based choices, or trying to eliminate them too soon.
6   *Quitting too soon*. A variety of pressures cause participants in a decision to rush to a premature solution with a false sense of urgency – an imposed time constraint that isn't real, or a sequenced decision process where it's presumed the next steps are awaiting the results of this stage of the process.
7   *Not knowing when to quit*. This is the flip side of quitting too soon. At some point the incremental gains are not worth the incremental effort and you need to satisfice. Keep in mind also that even the best management alternatives need to be implemented to be useful.

7.4   Case studies

In this section we present two case studies that help to illustrate some creative approaches to generating alternatives in a group deliberative setting. The first substantively concerns the development of a recovery strategy for endangered salmon using strategy tables. But the real

Table 7.3 Strategy table for Cultus Lake sockeye.

Cultus sockeye salmon exploitation rate (%)	Enhancement program	Freshwater projects
(a) Base alternative		
5	None: stop current enhancement activities	None: stop current programs to improve lake conditions
10	*Continue with current level of enhancement*	*Continue current programs to improve lake conditions*
20	Double current level of enhancement	Increase intensity of programs to improve lake conditions
30		
40		
(b) One possible alternative		
5	*None: stop current enhancement activities*	None: Stop current programs to improve lake conditions
10	Continue with current level of enhancement	Continue current programs to improve lake conditions
20	Double current level of enhancement	*Increase intensity of programs to improve lake conditions*
30		
40		

and transferable message is how effective this was in bringing together a group of people to work together productively on solutions when quite literally, it was said that absence of an agreement would lead to 'blood on the streets'. The second concerns a portfolio-building exercise. Although the context in this example is an energy planning process, we've used the same approach for a wide range of applications. If you're a city manager, then instead of portfolios to deliver energy services, imagine portfolios designed to deliver water from a variety of water supply sources, or portfolios designed to deal with solid waste – combinations of landfills, incinerators, or recycling programs to meet the projected solid waste stream, and so on.

7.4.1 Building strategies for Cultus Lake salmon

In early 2009, we assisted a multiparty advisory group charged with developing a precautionary management strategy for endangered populations of Cultus Lake sockeye salmon[8].

Various categories of actions were identified as potentially contributing to the development of an effective management strategy. In this simplified illustration, we're presenting three: the species exploitation rate[9], enhancement options[10], and freshwater projects[11] (see Table 7.3a and b). The top row names the categories. The items in each column are elements within the category that could be selected. Strategies were developed by highlighting selected elements, shown in the table by italicized text. Table 7.3a presents one possible strategy: continue with the current situation in terms of the previous year's exploitation rate (around 10%) and current levels of enhancement and freshwater project activity. Table 7.3b shows a possible alternative strategy: a higher exploitation rate (20%), reduced levels of enhancement, and increased freshwater project activity.

The elements shown in each category are not meant to cover all possibilities. For example, the discrete series of exploitation rates that is shown in the left-hand column (5%, 10%, 20%, 30%, etc.) could easily be augmented with other rates (15%, 18%, 35%, etc.). However, by covering a realistic range of exploitation rates using a representative set of discrete rates, the strategy table peeled away unnecessary complexity and focused decision makers on big-picture choices. Later refinements could address whether '20%' should really be '22%' and so on.

Here's how it worked in practice. Members of the consultative group met first to clarify the decision context and define management objectives and measures. Then they began to generate alternatives. In the first 'round' of alternative generation, the participants identified categories of management actions and the representative elements within each category. Then they developed a status quo alternative and several 'bookends' that maximized individual objectives (one focused on conservation, another on commercial fishing objectives, another on employment, and so on). In round 2, they developed and discussed a number of strategies that included creative ways to find a greater balance across objectives. If a higher exploitation rate were selected, for example, then a more aggressive set of enhancement activities could be proposed as an offsetting management response. If a more intensive level of freshwater predator removal was suggested, then it implied lowered enhancement activities. Each of these strategies could be compared with the status quo, and areas of relative gain or loss could easily be estimated. In the next round, they eliminated dominated alternatives, examined trade-offs, and proposed new and improved alternatives. In their fourth meeting, they moved toward selection of a preferred and broadly-acceptable win–win option.

The transparent and collaborative gaming nature of this strategy development process helped the advisory group to more fully understand the implications (i.e. strengths and limitations) of different combinations of management actions. Participants noted that creation of the strategy table helped to focus their discussions and to overcome a paralysis by complexity which, for several years, had prevented the development of new management initiatives. By structuring, in a simple way, the primary choices available to managers for meeting stakeholders' expressed objectives, the choices suddenly looked less confusing and more amenable to resolution through analysis and dialogue.

7.4.2 Building portfolios for long term energy planning

Sometimes there's a need to design a strategy to meet some predetermined design criteria. For example, you need to fund programs up to a budget cap, or develop supply options (water, wastewater, energy, etc.) to meet a specified forecast of demand, or meet a legislated emissions reduction target. There may be any number of criteria that have been fixed in policy (and are beyond your scope to question). In budget allocation decisions, there may be policy requirements that (say) at least one project be undertaken in each of six principal regions, or that funds be allocated equally to three main ethnic groups living within the target area. Sometimes groups decide that a strategy must contain a mix of both short- and long-term goals: it is essential from a biological perspective that funding be allocated to long-term restoration and research activities, but it also is essential for political and citizen-acceptance reasons that some visible short-term gains are associated with the program.

If there are firm targets or constraints that must be managed, then a useful alternatives generating process should acknowledge them. It's still possible to encourage the generation of creative value-focused alternatives within specified performance constraints. The following case study shows one example.

In this case study, we examine how an energy utility (BC Hydro) engaged stakeholders in an interactive process to develop strategies designed to specific performance criteria during its 2006 Integrated Electricity Plan (IEP) consultation process.

For the purposes of the IEP, a valid alternative had to meet the following constraints:

1 Describe a portfolio of resource options (e.g. wind, small hydro, etc.) that could be deployed in the next 20 years to meet forecast demand.
2 Meet targets for total energy supply (20000 GWh) and capacity (the amount of energy that can be reliably deployed at any give time: 2600 MW).

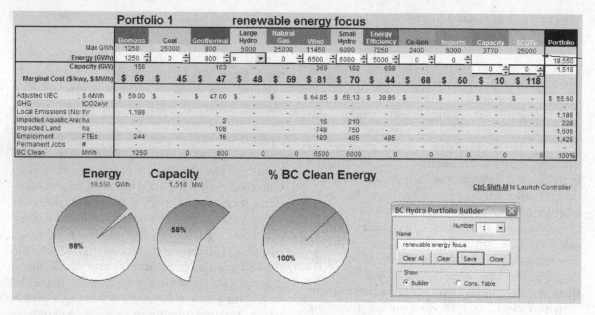

Figure 7.1 Portfolio building tool for the Integrated Electricity Plan (IEP).

3 Use only quantities of each resource option that are known to exist in British Columbia.

A spreadsheet tool was developed (in partnership with the client, Basil Stumborg) to allow stakeholders to build alternatives by assembling 'building blocks' of energy from a variety of resource options (see Figure 7.1). For most of the resource options the block size used was 500 GWh, but for others, such as geothermal and large hydro, the block size represented the actual projects available in British Columbia. For example, since there was only one potential large hydro site in the province under consideration at the time, participants could choose whether to use the one block of 5000 GWh hours of large hydro in an alternative or not.

Each time an energy block of a particular resource option was added to an alternative, the user was presented with a running tally of the implications of the alternative in terms of various indicators of interest to participants. These included the cost of the alternative, GHG gases emissions, local air emissions, jobs, the proportion of 'clean energy' (an indicator important to

provincial energy policy), and the area of land affected. Supply curves for each resource option were built into the tool, providing both the marginal cost of the next block of energy available of each type, and preventing the adding of quantities of a particular resource option that did not exist. A simple look-up table built into the tool contained emission factors and additional data required to calculate other impacts.

Before starting, the tool shows various resource options as columns in a table, with the indicators of interest shown as rows. Under each resource option's label is a reminder of the estimated maximum quantity of energy available in British Columbia of that resource (for biomass, 1250 GWh). Beneath this is the marginal cost of each resource, that is, the cost of the next block of energy of that type, in $/kWh.

For example, suppose we are interested in building an alternative that focuses on renewable energy. We might decide to start by adding one, 500 GWh block of biomass energy to our alternative (called 'Portfolio 1'). We do this by clicking on the up button of the 'spinner' control as indicated by the black arrow. By adding the first 500GWh block of biomass, participants can see that they have accounted for 3% of the energy total, 2% of the capacity target, have created 98 full time jobs and 475 tons/year of NOX, and that the alternative so far comprises 100% British Columbia clean energy. Since there are only 1250 GWh of biomass available, however, participants will soon need to start looking at other resource options to fill out the alternative. At this point, the alternative is close to meeting its target for energy but because of its reliance on intermittent resources, such as wind, has a way to go to meet its capacity obligations.

Once an alternative is completed, the data in it are sent to a consequence table for subsequent evaluation using trade-off analysis techniques. Later in the process, each alternative was modeled more completely by BC Hydro, with transmission, temporal issues, and other complicating factors overlaid. Using this technique allowed participants to sketch out alternatives interactively in a constructive, discursive environment. Recognizing how critical these kinds of tools are to helping decision makers make informed choices, we reiterate the importance of allocating time and resources to their development early in the process.

7.5 Key messages

The creative process of identifying good alternatives includes several key steps:

1 *Brainstorm a range of potential solutions.* This means returning to the objectives to ask 'how could we achieve this objective?' It means temporarily peeling away constraints to facilitate creative thinking. If you have a long list of ways to address the objectives then you may want to consider an approach to screening them, but always do so with caution. Screening involves some very significant value judgments – make sure they're explicit and made by those with legitimacy to make them.
2 *Organize individual actions or elements into complete, fully specified alternatives.* Use strategy

tables to help organize and clearly specify alternatives. There may be an infinite number of alternatives. Develop a few that are distinctly different and represent either fundamentally different ways to do things, or different value-based priorities. This will facilitate learning.
3 *Iteratively improve alternatives.* Once consequence estimates are available for the initial set of alternatives, eliminate alternatives that are outperformed by others and develop new ones based on the insights gained. This process will involve both technical and value-based judgments.

To help design good alternatives and know when you've got them, ask:

1 Are they complete and directly comparable – have different elements of a complex solution been combined? Do the alternatives have the same scope and underlying assumptions, i.e. are we comparing apples to apples?
2 Are they value focused? Do they address the fundamental objectives rather than means? Do they address all the objectives? Have they looked for creative and fundamentally different ways to achieve the objectives?
3 Are they fully specified – is what is meant by each alternative unambiguous? Would experts have enough information to evaluate their consequences and would different experts have the same understanding of the alternative?
4 Are they internally coherent – do they work and make sense together?
5 Are they distinct – are they really different? Have we done more than just tweak a single leading or favored alternative?

If alternatives have these properties, then they are well-placed to help decision makers accomplish the three key things alternatives are meant to do: explore creative and fundamentally new ways to achieve the objectives, expose and focus attention on fundamental value-based trade-offs, and give decision makers real, meaningful choices to make.

7.6   Suggested reading

Gregory, R. & Keeney, R.L. (1994) Creating policy alternatives using stakeholder values. *Management Science*, **40**, 1035–1048. Outlines the use of decision-structuring techniques for creating and evaluating improved alternatives that reflect expressed stakeholder values, based on a workshop in Malaysia that addressed participants' conflicting views concerning proposed future resource development priorities.

Nutt, P.C. (2004) Expanding the search for alternatives during strategic decision making. *Academy of Management Executive*, **18**, 13–28. This paper argues convincingly, based on the author's experience, that managers typically fail to devote sufficient attention to the creation of a comprehensive set of alternatives.

Raiffa, H. (2002) *Negotiation Analysis: The Science and Art of Collaborative Decision Making*. Belknap Press of Harvard University Press, Cambridge. The author helped to initiate the study of prescriptive approaches to decision making and negotiations; it's no surprise that this late-career book provides entertaining descriptions and thorough analyses of methods for developing innovative alternatives as part of public policy consultations.

7.7   References and notes

1   Some years ago, several newspapers in the southwest United States sought to save money by omitting their reporting on daily weather on other parts of North America. Their readers were incensed: it turns out that people living in Arizona or New Mexico like nothing better than reading, at some length, about the misery that snow and winter winds inflict on people living in northern states or Canada. It provides a reminder, on a daily basis, of how lucky they are to be somewhere else.

2   Derby, S.L. & Keeney, R.L. (1981) Risk analysis: understanding 'how safe is safe enough?' *Risk Analysis*, **1**, 217–224.

3   Hsee, C.K. (1996) The evaluability hypothesis: an explanation for preference reversals between joint and separate evaluations of alternatives. *Organizational Behavior and Human Decision Processes*, **67**, 247–257.

4   Nutt, P.C. (2004) Expanding the search for alternatives during strategic decision making. *Academy of Management Executive*, **18**, 13–28.

5   Failing, L., Horn, G. & Higgins, P. (2004) Using expert judgment and stakeholder values to evaluate adaptive management options. *Ecology and Society*, **9**, 13. Retrieved from http://www.ecologyandsociety.org/vol19/iss11/art13/

6   Schwartz, B. (2004) *The Paradox of Choice: Why More is Less*. Harper Collins, New York.

7   Our thanks to Mike Runge for helping us think more clearly about this; any remaining errors of logic are ours, of course.

8   Gregory, R., & Long, G. (2009).Using structured decision making to help implement a precautionary approach to endangered species management. *Risk Analysis*, **29**, 518–532.

9   A charming term used in fisheries management that essentially refers to the percent of fish that will be permitted to be caught.

10 Cultus sockeye salmon are being supported by a lake-based enhancement program that involves captive breeding of some of the fish. While improving numbers of fish, some are concerned about the longer term impacts of such programs on the genetic integrity of the fish.

11 These include culling Northern Pikeminnow, an invasive fish, and cutting back milfoil, an invasive plant that is thought to provide cover for these predators. These activities are labour intensive and of uncertain efficacy.

8
Characterizing Consequences

This chapter discusses some concepts and structuring methods to help groups build a common understanding of the anticipated consequences of management actions. In the first section we cover the basics of building effective consequence tables, including a discussion of the pitfalls of simple 'pros and cons' comparisons and the advantages of consequence tables. Subsequent sections address a variety of topics related to gathering the 'best available information' and working with experts, including:

1 How can we know what exactly is the best available information? How is our understanding of best available information changing and what are the implications? How should information from a variety of sources – including the knowledge of scientists, local community residents, resource users, and aboriginal communities – be collected and integrated?

2 Who's an expert? Societal trends toward the democratization of expertise remind resource managers that often there are a large number of individuals with relevant expertise. It's no longer a domain purely for trained scientists. What allows the experience or training of an individual to be viewed as providing credible inputs about consequences?

3 How can studies and models be designed that will be useful for decisions? It's easy to spend enormous amounts of both time and money filling information gaps. Are there guidelines available for scoping and prioritizing studies, as well as for developing models that are helpful to decision makers, while wisely using limited resources?

4 How can defensible judgments from experts be elicited? Although expert judgment methods can (and in some cases should be) quite sophisticated and resource-intensive, are there best practices that can easily be used without specialized training? To what extent is it helpful to encourage discussion among experts, or between experts and other participants as part of the decision process?

5 How can competing claims and hypotheses about consequences be examined critically to ensure high standards of information quality for decision making? The recognition of multiple sources of knowledge, the presence of uncertainty, and the desire (or, in some cases, legal mandate) for collaborative decisions all drive a critical need for approaches to exploring the quality of information that are rigorous, constructive, and support mutual learning.

Structured Decision Making: A Practical Guide to Environmental Management Choices, First Edition. R. Gregory, L. Failing, M. Harstone, G. Long, T. McDaniels, and D. Ohlson.
© 2012 R. Gregory, L. Failing, M. Harstone, G. Long, T. McDaniels, and D. Ohlson. Published 2012 by Blackwell Publishing Ltd.

8.1   Organizing information about consequences

What are we looking for at this stage? The bottom line is that a presentation of consequences should convey all of the information that is critical to understanding and comparing the alternatives. It needs to clarify the relative performance of each alternative – its relative advantages and disadvantages – in terms of the agreed-upon objectives and performance measures. It needs to convey differences in the uncertainty of the outcomes – are the consequences of some alternatives more uncertain than others? It needs to check for differences in the confidence of the respective experts regarding these predictions, identify the key trade-offs across consequences, and make clear how they're relevant to choices among the alternatives. And there it should stop. This point marks the separation of facts (what we know about the consequences of alternatives) and values (what we choose to do about it).

8.1.1   The role of consequence tables

Consequence tables have been introduced several times in preceding chapters. It's a pretty simple idea: a table with objectives in the rows and alternatives in the columns[1], with the cells providing estimates of the anticipated consequences for each alternative in terms of the specified objectives. You might think it's so obvious that it's not worth much additional space. But consequence tables play such a pivotal role in structuring both technical analyses and deliberative exchanges as part of an SDM approach that we're going to spend more time on them.

From an analytical perspective, the consequence table serves the role of focusing analysis on the assessment of consequences as defined by the expressed objectives and performance measures. This simple point has important implications. Baseline studies may be needed, but only in the service of estimating consequences. Risk assessments cannot end with the identification of hazards; they must go on to estimate impacts on the things that matter – as defined by the objectives and measures. Habitat mapping exercises cannot end with brightly colored maps and overlays of existing conditions; they must summarize the impacts of specific management actions on fundamental objectives. Access to the summary information conveyed in a consequence table profoundly changes the scope and nature of factual evidence that will be produced. This is not to say that habitat maps and models or risk assessments are not useful tools, yet it's important to realize that they only address a small part of the information needed for decision making. As noted in Chapter 3, environmental managers and advocates have a tendency to produce detailed studies on the biology of the resources and creatures they care about and leave the assessment of practical management alternatives to the end. Recognizing objectives and alternatives early on is key to an appropriate allocation of resources. Studies may be done, maps may be built, data may be collected, but these tasks all should be oriented toward obtaining the information that's needed to fill in the objectives-by-alternatives cells of the consequence table.

A consequence table also plays a key role from a deliberative perspective, particularly in the context of collaborative decision making with multiple stakeholders. It serves as a visible collective agreement about how consequences will be represented for the purposes of the decision at hand. The rows constitute agreement about what matters and what will be considered when choosing among alternatives. The columns represent agreement about which alternatives merit detailed evaluation. The predicted consequences entered in each cell represent a collective understanding of what is the best available information. They will have been estimated by people recognized as experts and representing a diversity of views about the anticipated consequences of the management actions under consideration. This collective understanding is essential to making environmental management decisions that are rigorous, inclusive, defensible, and transparent (remembering these guiding criteria from Chapter 1). Each individual participant in the

Table 8.1 Summary assessment, showing pros and cons.

Alternative 1: status quo	Alternative 2: conservation strategy	Alternative 3: moratorium on logging
Pro	*Pro*	*Pro*
• Does not require additional resources	• Government seen to be serious in taking the lead on conservation actions	• May win support from NGOs and result in additional funding opportunities
• Low impact to timber supply or forestry revenues	• Moderate risk that endangered plant species will become extinct	• Government seen to be actively leading conservation actions
• No change required to regulation or legislation	• Establishes legally enforceable measures for conservation	• Moderate risk that endangered plant species will become extinct
• Maintains high rates of recreational use	• Meets international treaty obligations	• Buys time until a more comprehensive strategy is developed
Con	*Con*	*Con*
• High risk that endangered plant species will become extinct	• Requires changes to policy, regulations	• May require compensation to the forestry sector
• Government will not be seen as responding to an emerging issue	• Continued complaints and NGO pressures	• Impact to timber supply and forestry revenues
• Moderate risk that endangered plant species will become extinct	• Potential impact to timber supply and forestry revenues	• Will require substantial additional resources
• Is a deferral, not a resolution, of the issue	• NGOs will likely continue to pressure government for greater conservation	
Will likely result in continued legal challenges and NGO pressures		

process may have their own views about the best model, data, or expert to employ. But the phenomenon of dueling experts is not conducive to collaborative decision making. Instead, participants can use the visual vehicle of a consequence table to collectively define what will be considered the 'best available information' for addressing the decision at hand. This collective agreement is possible only when decision-relevant objectives, performance measures, and alternatives have been agreed upon and, in the presence of uncertainty, when key aspects of uncertainty have been characterized and reflected in the consequence table.

While consequence tables aren't the only way to present information, they are a pretty universally useful one. The following sections provide guidance to help ensure that the presentation of consequence information in a consequences table is useful to resource managers, to the stakeholders they serve, and to the decision makers who ultimately will be responsible for making a decision about management strategies.

8.1.2 *The pitfalls of pros and cons*

Imagine you are a government decision maker charged with approving a controversial conservation strategy to protect a rare and endangered plant species that is under threat from forestry and logging operations. Your staff have just sent you a briefing note with what is referred to as a summary assessment of alternatives. In many cases this will be in the form of a rambling prose account, which means that the hard-working manager in question will need to tease out the relevant pros and cons for each alternative. Suppose, however, that you are fortunate and the summary note sent to you is similar to that shown in Table 8.1 (which is based closely on an actual decision note prepared by government managers).

We suggest there are at least six problems with this presentation; mistakes that are commonly made in government or corporate analyses and briefing notes that use lists of pros and cons to summarize the predicted consequences of a proposed action:

Table 8.2 Example consequence table.

Objectives	Evaluation criteria	Alternative 1	Alternative 2	Alternative 3
Conservation	Probability of plant species becoming extinct within the next 10 years	30%	10%	10%
Financial	Average annual additional costs to government ($000)	10	20	35
	Average annual loss of government revenue (taxes, license fees) ($000)	60	62	57
	Average annual forestry sector revenue ($000)	0–20	20–140	120–180
Social	Public and NGO support	Widespread complaints and organized legal challenges	Widespread complaints	Moderate support by most NGOs
	Recreation opportunities (additional visitor days)	300	175	400
Cultural	Altered archaeological resources (no. of culturally modified trees affected)	40	25	80
Strategic	Government leadership via legislative change	Government seen as unresponsive	Government seen as competent	Government seen as innovative leader

1 *Inconsistent treatment of objectives or concerns.* We know the implications for meeting international treaty obligations for Alternative 2, but no information is provided on this concern for the other alternatives. Similarly, effects on recreational opportunities are noted only for Alternative 1.

2 *Incomplete set of consequences.* Cultural resources, for example, are normally an important consideration in decisions of this type, but they do not appear. Does this mean that the potential impacts: (a) were minimal or unimportant; (b) did not vary across the alternatives; (c) could not be estimated or were associated with too much uncertainty; or (d) were simply forgotten?

3 *Vague verbal descriptions.* What is a 'high' versus a 'moderate' level of risk? What is the difference between 'impact' and 'potential impact' on forestry revenues, or the significance of moving from a high to moderate risk that the plant species will become extinct?

4 *Hidden or arbitrary value judgments.* For example, moderate risk of plant extinction is stated as a 'pro' for Alternatives 2 and 3, which may reflect the perspective of a biologist charged with preparing the brief. A different decision maker, however, may well see this as a 'con' given the associated costs and political pressures from industry groups. So there is an open question regarding 'whose judgments were used' to fill in the summary table.

5 *Confusion between means and ends.* 'Establishes legally enforceable measures for conservation' is a means to an end, where the end presumably is conservation of the species – yet this is already reflected in the upgrading from high to moderate risk of extinction.

6 *Double counting.* 'Continued complaints and NGO pressures' is probably redundant with 'NGOs will likely continue to pressure government for greater conservation'.

Table 8.2 shows the same decision in consequence table format, using a mix of natural, proxy, and constructed performance measures (as described in Chapter 5). We contend that Table 8.2 provides a more complete and consistent summary of the anticipated consequences and would better inform decision makers. The best news is that turning lists of pros and cons into consequence tables can always be done: it's a low effort, high value activity that helps to eliminate sloppy thinking and clarify key messages.

8.1.3 Building and using consequence tables

Much of what constitutes best practice in estimating and communicating consequences has already been discussed in other chapters:

1 Consequences are meaningful only if they link to a well-structured set of objectives and performance measures that are complete and concise, unambiguous, understandable, direct, and operational.
2 Good alternatives are complete and comparable, value-focused, fully specified, internally coherent, and distinct. A well-developed set of alternatives will expose the fundamental value-based choices facing decision makers.

If you've done all these things, you're well on your way to a good expression of consequences. In this section we address several additional considerations, each of which we've encountered at some time as a stumbling block to the development of effective consequence tables:

1 Separate facts and values.
2 Report consequences in consistent terms and units.
3 Focus on relative performance.
4 Expose key trade-offs and uncertainties.
5 Provide context for interpreting the significance of consequences.
6 Refine consequences iteratively.

Separate facts and values

There are two sides to this simple but powerful advice (a recurring theme, we know). First, entries in the consequence table are fact-based and should be provided by experts. Regardless of

how inclusive the process is, don't let it be suggested that everyone should participate in estimating consequences. While the nature of the relevant expertise will vary, there will, in virtually all cases, be a set of people who have specialized knowledge that should be used in estimating consequences. Second, choices based on the information in the consequence table are value-based and should be made by decision makers. Don't let experts make choices. They, and the consequence table, should say nothing about which objectives are most important or which alternatives are preferred. We note that the simple act of estimating the consequences of *multiple* alternatives helps avoid hidden value judgments. We've seen plenty of situations where the staff's task is to develop 'a plan' or 'a strategy' (with emphasis on the singular). Staff then make (and must make, to complete the assignment) all kinds of value judgments to come up with a recommended plan. This procedure confuses the role of staff and decision makers: staff need to create, investigate, and evaluate realistic alternatives and weed out the clear losers among them in order to present decision makers with high quality alternatives. They must then turn the decision over to those with legitimacy to make difficult value judgments, which usually means either elected officials or their designates. If staff are tasked with producing a consequences table with multiple good (and mutually exclusive) alternatives in it rather than a single recommended plan, then the tendency for this mix up of roles will be reduced.

Report consequences consistently across alternatives

Obviously our hope is that a consequence table will be populated with well-designed performance measures, as outlined in Chapter 5. But at the most basic level, the key to a useful consequence table is ensuring that consequences are reported consistently across the alternatives. Each type of consequence may be reported differently of course – you won't report the employment effects of an alternative in the same units as its impacts on an endangered species. However, for each objective, the consequences should be reported consistently. You can't make informed trade-offs if you don't have a consistent characterization of consequences.

Focus on relative performance

A good consequence table focuses the attention of analysts and decision makers on the *relative* performance of alternatives. In part, this advice simply reflects how people make sense of information. Recall from Chapter 2 that people tend to be more sensitive to changes in performance or consequences than to the absolute value of consequences. Recall also the experiments (by Chris Hsee and others) described in Chapter 7: perceptions about the value of one alternative are strongly influenced by its comparison with others.

Another reason to focus on relative performance is that it can substantially simplify the estimation of consequences. This happens for two reasons. The first is that often the initial condition or base case is hard to define and quantify. Consider regulatory costs. Estimating regulatory costs under a reference or base case can be quite complex. However, estimating the *incremental* costs associated with a new regulatory regime can be estimated much more readily. The second reason is that sometimes it's difficult to estimate the absolute value of an environmental parameter with confidence. However, if this parameter affects all alternatives equally (which is often but not always the case), then it's possible to estimate the relative performance of alternatives accurately even without being accurate about the absolute value.

This emphasis on relative performance is useful if the focus is on incremental change. Of course, the value of an incremental change depends on what you start with. As Hammond *et al.*[2] put it, 'it's not enough to look at the size of the slice; you also need to look at the size of the pie' (p. 149). This takes us to the critical task of defining the 'base case.'

One implication of the emphasis on relative performance is the need to carefully define the reference case (or base case) that will serve as the standard by which comparisons with other alternatives are made. In some cases, the reference case will be the current or status quo situation, in other cases it might be a historic situation. Given the sensitivity of people to losses and gains, and the different ways we value them, the choice of a clear reference case can be important to the construction of alternatives and, in turn, to the identification of consequences.

Expose key trade-offs and uncertainties

The single most important task for a consequence table is to expose and focus decision makers on the relevant aspects of performance of the alternatives under consideration. This includes both the key value-based trade-offs and trade-offs involving uncertainty. The expression of consequences should include a clear and honest representation of uncertainty – not just an expected outcome but the range of possible outcomes – with emphasis on exposing different risk profiles.

Uncertainty can be difficult to express in the cells of a consequence table in ways that meet both analytical and deliberative needs. Recall the discussion in Chapter 6 about showing the uncertainty associated with different management alternatives. As summarized in Figure 8.1, proposed management Action A is expected support an abundance of 60 individuals, whereas the best estimate for Action B is an abundance of 80. Clearly, with other things equal, Action B would be preferred. However, assume there is uncertainty in the estimates of potential losses: under some plausible conditions Action B could lead to substantial losses (so that the lower bound of the experts' estimates of abundance drop to zero) whereas the worst plausible estimated performance for Action A would be an abundance of 50. In this case, the alternatives involve different degrees of risk, and different decision makers might make different choices depending on their risk tolerance. When this is the case, it's important for the performance measures and consequence table to provide concise information on the distribution of possible outcomes.

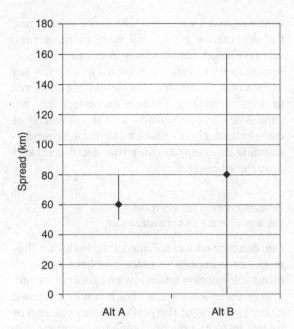

Figure 8.1 Alternatives with different risk profiles.

How best to capture this in a consequence table depends on the specific decision context. One possibility is to calculate the expected value (EV). Recall though that the use of expected value assumes that decision makers are risk neutral. In some cases – especially repeated, routine decisions with no real catastrophic potential – use of an EV may well be warranted: the goal is good average outcomes over time and individual bad outcomes are acceptable. In other cases – high stakes, one-off decisions where a bad outcome could mean a catastrophic or irreversible loss (species extinction, contamination of drinking water, dam failure, invasive species introduction, and so on) – decision makers are anything but risk neutral and use of the EV hides important information. How then to report outcomes concisely?

One commonly used way is simply to report the range as a confidence interval, as in Table 8.3(a). This is easy, intuitive, and pretty useful, but there are a few disadvantages. First, no information is provided about relative probabilities – how much more likely is the best guess than the upper or lower bound? Is the probability of a bad outcome

the same as the probability of a good one? Second, our experience is that many people (including experts) assume that all outcomes within the range are equally likely. And third, ranges are hard for people to compare and, if quantitative approaches to trade-off analysis are to occur later in the process, hard to assign weights to.

A useful way to summarize information from a distribution is to select a specific part of the distribution to report on. This will make it more accessible to a non-technical audience, but of course information will be lost. So it's critical to understand which part of the distribution is most central to decision making. Often it is the consequences of being wrong that are most important to decision makers. When this is the case, a useful approach is what we call 'downside reporting,' which involves reporting the best estimate as well as the consequences associated with the 'worst plausible' scenario. This is consistent with the 'mini-max' decision rule (which is short for 'minimize the maximum regret') and low regret decision strategies made popular by Stokey and Zeckhauser[3] and others in the 1970s. In lay terms, downside reporting answers the question 'what's the potential downside of the alternatives?' This approach is illustrated in Table 8.3(b). The most likely abundance under Alternative A is 60 but there is a 5% chance that it will be as low as 50. Under Alternative B, the most likely abundance is 80, and there is a 5% chance that it will drop to zero.

The choice between these alternatives is value-based and depends on the risk tolerances of decision makers. It can be a tough decision. The higher best guess in Alternative B is associated with the largest downside risk, whereas for Alternative A the best guess is lower but much of the downside risk is avoided. Some decision makers will choose to accept the downside risk in order to achieve the higher expected benefits of Alternative B as compared with Alternative A; others will not.

The example in Table 8.3(c) is a reminder that all this has to fit into a multiattribute framework. Beware the tendency to be comprehensive in

Table 8.3 (a) Consequence table showing best guess and 90% confidence interval.

	Alternative A	Alternative B
Species abundance		
Best estimate	60	80
90% confidence interval	50–80	0–120

Table 8.3 (b) Consequence table showing best guess and downside potential.

	Alternative A	Alternative B
Species abundance		
Best guess	60	80
Downside potential (5%)	50	0

Table 8.3 (c) Consequence table showing multiple objectives with best guess and downside potential.

	Alternative A	Alternative B
Species abundance		
Best guess	60	80
Downside potential (5%)	50	0
Recreational user days		
Best guess	400	600
Downside potential (5%)	350	540
Management cost ($000s)		
Best guess	$1500	$2400
Downside potential (5%)	$1800	$3000

characterizing uncertainty relating to the consequences of actions. If you include complex information about uncertainty – probability distributions, sensitivity analyses, and so on – on every cell in the table (as in Table 8.3[c]), you may quickly overwhelm even your most numerate participants. The job of the analyst is to judiciously select those aspects of uncertainty that best expose risk profiles and allow decision makers to express their risk tolerance. Although a variety of presentation formats can aid participants' understanding, there are important trade-offs between the amount of information presented and people's

ability to make sense of it[4]. Both the analyst and the deliberative group will need to make (and revisit) choices about which information to present in order to capture most accurately the key aspects of performance that should be considered in decision making. There is no perfect way, but careful design of measures and the provision of complementary information to improve understanding can help to ensure that decision makers have the whole picture.

Provide context for interpreting the significance of consequences

One concern with consequence tables is that they might oversimplify complex problems by force-fitting information into a row and column format. This is a relevant concern but it can be mitigated either by design of the performance measures or by the introduction of complementary ways to present relevant information. For example, if a performance measure for aquatic productivity is 'juvenile salmonid density' (the density of young salmon in the river, as estimated in kg/m^2), it may be useful to remind participants how this river compares with others in terms of the salmon density. If it is in the upper 10th percentile of all rivers in the country and is projected to remain so under all proposed alternatives, this may affect how decision makers value small changes in salmon density across alternatives. At other times, or with other participants, it may be helpful to show pictures of what the river would look like if this salmon density were to occur. Such contextualizing information is critical for helping decision makers make informed judgments about the meaning of the stated consequences and how much their relative importance.

There are a variety of ways to introduce this contextualizing information:

1 *Show and tell*: add complementary data alongside the consequence table and talk about it. This could include graphs and charts that illustrate risk profiles or information about the condition of this river, species, airshed, or local economy relative to others. It could also include maps and stories – recall from Chapter 5 the discussion of a performance measure for navigability that was accompanied by maps to show which parts of the river would be affected. Stories can also play a role – real people talking about how changes in measures have or are expected to change their lives or their experience of their environment. Some people will say (correctly) that the introduction of stories may bias choices – introducing emotional or affective responses, availability bias, anchoring on potentially irrelevant details, and the like. The counter point is that quantitative analyses may omit important considerations and, for some audiences, be perceived as 'cold' or difficult to understand, thus creating a barrier to meaningful dialogue. We recommend multiple framings – different ways of examining and expressing impacts – along with a reminder to allow time for participants to reflect and to reach relatively stable conclusions.

2 *Smart performance measures*: as we noted in Chapter 5, performance measures can be designed to reflect relevant thresholds or benchmark information. For example, instead of a measure that reports the 'juvenile salmonid density' in kg/m^2, it could report the percentile corresponding to that estimated density (e.g. under this alternative, the river would fall into the 85th percentile for juvenile salmonid density).

Refine consequences iteratively

A useful consequence table needs to expose key trade-offs and uncertainties. If it doesn't, then it may be necessary to go back and redefine objectives and/or performance measures. This isn't uncommon; SDM is an iterative process at every stage. The goal is to get information that's good enough to make an informed choice.

You may also need to go back and improve the estimation of consequences. A key question is

when is a more rigorous and/or greater exploration of uncertainty warranted. This is a complex question, but as general guide, triggers for further analysis may include:

1 Sensitivity of an estimated consequence to an uncertainty.
2 Proximity of the estimated consequence to a standard or target.
3 High cost of being wrong.
4 High degree of disagreement among experts.
5 Demonstrated sensitivity of the decision to the uncertainty.

The last point is probably the most important (and largely serves as a synthesis of the other points). Many performance estimates are very uncertain but don't turn out to be important drivers in the final decision. However, if a performance measure plays a central role in the deliberations *and* is very uncertain, then it may be appropriate to pause and, if resources allow, undertake initiatives designed to refine the data, models, or analyses on which the estimates are based.

8.2 Understanding the best available information

Now that we've discussed how to set up a consequence table, the burning question is: what goes in its cells? In this section we explore emerging perspectives on what constitutes the best available information, with emphasis on sources of knowledge and expertise that must be tapped into. In the next section we explore three broadly applicable ways for exploring information about consequences.

8.2.1 What is meant by the 'best available information'?

A first point is that our use of the word 'information' is not accidental. Conventionally, environmental managers have relied on best available 'science' to inform decisions. To most modern observers, the scientific process implies properties that include: a systematic and repeatable methodology; an emphasis on scrutiny, skepticism, re-evaluation and learning; professional independence, objectivity, and accountability; quality control by peer review; and transparency. However, the reliance on 'science' is changing – blind trust in scientific experts and institutions is waning and other kinds of knowledge are increasingly viewed as bringing legitimate insights to decision making[5]. This shift is perhaps demonstrated most dramatically in Europe, where public reaction to science in policies related to mad cow disease, genetically modified organisms, nanotechnology, and other issues has evolved into nothing short of a crisis for scientific and political institutions[6]. Similar controversies are increasingly prominent throughout the world and are a major theme in environmental risk management. This trend is occurring at least in part because the nature of the questions science is called upon to answer has changed, while the nature of scientific inquiry and its presentation as part of a public policy-making process has not – or at least has not kept pace.

With respect to the management of risks to natural resources, human health, and the environment, science now finds itself routinely operating in an arena of apparently irreducible uncertainties and irresolvable value conflicts, a situation some have labeled 'post-normal science'. In the face of these challenges come calls for broadening public involvement, democratizing expertise, and establishing what has been called an 'extended peer community' to aid the development and peer review of scientific results[7]. Key trends include acknowledging the limits of conventional science, addressing the values basis for decision making, understanding the implications of uncertainty and the limitations of experts, acknowledging the role of local and traditional knowledge, and improving accountability and communication between scientists and the public.

This democratization of expertise and science leads to important questions about the extent to which 'right' information for a decision really exists. One perspective is that determining the knowledge to use in decision making is less a search for truth than it is a negotiation where, in essence, a group agrees to reach a common understanding or interpretation of the truth for the purposes of *this* decision, with *this* group of participants, and given *this* mandate by decision makers. This context-dependent perspective correctly recognizes that knowledge about consequences needs to be responsive to the ecological, social, and legal environment within which decisions occur. However, it also raises worrying possibilities: that the distinctions between expertise and lay knowledge become so blurred that important insights are lost; that sham information, based on misunderstandings or insincere motivations, is used alongside high quality information; or that a false convergence is promoted, so that the uncertainty in key sources of information about consequences is hidden. From a decision quality perspective, the concept of working toward a mutually agreed knowledge base for a particular decision is a reasonable and defensible goal if – and only if – uncertainty is both adequately characterized in the presentation of consequences and reflected in the development of alternatives[8]. The former involves rigorous but understandable treatment of consequence estimates, with the involvement of a diversity of experts. The latter involves a commitment to explore alternatives that are robust (perform well under a range of uncertainties), precautionary (based on conservative estimates of impact on sensitive resources), or adaptive (designed to promote learning over time and build in review mechanisms to incorporate that learning in future decisions).

In our work, and in this book, we emphasize approaches that we believe strike a balance between drawing on the specialized knowledge of experts – and the very real insights such expertise provides – and drawing on the experience, observations, and common sense of lay people to interpret and use that knowledge in ways appropriate to the decision at hand.

8.2.2 What sources of knowledge can be used?

Different sources of knowledge

Science has long been the most widely acknowledged source of credible information about consequences for public decision making – and not without some good reasons: the principles noted above are exactly what has made science so successful over the past 200 years. When we talk of science as a source of knowledge for decision making, we're referring not just to natural or biological sciences, but also to economics, the applied sciences (engineering and technology), and the social sciences, including among others, anthropology, sociology, psychology, and decision sciences.

Each of these disciplines is characterized by commitments to the accepted scientific method (defining a hypothesis, making explicit predictions based on that hypothesis, recording refuting and corroborating observations, etc.) and to a set of institutions (academic journals, peer review, etc.). SDM approaches seek to broaden this usual frame of reference when thinking about consequences by making use of a variety of information sources, including the knowledge of local residents and resource users and the traditional knowledge of aboriginal communities, and addressing the deliberative needs of a multiparty decision process. As a result, the role of science remains critical but also specific and limited (see Text box 8.1).

In the context of estimating consequences, a key consideration is the need to incorporate local and traditional knowledge meaningfully into the decision-making process. What do we mean by local or traditional knowledge? While there are various interpretations, three characteristics of local knowledge have been widely noted. First, it is typically experience-based, relying more (but not exclusively) on personal observation than on quantitative data and controlled experimentation. Second, local knowledge tends to be expressed in ways that are more holistic (often reflecting eco-systemic properties) and less reductionist than that of western science. Third, local knowledge is usually anchored firmly in the experience of place, and as such it tends to deal with particular concerns rather than more general categories, and with time- and context-specific observations and conclusions rather than fixed or generalizable rules. Traditional ecological knowledge is a particular type of local knowledge. It is seen to have something to teach western science in terms of ways to respond adaptively to natural variation and to incorporate concepts such as flexibility and resilience as part of management strategies.

There are difficulties in all these distinctions: 'traditional' tends to imply that knowledge is in a static historical condition, whereas in fact it is dynamic, continuously adding insights to a pool of knowledge. 'Western' science implies that the scientific paradigm is limited to western cultures when clearly it is not. 'Ecological' implies a body of knowledge that is limited to specific knowledge about ecosystem properties. We agree with Berkes, Colding, and Folke[9] who note that 'whether a practice is traditional or contemporary is not the key issue. The important aspect is whether or not there exists local knowledge that helps monitor, interpret, and respond to dynamic changes in ecosystems and the resources and services that they generate'. From the perspective of decision quality, the important requirement is that inputs from alternative knowledge sources receive equitable treatment as part of the decision process and are evaluated on the basis of their relevance and merit to the decision at hand[10].

Integrating different sources of knowledge on decision making

Attempts to 'integrate' different sources of knowledge into regulatory or public decision-making processes generally have been disappointing. Many resource managers continue to voice concerns that stakeholder-based processes will compromise the integrity and importance of science as a guide to risk management. Many resource users and aboriginal communities, on the other hand, continue to worry that integration is a euphemism for decision makers to continue sidelining or ignoring valuable lay and

Text box 8.1:   The limits of science

The primary source of expertise for most resource-management decisions comes from the domain of science. Now is a good time to remind ourselves that the role of science is critical but specific and limited. Science can alert us to problems that may be potential targets of management action, but cannot tell us what to do or who should be involved. Science has a role to play both in directly identifying candidate actions and in acting as a test of relevance for actions proposed by non-technical stakeholders – are they technically feasible? do they have a realistic possibility of improving key concerns? – but it cannot tell us which of the candidate actions is preferred. When addressing consequences, science is often uniquely designed to identify the potential effects of proposed actions on the expressed objectives. But as noted by the National Research Council[11], scientific expertise cannot directly address the social, cultural, or economic importance of these effects (which often serve as the initiating reasons for considering an action). At the latter stages of a decision process, the technical information provided by scientists can help decision makers put trade-offs in context, but this information rarely is sufficient to make choices[12].

Thus science, as it is normally conceived, is essential to the completion of only a portion of environmental risk management tasks – those involved with the identification and characterization of risks and estimation of the consequences of risk management actions. In contrast, science provides relatively little help in the initial structuring stages of risk management, or the final balancing stages of risk trade-offs and decision-making. Whenever public agencies claim that their decisions are 'science-based', then decisions that should be made on the basis of the values and trade-offs of stakeholders are instead being turned into technical debates. As noted by Coglianese and Marchant[13] (p. 1258), 'agencies need to explain their decisions by reference not only to scientific evidence but also to policy principles that speak to the value choices inherent in their decision making'.

cultural knowledge[14]. Although we don't explicitly use SDM as a way to facilitate meaningful integration of different knowledge, we have observed that it's been a useful way to support that goal. We offer five suggestions based on our observations:

1 *Focus on the decision at hand*. People are divided by debates about which source of knowledge is more right than another, but they are united by a desire to find creative solutions. By engaging people in a process centered on decision making, many of the intractable debates about consultation protocols, intellectual property rights, and power imbalances become not less important, but either less relevant to the problem at hand or more focused on salient aspects. Further, questions about the quality and relevance of information, as

we will discuss later, cannot be determined without reference to the decision at hand.

2 *Separate facts and values*. The knowledge brought to a decision-making process encompasses both knowledge about 'what is' – conditions or predictions about the resources – and about 'what should' – what actions should be undertaken. Although some argue that this distinction cannot legitimately be made, we have found that in a decision-making context the distinction can and should be clarified because facts and values warrant different treatment (see the more detailed discussion in Chapter 2). It is the failure to separate them and to find a legitimate place for values in decision making that leads to friction between science-based decision processes and indigenous community members. Yet resource managers should recognize that the topic is sensitive and

choose their language carefully. A common and legitimate objection to western science and conventional decision processes is that they compartmentalize and de-contextualize traditional ecological knowledge, extracting relevant fact statements but removing from it the specific practices, events, or culturally derived norms that make it relevant and meaningful[15]. For this reason it can be controversial to talk about separating facts and values. However, the *practical* consequence of this separation is typically not controversial: hypotheses or beliefs about consequences need to be treated differently than judgments about how important they are and what to do about them. This makes good sense to people, and after exposure to the practical demands of an environmental decision-making process we have yet to meet a person who doesn't agree.

3 *Reflect fact-based traditional knowledge as part of hypotheses linking actions and outcomes.* Typical approaches to traditional knowledge often focus on documenting traditional use areas. While this is a useful and often essential starting point, we find it's often insufficient, as its relevance to decision making is not always apparent. Given the prevalence of uncertainty in environmental decision making, we've found that a useful approach is to define traditional knowledge in the form of hypotheses about cause and effect, in particular about the pathways of effect from a proposed action to an outcome[10]. Such hypotheses become important factors in identifying candidate management alternatives, designing impact models, interpreting the significance of uncertainty, and developing modeling programs. All of these activities are central to decision making, thus providing a clear basis for integrating traditional knowledge in the process.

4 *Reflect value-based traditional and local knowledge in objectives and trade-offs.* Providing an explicit place for value judgments in a decision process – by defining objectives related to traditional values, for example, or identifying participants' value judgments about how much risk is acceptable – is a critical way to demonstrate the incorporation of traditional knowledge in the decision process. Having clearly defined steps in the decision process helps. In particular, knowing that questions related to values are explicitly addressed at the objectives step and the trade-offs step, makes it easy to agree on appropriate ways to address questions of fact at the consequences step.

5 *Treat competing fact-claims equitably.* When there are competing hypotheses, all need to be subjected to the same lines of questioning about their relevance and reliability, regardless of whether they stem from a scientific or traditional source. Conventional scientific concepts such as replication and peer review have their parallels in traditional cultures – not all knowledge emanating from a traditional community is viewed by its members as equally reliable. Later in this chapter (section 8.3.2) we outline some considerations for exploring the quality of competing claims, in ways that can be meaningfully and respectfully applied to all knowledge regardless of its provenance.

8.2.3 Who's an expert?

Expert input occurs at each phase of an environmental management process. In most cases, an expert is considered to be someone who by virtue of their training, experience, or professional standing has specialized knowledge not available to most people. Yet as we've noted, the definition of an expert is changing, particularly in the realm of deliberative decisions: professional training and academic expertise no longer mean that an expert's testimony will automatically be accepted. Not only is lay knowledge grounded in real-world experience and local realities increasingly recognized as valuable but the vulnerability of expert judgments to the damaging influences of cognitive biases, conflicts of interest and value judgments is widely acknowledged. As a result, the acceptability of contributions by experts is

Text box 8.2:   Integrating hypotheses from scientific and traditional knowledge sources

As part of a water use planning process on the Bridge River in British Columbia, a fisheries technical working group developed an influence diagram to document hypotheses about the major factors limiting fish populations[16]. Both agency fisheries biologists and traditional knowledge holders identified hypotheses about the effects of changes in reservoir operation (volume, timing, elevation) on the fundamental objective of maximizing fish abundance. Each of the pathways in the figure represents a hypothesis about an impact pathway.

The identification of spawning tributary access as one of these factors stemmed primarily from aboriginal participants. A performance measure defined as 'spawning success' was developed that combined estimates of spawning success in each tributary (a function of water elevations associated with each proposed management alternative) and the tributary's importance weighting (a judgment of the contribution of each tributary to overall juvenile recruitment in the reservoir). The estimate of spawning success as a function of water levels was developed partly from scientific insights and analysis (how long and at what depth can eggs survive inundation) and partly from aboriginal knowledge (observations of how flexible fish are in their spawn timing). Tributary weighting was based on local and scientific knowledge of the relative utilization of the tributaries.

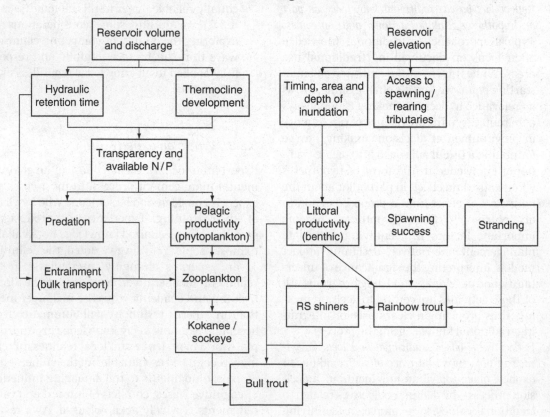

Integrating scientific and traditional knowledge.

By framing knowledge about what was affecting fish abundances as hypotheses, the input of both agency fish biologists and local community members was accommodated in the decision process. The emphasis here was on treating competing hypotheses consistently and equitably. Discussion included: what is the hypothesis about how this factor influences fish abundance?; what evidence do we have in support of this hypothesis?; what evidence do we have against it? Ultimately, this process led to breaking out the overall fisheries objective (maximize the abundance of fish) into five sub-objectives:

1 Minimize entrainment (in the turbines).
2 Minimize stranding.
3 Maximize spawning success.
4 Maximize littoral productivity (growth rate factors in and around shallow edges of the reservoir).
5 Maximize pelagic productivity (growth rate factors in the open, deep water).

Field studies and models were executed to facilitate the selection and calculation of mutually agreeable performance measures and to clarify the relative importance of each sub-objective as a limiting factor. Local communities' hypotheses about the factors affecting fish abundance were visibly reflected in the definition of sub-objectives, the selection of performance measures, the prioritization of field studies, and ultimately, the evaluation of alternatives.

increasingly subject to review, with their credibility depending strongly on the decision context[17].

Different types of expertise

When estimating consequences, it's typical to look for experts with substantive expertise in the subject matter – economists for estimating employment effects, marine biologists for estimating effects on whales, and so on. However, a number of other types of expertise also are important to consider, including analytical, interactional, predictive, and implementation expertise (adapted from Burgman[18]). Getting the best available information on consequences for use in a deliberative setting may require eliciting information from a set of experts with many or all of these types of expertise.

1 *Substantive expertise.* This involves the selection of experts based on their training, experience, and problem solving skills in the domain of interest. Substantive experts are expected to be both effective and efficient – in other words, they should be able to get the right

answer and get it quickly – and they should know the limits of their own knowledge. Relevant substantive expertise could be either context-specific or broader and more systemic (acquired in other contexts and reapplied here); it also could be rooted in academic training or experiential training acquired over a long period of resource use.

2 *Analytical expertise.* This involves the selection of an analyst (sometimes called a technical facilitator or integrator) to guide the expert judgment process; someone who understands how decisions are made under conditions of uncertainty and knows about the pros and cons of different elicitation and aggregation procedures. A detailed knowledge of the subject area is neither required nor expected, but an overview is helpful in order to frame issues and introduce follow-up or clarifying questions.

3 *Interactional expertise.* This includes the ability to communicate technical information to lay persons and to work effectively with groups. While it would obviously be counter-productive to reject good information solely because

its carrier was not eloquent, the days when experts can launch advice from the safety of their labs is long gone. There will be a need for experts to exchange knowledge; there will be a need for and expectation that non-specialists will have an opportunity to examine critically and understand complex technical judgments.

4 *Predictive expertise.* With decision making, if we are asking for predictions about the consequences of management actions – perhaps 15–20 years away – even the most skilled experts often will be operating in an area well outside their usual area of practice. Ask yourself: 'who would constitute the better expert: one trained in the subject matter at hand but with little experience in making predictions, or one skilled in the art of making forecasts but with less-substantive knowledge'? Or perhaps both types of expertise are needed; in setting up a panel of experts to explore the implications of uncertainty on estimates of consequences 50 or 100 years into the future, one can imagine substantial insights coming from the dialogue between a subject expert and forecasting expert.

5 *Implementation expertise.* With SDM, we're typically interested in judgments leading to the predictions of consequences of policy or management actions. The expertise about the effectiveness of actions themselves (i.e. the impact of a market instrument, uptake of a code of practice, etc.) is very different from expertise about the affected resources (e.g. species or aquatic ecosystems). Further, a variety of institutional considerations could affect the probability that a proposed action continues to remain a management or political priority. This implementation expertise may reside in quite different people than the biologists or ecologists who often are consulted as substantive experts.

Implications for populating consequence tables

Who should populate a consequence table depends on the decision context – it should be the individual, or more commonly the individuals, who are recognized as experts in the subject matter. If the decision is about a proposed prescribed burn in a park near to an urban area and the performance measure in question is 'quality of the recreational experience', then non-technical stakeholders or resource users may in fact be the relevant experts who should assign consequences to the alternatives. If instead the effect in question concerns possible changes in harvesting local foods, then traditional knowledge holders in an aboriginal community may be the relevant set of experts. For estimates about the health impacts from air emissions, then the experts will comprise epidemiologists and risk analysts, who may be called upon to develop (and clearly explain) probabilistic air dispersion models. The point is that there is a body of expertise – specialized knowledge not available to just everyone – that will aid in informing the decision. Adopting a collaborative decision-making approach should not mean that the knowledge of everyone is considered equally reliable. Instead, it is the task of the technical facilitator to ensure the use of the most relevant and highest quality expertise, regardless of the source, and the documentation of the rationale for that determination.

8.3 Gathering information about consequences

Only rarely is information about consequences readily available for use in decision making. Baseline studies and mechanistic models of ecological systems sometimes exist, but almost always they will need to be augmented with new decision-focused analyses. Information about consequences comes from three main sources: (1) studies and predictive modeling; (2) expert judgment and; (3) group deliberations or peer review.

8.3.1 Designing studies and predictive models for estimating consequences

Scoping studies and building predictive models that support decision making isn't difficult, it's just different.

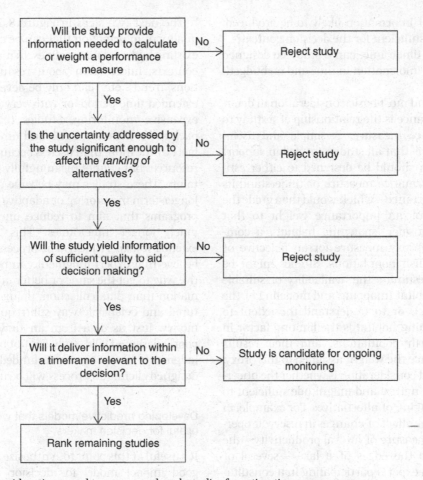

Figure 8.2 Considerations used to screen and rank studies for estimating consequences.

Scoping and prioritizing studies

The first challenge may well be that the list of studies identified for estimating consequences exceeds available budgets, and there is a need to identify, scope, and prioritize them. As usual, a useful starting point is sketching the decision and developing a solid objectives hierarchy. Develop a conceptual model that links candidate alternatives to the objectives and performance measures, and from this, identify information needs for calculating performance measures for each alternative. If a proxy measure has been selected, identify whether there are information needs critical to interpreting the importance of the proxy – that is, the relationship of

the proxy to the fundamental objective. Now you're ready to design studies to address the unknowns.

By studies, we mean any investigations required to gather information for decisions. They could include literature reviews, analysis of existing data, collection of new data, the development of impact models for estimating consequences/ impacts, and eliciting or synthesizing expert judgment. Once the set of candidate studies is developed, three general criteria (shown in Figure 8.2) are useful for prioritizing them:

1   Relevance to decision making: will resolution of the uncertainty affect the choice of alternative?

2 Quality of information likely to be produced: will it be sufficient for the decision context?

3 Cost and timeframe: can a study be designed to deliver information in time and on budget?

The first and most basic consideration in determining relevance is the relationship of a study to a performance measure. A simple and useful scoping rule is that all studies funded in support of a decision should be designed to either estimate a performance measure or understand its relative importance – which would then guide the assignment of an importance weight to that measure. Consider 'spawning habitat', a commonly used proxy measure for an objective of maximizing fish populations. Studies might be designed to estimate the availability of suitable spawning habitat (mapping and modeling of the affected river), or to understand the extent to which spawning habitat is the limiting factor in increasing fish populations, and their results could influence the weight assigned to this proxy.

The second consideration is whether the uncertainty is of a nature and magnitude sufficient to affect the ranking of alternatives. For example, in estimating the effect of changes in reservoir operations on a measure of littoral productivity – the area close to the edges of a lake – several of the technical experts participating in a consultative group thought it necessary to make an assumption about the depth of light penetration in the lake, which was not known with any confidence. A proposed study to obtain more accurate estimates of light penetration was rejected, however, because it was recognized that the 'depth of penetration' wouldn't change across the alternatives. Although the absolute value of the estimated littoral productivity might change with a more accurate estimate of light penetration, the relative performance (and thus the ranking) of alternatives would not. Any reasonable assumption about light penetration (for example, based on similar lakes with known light penetration) was sufficient. So this uncertainty did not affect the decision and the study was not a candidate for funding[19].

The next two steps in Figure 8.2 concern the quality of information that can be collected given existing budgets and timelines. In many ecological contexts, information about resources (populations, trends, etc.) can only be developed over an extended time period, or with very extensive and expensive monitoring. Studies that are scaled back to be within budgets and timelines may in fact be little more than useless because they use up resources but cannot meaningfully reduce uncertainty. These studies may become candidates for longer-term monitoring or adaptive-management programs that aim to reduce uncertainty over much longer timeframes. This consideration extends to expert judgment processes. While we believe that expert judgment can be a cost-effective way to get the same or better quality of information than data collection, doing so in a structured and defensible way still requires time and money. Just as data from an underfunded and poorly designed field study cannot be relied upon, judgments from an underfunded and poorly designed elicitation process will be unreliable.

Developing predictive models that will be useful for decision makers

It's useful at this point to synthesize what makes a good impact model for decision making. The range and scope of impact models used in environmental management are dizzying and their details are discipline-specific. Key considerations in designing models to estimate economic or employment effects for example, are different from those for estimating biological responses. However, there are a few general guidelines for developing decision relevant models that can be applied across disciplines.

1 Develop a conceptual model that illustrates the causal relationships between proposed management actions and the measures that will be used to assess performance. These might take the form of means–ends networks, influence diagrams, impact hypothesis diagrams, probability networks, Bayesian belief networks, event trees, decision trees, and so on. Models should be designed to estimate the consequences of management actions on the objectives, using the agreed performance measures to report these consequences.

2 Ask whether meaningful prediction is possible. In some systems, if uncertainty is deep and irreducible, it may not make sense to invest in predictive modeling – climate change mitigation and adaptation applications are a good example.

3 As a general rule, plan to report consequences probabilistically. A good case can often be made that decision makers will be better informed by models that produce probabilistic information about the thing that matters (fundamental objectives) than by detailed mechanistic models that produce precise information about proxy measures.

4 Ensure that models are only as complex and as accurate as they need to be to inform decision making. If a parameter doesn't help to clarify preferences among alternatives, it should probably be omitted. In most cases, models need only be able to distinguish *relative* performance differences across alternatives. It's useful to design models so that they can be iteratively improved. Start with something simple, and enhance it when you need to.

5 Design models so that they can be used interactively in a deliberative environment to facilitate the exploration and iterative refinement of different alternatives. It's usually important to design them to facilitate the exploration of the implications of uncertainty via sensitivity analysis.

6 Because of the importance of having models that can be iteratively improved and used interactively, the most useful models are usually those that can be used and updated by people without special model building expertise or proprietary software.

Decision support is not of course the only reason to perform studies or build models, but in SDM the goal associated with data collection is to make better decisions. This means it is focused on two things: gaining a better understanding of the potential consequences of proposed actions and identifying better alternatives. If the potential reductions in uncertainty won't help managers to make a better choice in the context of the specific management issue under consideration, then the value of the information to the decision at hand is negligible. But of course, improving biological and scientific information may be valuable for many reasons – establishing a base-line inventory, creating a comprehensive listing of resident species and their preferred habitats, understanding key ecosystem functions. Or from a process perspective, data collection may create a means to involve community stakeholders who up to this point in time have not fully participated.

8.3.2 The role of expert judgment in estimating consequences

If reliable data and models are not available or appropriate to the decision at hand, then a good approach is to elicit expert judgment. Expert judgment involves asking a range of recognized experts to synthesize the full range of evidence and theory and provide judgments. The judgment may involve assigning discrete probabilities (what's the probability of a particular event or set of events?), assigning consequences using a defined impact scale (see Chapter 5) along with associated probabilities, or developing a subjective probability distribution.

Of course experts are often asked for input on a wide variety of issues. Here we restrict ourselves to the elicitation of judgments leading to estimates of consequences. Protocols for eliciting such judgments have been extensively examined by researchers in decision analysis and applied behavioral decision making. What has emerged is a fairly clear convergence on best practices. Small differences in how procedures are described by different practitioners are dwarfed by the broad consensus on approach. In this section, we summarize these best practices. In some situations, eliciting defensible judgments will require a technical facilitator with specialized expertise in decision analysis and elicitation methods. However, many of the practices can be usefully applied in relatively routine decision situations without specialized expertise, with important improvements in the quality, transparency and defensibility of the resulting judgments. Our goal in this section is to provide enough guidance to allow the unaided use of best practices in these routine situations, while recognizing the role that externally facilitated elicitations could play in more complex or higher stakes situations. A variety of texts provide detailed 'how-to' guidance; we provide a list of further reading at the end of the chapter.

The promise and pitfalls of experts

Estimating consequences in environmental management unavoidably involves incorporating the judgments of experts. Yet in many cases, we see environmental managers 'resorting' reluctantly and apologetically to expert judgments, either because of short timelines or limited budgets. We propose that expert judgment is more than just an unpleasant compromise. A well-structured elicitation brings the principles of good science to the practice of eliciting probabilistic estimates through the use of systematic and repeatable methodologies, careful documentation of assumptions and results, and quality control through peer review. Done well, it can in some cases provide better information for decision making – higher quality information in a more relevant form and at lower cost – than additional data collection or modeling exercises. Done poorly – as unfortunately, it often is – it is indeed an unpleasant compromise.

The reliability of expert judgments depends on many things; most notable are the nature of the judgment and the method of interrogation. Experts are most reliable when they make frequent, repeated, and easily verified judgments that allow them to learn through feedback. Weather forecasters come to mind. Unfortunately, most environmental management problems exhibit none of these characteristics. Judgments are typically novel, infrequent and untested, so that monitoring and feedback are rare. In such circumstances, unrealistic expectations on the part of decision makers, coupled with what Burgman[19] calls 'unquenchable expert optimism', create a growing dependence on 'widespread, unreliable expert opinion that wears a mantle of scientific reliability'. These are tough words from one of the world's leading researchers into the methods and uses of expert judgment.

The good news is that there is a well-established body of literature on decision-focused approaches to eliciting expert judgment, demonstrating a consensus on best practices related to expert selection, elicitation protocols, bias avoidance techniques, treatment of uncertainty, documentation and peer review[20]. The bad news is that in many resource-management agencies, there is a disturbing lack of uptake of even simple structuring tools and methods that have been shown to improve the quality and relevance of judgments for decision making.

A practical elicitation process

The elicitation process can be thought of as three major steps: preparation, elicitation, and synthesis. The steps can be conducted to various degrees of rigour depending on the nature of the decision. In some cases, the elicitation may consist of calling a couple of trusted colleagues or staff members into your office for an hour or two of discussion. In such cases, the preparation stage is relatively trivial but good structuring can still help to make sure you get the best possible information for your decision. In other cases, when a more formal approach is called for, a practical and defensible approach usually will involve a mix of individual elicitations and group workshops in order to benefit from both independent thinking and the insights that come from sharing and deliberation.

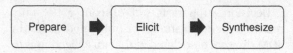

Step 1: prepare

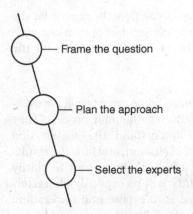

- Frame the question
- Plan the approach
- Select the experts

Frame the question

This task involves defining the technical question or questions that need to be addressed, the reason the information is needed, and how exactly it will be used in the decision-making process. Questions must focus on technical judgments (i.e. those requiring specialized technical knowledge) rather than value judgments about trade-offs or the 'acceptability' of risk. We know you can't completely separate them, but there are questions that are fundamentally about facts (such as what will happen to sea turtle populations if we protect their habitat) and others that are fundamentally about values (such as how much should we spend to restore sea turtle habitat). To ensure their relevance to decision making, the questions should be framed in response to a decision sketch, so that the objectives, measures, and alternatives are understood, and a clear need has been identified for the information. In some cases experts may be asked to estimate consequences directly – that is, estimate the values of performance measures under different alternatives. In other cases, they may be asked to estimate some quantity that will be needed to estimate a performance measure.

Plan the elicitation approach

At this stage the plan for the elicitation process is established. For routine problems, this could be as simple as meeting with colleague and talk-ing for an hour to determine areas of agreement and disagreement between the two of you. In more consequential (and hence more formal) applications, key activities include:

1 *Select an analyst and an in-house subject matter expert.* The analyst could range from an independent technical facilitator for complex high profile cases to an in-house resource for more routine cases. The 'in-house' expert is a subject-matter expert who will help to structure and pilot test the elicitation protocol.
2 *Sketch a preliminary conceptual model.* A mean–ends network or influence diagram illustrates how the judgment being elicited relates to the alternatives under consideration and the fundamental objectives that will be used to evaluate them. If the judgment involves estimating a performance measure or a quantity that will be used to estimate a performance measure (as is commonly the case), the influence diagram will show the relationship of the quantity to be elicited to the calculation of the performance measure.
3 *Develop the elicitation approach.* For example, the degree of quantification desired, the number of experts, group versus individual elicitations, the method of aggregation. This must be done in consideration of budget and time-line considerations as well as the nature and stakes of the decision.
4 If the question involves directly predicting the consequences of alternatives, make sure the alternatives are unambiguously specified (see Chapter 7).
5 *Pilot-test the elicitation protocol internally prior to asking external experts to use it.* This helps to ensure the context is clear, to test that the questions and response modes are practical, and it will help to identify relevant data and literature to provide to experts.

Select the experts

Much has been written about selecting experts. It's important but it doesn't have to be complicated:

1   Write up your criteria for selecting experts – what expertise do you need and why? – and the process for selecting them. Sometimes it will be useful to have a group of people make the selection.

2   Select multiple experts, representing a diversity of affiliations, types of expertise, technical perspectives and knowledge sources. While the number will depend on the application, some practitioners suggest that three to seven experts is sufficient[21]. The number might rise if relevant knowledge must be elicited from both scientific and local/traditional knowledge experts.

3   Experts selected should be widely recognized as experts in the subject matter. At least some of them should be skilled at communicating with experts and non-expert stakeholders.

4   Make an effort to ensure impartiality. Experts with obvious conflicts of interest or motivational biases should be avoided. However, no one is completely impartial, and compromises may need to be made when the pool of available experts is limited. If experts are not impartial, balance that by having experts with a diversity of affiliations.

5   Document the rationale for selection, including notation of any known conflicts of interest or other biases.

Step 2: elicit the judgment

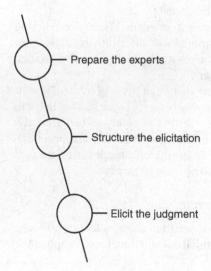

- Prepare the experts
- Structure the elicitation
- Elicit the judgment

Prepare the experts

This step is designed to prepare the experts for the elicitation. It can be done either one on one with individual experts, or in a group with all the experts involved.

1   *Clarify the context.* Some experts will have reservations about an unfamiliar process. They will need to understand the reason and method for the elicitation, and how the results will be used. Often it's important to clarify that uncertainty will be explicitly characterized – the goal is to expose and understand uncertainty, not to hide it.

2   *Build awareness of common biases.* Although awareness doesn't eliminate biases, it can help to mitigate them, or at least to explain why some parts of the elicitation are done the way they are. Common biases include overconfidence, anchoring (with insufficient adjustment), and availability (see discussion in Chapter 2).

3   *Provide a diversity of background information.* This is the task of providing experts with a range of relevant information and benchmarks on the subject at hand. It helps to reduce biases related to anchoring and availability (where experts over-rely on studies they've recently conducted or that tend to confirm their pre-existing views).

4   *Provide probability training if necessary.* If you are asking for probability judgments, it's important to know what level of familiarity your experts have with basic probability concepts. Some training may need to be provided, depending on the background of the experts and the decision context within which the elicitations are being conducted.

5   *Provide an opportunity for practice and feedback.* Practice and feedback is particularly important for countering overconfidence. Give experts some simple questions in their field for which the correct answers can be determined. Let them make some estimates and then compare their answers to the right ones[22]. Formal calibration exercises have been shown in some

cases to improve the quality of judgments, but if you don't know how to do these, don't let that stop you from doing the basics.

Structure the elicitation

Structuring and decomposing reduces the potential for error and bias. It also improves comparability among experts, transparency, and the accessibility of the judgment to peer review. It is at this step that the stage is set for building an audit trail that can be easily reviewed and understood by others. A key task is developing and documenting the conceptual model that reflects the expert's thinking about cause-and-effect relationships. From this model, the specific quantity(ies) to be elicited can be identified in relation to other factors. Sources of linguistic uncertainty (ambiguity and underspecificity) should be eliminated.

An important question is whether the conceptual model should be developed independently (each expert has their own) or whether experts work together to develop a common conceptual model and then provide independent judgments of a quantity within that structure. Because of the inevitable effects of group think, the collective approach will almost certainly result in a narrower representation of uncertainty than an independent approach. On the other hand, in a management context it has two important benefits. First, it streamlines the ability to aggregate across experts. If experts have different models and underlying assumptions, then making sense of them will take more time and resources. Second, a common conceptual model builds a common understanding of the system being managed and can facilitate better communication among managers on an ongoing basis. So there may be compromises between accurately characterizing the full range of uncertainty and using the process to facilitate more effective management and communication. The important thing is to be aware of the effect of group work on the characterization of uncertainty: groups will – rightly or wrongly –

tend to converge on particular models or particular values of uncertain quantities, and become overconfident in them.

Elicit the judgment

At this stage the expert is either interviewed in person or provided with a questionnaire to complete. The objective is to derive an unambiguous description that best reflects the expert's beliefs about the range of probable outcomes (or hypotheses or states of nature, depending on the context) and their likelihood.

The prevalence of biases in expert judgments has led to the development of prescriptive procedures to counter them; overconfidence has received particular attention. As a general rule, overconfidence is countered by asking experts to: (a) focus on extreme situations first; and (b) explicitly consider and find explanations for why their judgments might be wrong.

The exact form of an elicitation depends on what's being elicited. Here's an example of a broadly applicable protocol for eliciting a subjective probability distribution for an uncertain quantity:

1 Confirm the quantity to be elicited, being careful to eliminate ambiguity and underspecificity.
2 Remind experts that their judgments will be open to peer review and that they will be accountable for them (the process will not be anonymous).
3 Minimize overconfidence by using a four-point interrogation procedure that elicits a 'free' interval[23,24] (see Text box 8.3).
4 Elicit a plausible upper bound.
5 Elicit a plausible lower bound.
6 Elicit the most likely value.
7 Assign a level of confidence to the interval.
8 Ask experts to consider the reasons why they may be wrong: 'what reason could explain how the true value could turn out to be much larger or smaller than your extreme value?'
9 Document conditionalizing assumptions, along with their rationale. This both improves

transparency and facilitates comparison across experts.

10 If multiple probabilities have been elicited, ensure the judgments obey basic probability laws (e.g. summing to one, etc.)

11 Iterate and revise the judgment if necessary.

If experts are being asked to assign consequences using a pre-defined impact scale (Chapter 5), the same approach can be followed – see the example in Chapter 9.

As a general rule, face-to-face interviews are preferred. This ensures that the respondent is giving full attention to the task, has an opportunity to ask questions, and has access to relevant technical resources.

Step 3: synthesize

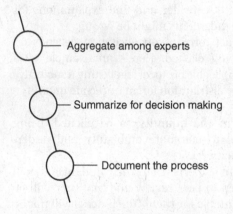

- Aggregate among experts
- Summarize for decision making
- Document the process

Aggregate across experts

If multiple experts have provided judgments, it will be necessary to aggregate their answers in some way to develop a clear picture of the range of beliefs. There are two basic ways to aggregate: numerical and deliberative (many texts call the latter 'behavioural'). Numerical methods use quantitative methods to arrive at a combined estimate. There are a variety of both probabilistic and non-probabilistic methods (e.g. average and weighted averages, evidence theory, fuzzy sets, and so on). Beyond simple averaging, which can

---

Text box 8.3: Using a four-point process to elicit a free interval

Andrew Speirs-Bridge and colleagues[23] report a substantial decrease in experts' overconfidence using a four-point elicitation procedure rather than the more conventional three-point (lower limit, upper limit, best guess), with the experts assigning their anticipated confidence in the interval produced as part of the fourth step in the expert judgment elicitation process. Here's what the four point elicitation protocol looks like in terms of specific questions asked of an expert. This example is based on the description given by Speirs-Bridge *et al.*[23] of the elicitation of a 'free interval' from experts about rates of infectious diseases.

1 Realistically, what do you think the lowest rate of this disease could be for the 12 months to September 30 2007?

2 Realistically, what do you think the highest rate of this disease could be for the 12 months to September 30 2007

3 Realistically, what is your best guess (i.e. most likely estimate) of the rate of disease for the 12 months to September 30 2007?

4 How confident are you that your interval, from lowest to highest, could capture the reported rate for this disease? Please enter a number between 0 and 100%.

---

be appropriate in some cases, we generally do not advocate these methods for three reasons. First, they require special expertise that's usually unavailable. Second, the assignment of weights to the expertise of different experts is controversial and may distract resources and attention from more important considerations. Third, and most important, it doesn't usually help. If there is a high degree of overlap among experts' distributions, this is generally obvious once they're plotted, and research shows that the choice of weighting and aggregation method makes little difference to the results. If, on the other hand, there is a high degree of disagreement among experts (little overlap among their distributions), then decision makers need to know more about the nature and reasons for that disagreement in order to understand the implications for decision making, and mathematical aggregation won't help with that.

We, therefore, advocate a deliberative approach, which basically means talking and thinking critically about what the results mean for the decision at hand, with an emphasis on clarifying areas of agreement and disagreement among experts. Some texts/practitioners emphasize consensus-based approaches (such as the Delphi method). However, professionals working in this field generally concur that consensus is not generally an appropriate goal – or certainly not the primary one. The goal is understanding and accurately characterizing uncertainty, including the key disagreements among experts that remain after the arbitrary noise of linguistic ambiguities and underspecificity have been eliminated.

A useful deliberative approach to aggregation involves eliciting independent judgments, convening the experts to clarify key areas of agreement and disagreement, and then allowing experts to revise their judgments before synthesizing the responses. Useful discussion questions at an aggregation workshop could address:

1 *Differences in conceptual models.* Why are the models different? Do these differences result in significantly different judgments about the quantity under consideration?
2 *Differences in source data and related studies.* What sources did experts rely most on and why? Which did they dismiss as irrelevant and why?
3 *Low-probability events.* What specific conditions or scenarios did experts consider when thinking about low-probability events?
4 *Differences in conditionalizing assumptions.* How would the experts' judgments change under different assumptions – that is, those used by other experts? Where differences in assumptions are arbitrary, is it possible and useful to agree on a common set?

5 *Differences in specific judgments.* Encourage outliers to describe the rationale for their judgments. Or encourage others to try to come up with plausible rationales for these judgments, even if they are not their own.
6 *Alternative ways to present information to decision makers.* Is it reasonable, given residual differences among experts after efforts to reconcile assumptions and data sources, to develop a single probability distribution that captures the views of all experts? Would this be useful to decision makers? Would doing so clarify or obscure key messages about the state of agreement or disagreement among experts? Would it be better to summarize findings using the individual distributions?

As a minimum, the discussions should highlight key differences in judgments and the main reasons for them. Depending on the degree of agreement or disagreement and the anticipated uses of the results, the group may work together to produce a single consensus judgment that encompasses the views of all participants, produce a majority judgment with documentation of dissenting opinions, or simply present a set of individual judgments. As a general rule, unless there are very good reasons for doing so, combining very different probability distributions is to be avoided.

Tremendous insights can come from group work. Many differences in judgments stem from easily resolvable differences – relatively arbitrary assumptions that could easily be changed without loss of information. Or one expert learns that another has a much higher quality data set to draw on and reworks his/her judgment using the new data. And so on. Keep in mind, however, the problems associated with group think: the tendency of a few individuals to dominate (even though they don't have greater expertise), a general desire of participants to avoid conflict, and various motivational and cognitive biases can lead to a false convergence of opinion and overconfidence in a consensus opinion.

Summarize and present implications
for decision making

At this step, areas of agreement and disagreement are documented, with emphasis on making the results relevant to decision makers' needs. In an SDM context, this will normally mean showing how the results affect the comparison of proposed alternatives. The expert judgment process is not intended to recommend alternatives; it simply clarifies their possible consequences, and the extent to which there is agreement about those consequences. It may help to identify which alternatives are more robust than others, and which uncertainties are more or less significant. Experts may also identify cases where further analysis may be useful and suggest approaches to reduce uncertainty over time.

The format of the presentation of results depends on the nature of the elicitation process and the nature of the decision. Commonly, it will take the form of a completed consequence table, with expert judgments forming the basis for the estimation of consequences on one or more of the evaluation criteria. As with any other method of estimating consequences (e.g. modeling or other analysis), things that cause the decision (or choice of preferred alternative) to be sensitive to the uncertainty may trigger further analysis.

Document and review

The entire process and the results should be documented and available for scientific and public review at any time. Anyone with a stake in the outcome should have an opportunity to question and criticize the judgments. Experts should be accountable for their judgments and prepared to explain them.

While expert judgment processes are not necessarily repeatable in the sense that data collection or experiments are, they can be systematic and transparent. The conduct and the documentation of the process should create an audit trail that clearly outlines the basis for the judgment – which experts were used and why, what data/models/studies were considered, which conditionalizing assumptions were made and why, what possible scenarios were considered in setting bounds for possible outcomes. This audit provides the basis for open peer review, and increases the confidence others will have in the analysis. As in other steps, the scope of documentation should be consistent with the importance and complexity of the issues under consideration.

8.3.3 The role of the deliberative group: a new kind of peer review

At this point you may be getting the impression that all will be solved so long as experts are asked for their input in neatly structured ways. Not so. While formal expert elicitations help, the reality is that much of the time, the examination of information quality occurs right in the deliberative setting. It's a messy place to do business. It's populated by a mix of expert and non-expert participants. Competing hypotheses are put forward by multiple parties with different technical and political world views. There's a complex mix of value judgments and technical judgments at play. Uncertainties are irreducible. There's no blind trust in experts, and even if there were, the experts don't agree.

In the conventional approach, different parties identify different experts, studies are commissioned and sent out to external experts, contradictory reports are produced and then distributed for peer review. Reviewers offer useful and important caveats or commentary on the conclusions, but provide little resolution of fundamental uncertainties. Parties remain polarized. Conventional peer review is further hamstrung when knowledge stems from different sources – peer review processes seen as relevant to science-based knowledge are not seen as relevant for exploring local or traditional knowledge. In such an environment, what's needed is a fundamentally new approach to exploring the quality of information underlying competing claims.

Text box 8.4:   Exploring the 'best available information' in harvest management

As is the case in many other jurisdictions, the Ministry of Environment in the Province of British Columbia is responsible both for protecting environmental values and for managing their 'sustainable use'. In the context of mule deer – a common ungulate in many parts of the province – Ministry staff are mandated to seek a balance between ensuring mule deer are sustained in healthy numbers, with suitable demographic characteristics (i.e. age distributions, gender distributions, etc.) and providing suitable hunting opportunities.

The Ministry was interested in evaluating alternative harvest management strategies. This required understanding the relationship between regulatory alternatives with assumed hunter behavior in order to better understand the likely impacts (ecological, economic, and hunter satisfaction, etc.) of regulatory alternatives.

The figure below is a simplified representation of the Ministry's existing population model that we expanded to reflect the wider decision context. Very simply, for any given year the model makes assumptions about natural conditions (e.g. how harsh was the winter?), combines this with the status of the previous year's population[25], considers hunter activity and figures out how many of what animals are actually harvested that year. It then loops back and simulates many years of conditions to establish long-term averages. Hunter activity is determined by regulations, which set the context for the alternatives under consideration, and outputs are estimated in terms of multiple accounts that include ecological, economic, and social (hunter satisfaction) concerns.

One area of particular concern to the Ministry concerned the uncertainties around assumed hunter efficiency. A first step in exploring this was to use an expert judgment process to describe how efficient hunters were (in terms of 'days required to harvest a buck') in the target management area under specified conditions.

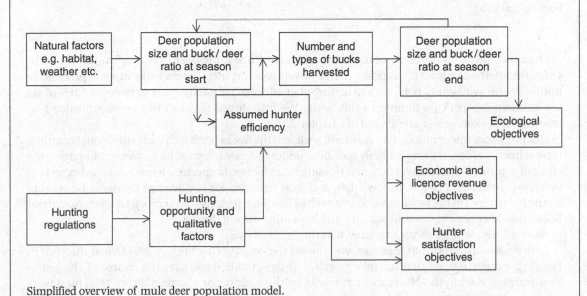

Simplified overview of mule deer population model.

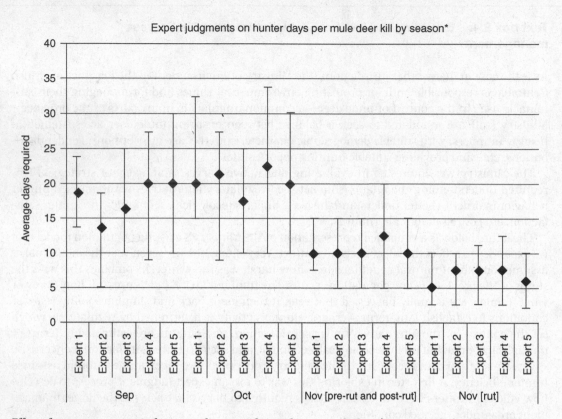

Effect of season on expert judgments of average days to hunt a mule deer. *For a specific location and buck-to-doe ratio.

Over the course of an afternoon, we led a group of five mule deer experts through the steps we've outlined – at a level appropriate to a relatively routine, internal decision support process. To minimize the availability bias, we had arranged in advance for each expert to be sent copies of six key academic papers pertaining to this issue. We introduced them to the overconfidence bias using several exercises– as described in Chapter 2.

Holding other important factors constant, we asked the five experts a series of questions regarding typical hunter efficiency under highly specific situations. Among several factors we explored was the effect of season on the hunter efficiency. Holding other factors at specified levels, we asked experts in individual interviews to give us low, high, and 'best estimate' for the average number of days that would be required to harvest a mule deer in each of four time periods – September, October, November before and after the rut (mating season), and November during the rut.

Some of the data we received is shown in the figure above.

Late on in our afternoon together, we showed the experts this finding and talked through it. People generally agreed that the relative ratios of days made sense over the course of the year – that hunting during the November rut would be the easiest, for example. From this, managers were able to use quite defensible ranges of values for these parameters for subsequent modeling exercises. Although the ranges of uncertainty for each expert were quite wide, particularly in the earlier months of the season, the estimates of ranges in the critical November period were much tighter – and certainly more consistent than managers had expected to see. For this reason, they concluded that the uncertainty about hunter efficiency in this context wasn't necessarily the most important unknown in this situation, so attention could turn elsewhere.

Although we undertook several judgment rounds, we were able to gather much valuable information in one afternoon. Imagine the cost of undertaking a field study to obtain this same information. Would the (presumably) better quality information of a properly controlled field experiment be worth the additional cost?

In a structured decision-making process, we put everyone in the same room. We use structuring tools to build common understanding and we use structured lines of questioning to make sure every hypothesis receives rigorous and equitable treatment. It's a review by an extended group of peers, as some would call it, all motivated by a desire to take some responsible action, leading to agreement on the 'best available information' for decision making. We're not proposing it's perfect, but it's better than the alternative.

In constructively debating the strengths and weaknesses of a hypothesis about consequences, it's useful to talk about three criteria: relevance, reliability, and sufficiency[26]. The *relevance* of a hypothesis is examined by clarifying how it affects the decision. Does it help in clarifying objectives or defining the range of potentially affected interests? Does it help to identify alternatives? Does it help to estimate the consequences of alternatives on the criteria that participants have identified for evaluating alternatives? Does it help to clarify the importance of losses or gains likely to result from choices? If not, it may not be a priority for discussion.

The next question is: what is the *reliability* of the information? Is the hypothesis of a type and quality reasonably relied on by experts in the particular field in forming opinions or inferences? Which hypotheses are most plausible? A useful approach involves:

1 Defining the hypotheses.
2 Sketching an 'impact hypothesis' diagram.

3 Examining the evidence for and against the hypothesis using a consistent line of questioning.

What constitutes *reliability* is something that needs to be determined and then consistently used by the group. In general they will refer to qualities of the evidence itself (is there a plausible mechanism for the hypothesized impact?; are supporting data/studies representative, replicated, or corroborated?, etc.), the process by which it was developed (was it a recognized systematic, transparent, peer reviewed methodology?), and the source (does it stem from a source with recognized expertise and some degree of impartiality?). Some care should be taken to make sure the questions are framed in a way to be relevant: (a) across disciplines – so that the process can apply to hypotheses about biological impacts as well as hypotheses about social or economic ones; and (b) across sources – so that it applies to hypotheses stemming from both local and scientific knowledge. This is isn't as hard as it sounds. The question 'is there a plausible mechanism for this impact?' can apply to hypothesized ecological, social, and economic impacts. 'Have the findings been replicated or corroborated?' can apply both to scientific analyses and field observations. 'Does the source have expertise that's recognized by a community of peers?' can apply to both scientific and local knowledge holders. If you make this a habit, it becomes just the way of doing business, rather

---

Text box 8.5:   Case study: working with experts to explore hypotheses

All these approaches tend to merge slightly in practice. We've examined competing hypotheses with expert panels and we've used structured elicitation methods with an expert sub-set of a multi-stake-holder committee. Here we describe a case where we explored competing hypotheses with a panel of highly trained scientists. Typical of many situations we find ourselves in, the group spent a lot of time in technical debates over the strengths and weaknesses of various hypotheses, but had so far failed to make much progress. Like many other groups, they experienced:

1   Problems trying to clearly link the outcome of discussions about hypotheses to the management requirements.
2   Problems trying to effectively expose fundamental differences in experts' assumptions and perspectives because of the complexity and nuance of the issues.
3   Problems trying to distinguish consistently research needs from the management actions that enjoy widespread acceptance and might be undertaken immediately.
4   Problems trying to structure group discussions and, as a result, inefficiently using management time.

In 2007, a joint United States–Canadian Technical Working Group (TWG) was tasked with developing recommendations in support of an official recovery plan for Upper Columbia River white sturgeon. This giant fish is formally listed under endangered species legislation in both the United States and Canada, and has suffered almost complete 'recruitment failure' – that is, failure to reproduce – for the last 20 years or more (sturgeon often live to be more than 100 years old). Clearly the construction of dams along the Columbia River had a major detrimental effect on species in previous decades, but the question of which specific mechanisms are causing ongoing recruitment failure is a subject of considerable uncertainty and controversy. Without a clearer understanding of the mechanisms that are causing this continued failure, the social impacts of potentially helpful management interventions, (which would often be in the tens or hundreds of millions of dollars per year), are difficult to justify.

The group was composed of members who in general, after working together for several years on issues related to management of sturgeon in international waters, were familiar with each others' general perspectives. Yet there was a growing sense of frustration with the group's inability to reach agreement on the causes of ongoing failure, exacerbated by the knowledge that some form of action was likely urgently needed if Upper Columbia River white sturgeon were to avoid extirpation.

Our role, as analysts and facilitators, was to help the TWG develop and critically examine a set of plausible hypotheses that could explain the recruitment failure, and ultimately to help members to identify and agree on useful near-term management next steps. One early emphasis of this work was to reduce the ambiguity in the definitions of the competing hypotheses themselves so as to improve the quality of discussions about the strengths or weaknesses of the evidence regarding each one.

After discussions with the experts, we realized that the competing hypotheses were best represented as pathways of effects that linked ultimate ongoing causes with recruitment failure. Working with the TWG, we developed the generic pathway structure shown in the figure below on which one of the candidate hypotheses (labeled 'LC1') is highlighted for illustration.

The 'hypothesis'[27] highlighted in the figure encapsulates the view that changes in flow regulation (i.e. the presence of a dam) have caused a change in the water flow patterns that, most significantly for white sturgeon, have created chronic problematic conditions for feeding and hiding in 0–40 day-old sturgeon, leaving them vulnerable to predation from other species and therefore causing ongoing recruitment failure. Secondarily, the hypothesis notes the more minor roles of other related factors, such as reduced food availability and impacts on growth.

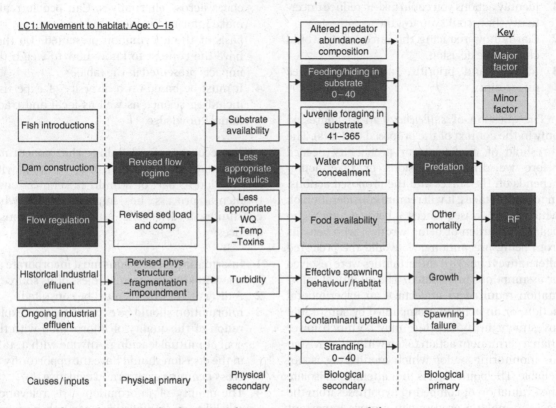

Impact pathways for Upper Columbia River white sturgeon recruitment failure.

For each of the hypotheses they developed, the expert participants discussed:

1 What is the hypothesis?
2 What is the evidence for and against this hypothesis?
3 What is the quality of this evidence?
4 What are the key information gaps and sources of best available information?
5 What are the estimated cost and time requirements associated with potential reductions in uncertainty)?

By structuring the hypotheses effectively, the group was able to (a) focus on key technical issues rather than getting caught up in the ambiguities of language and (b) critically examine and document the evidence for and against competing hypotheses.

In Chapter 10, we discuss further how we helped this group decide what next steps to take with respect to each of the identified hypotheses.

than a personal challenge to an individual's credibility or expertise.

In some cases it may be useful to go on to:

1   Identify actions you could take to reduce uncertainty (how could you test the hypothesis?).
2   Clarify how reducing the uncertainty would change the decision.
3   Identify and prioritize actions to reduce uncertainty.

The question of *sufficiency* can be addressed only in the context of the proposed action(s). The threshold of confidence or probability needed before we draw conclusions or take action depends on the stakes and the proposed actions. In terms of stakes, it will require consideration of: what are the consequences of being wrong?; who bears the burden of being wrong?; who benefits from being precautionary?; would the preferred alternatives change if uncertainties were resolved or assumptions invalidated? The quality of information required to undertake an experimental action, or an action accompanied by substantial investment in monitoring, may be much lower than a permanent action taken without provision for monitoring or for which monitoring is not reliable. The point then is that attempts to isolate the evaluation of competing hypotheses from the decision-making process can remove important context that would aid the deliberations.

8.4   Key messages

In this chapter we introduce three main issues critical to the success of SDM analyses and deliberations concerning the anticipated consequences of management actions. On the first topic, consequence tables, best practice means:

1   The presentation of consequences must allow consistent comparisons of relative performance across alternatives.
2   It should expose key value-based trade-offs and key uncertainties – that is, are some alternatives more risky or less well understood or less reliable – than others?
3   It must do so in a way that allows decision makers to interpret the significance of differences across alternatives. Can people really make informed value-based trade-offs on the basis of the information presented? Do they have the context to know how to weigh the impacts presented in the table?
4   It must be based on a diversity of expertise, including science as well as local and traditional knowledge.

The second topic, defining what constitutes 'best available information' and working with information as part of a multi-stakeholder environmental process, has implications for what constitutes a defensible estimation of consequences:

1   Information for decisions must incorporate all relevant knowledge, regardless of its source.
2   A diversity of experts must be consulted.
3   Information should be exposed to open exploration of the quality of information with the goal of mutual learning. Anyone with a stake in the decision should have the opportunity to cross-examine experts and evidence.
4   The quality of information – its relevance, reliability and sufficiency – can only be examined with reference to a particular decision. The best available information is information that is 'good enough' for the decision at hand.

The third topic is the gathering of information about consequences, for which there are three primary sources: predictive models, expert judgments, and the deliberative group itself:

1   An emphasis on the decision can help to make models and data collection more focused and streamlined. Models that are useful for decision makers will tend to be those that link actions to objectives, produce probabilistic estimates of performance measures, facilitate sensitivity analysis, and can be used

interactively to explore the implications of different alternatives.

2   All environmental management decisions will involve expert judgments. There are established best practices for eliciting expert judgments. They don't have to be onerous to use. In some cases, they can be applied in a few hours to improve the quality of information available for even relatively routine decisions.

3   Both expert and lay judgments are subject to a host of common biases and frailties. Structured elicitation procedures help to mitigate them. These procedures are well-established and there is no excuse for not using them. They should be adopted as a matter of policy.

4   Competing hypotheses should be subject to constructive and consistent scrutiny. It's useful to establish clear and common expectations about how competing hypotheses will be examined – lines of inquiry that will apply to all claims or hypotheses regardless of their provenance. Groups should make a habit of using them, so that questioning is just the way of doing business, not a challenge to an individual's credibility.

5   Much of the information that will be used for decision making is developed in the deliberative environment, among a mix of expert and non-expert participants. One of the key distinguishing features of an SDM process, and we believe a critical factor in its success, is the creation of an environment that emphasizes and supports disciplined, collaborative learning.

8.5   Suggested reading

Berkes, F. (1999) *Sacred Ecology: Traditional Ecological Knowledge and Resource Management*. Taylor & Francis, Philadelphia. An excellent introduction to the role of multiple knowledge sources in resource management, traditional ecological knowledge, and its relation to conventional science.

Burgman, M. (2004) Expert frailties in conservation risk assessment and listing decisions. In: *Threatened SpeciesLegislation: Is It Just An Act?* (eds P. Hutchings, D. Lunney & C. Dickman). pp. 20–29. Royal Zoological Society of New South Wales, Mosman, Australia. A highly readable and personal review of reasons why expert judgments, especially of rare events in novel circumstances (which is frequently when experts' input is most needed), are fallible and prone to error.

Failing, L., Gregory, R. & Harstone, M. (2007) Integrating science and local knowledge in environmental science and local knowledge in environmental risk management: a decision-focused approach. *Ecological Economics*, **64**, 47–60. This article uses several SDM case studies as entry points to discuss ways in which science and local knowledge, including the perspectives of aboriginal populations, can be integrated as part of resource planning and evaluation studies.

US National Research Council (2009) National Academies Press, Washington DC. A collection of papers by different authors, *Science and Decisions: Advancing Risk and Assessment* provides a constructively critical review of human health and ecological risk assessment concepts and practices used by the USEPA, with good discussions of uncertainty, cumulative risks, and risk-based decision-making frameworks.

Yearley, S. (2000) Making systematic sense of public discontents with expert knowledge: two analytical approaches and a case study. *Public Understanding of Science*, **9**, 105–122. Applies both theory and practical experience to understanding why the knowledge of experts is often met with criticism and mistrust by members of the public; good background reading for any scientist about to participate in their first multi-stakeholder environmental decision-making process.

Meyer, M.A. & Booker, J.M. (1991) *Eliciting and Analyzing Expert Judgment: A Practical Guide*. Academic Press, San Diego. One of many texts on the elicitation of expert judgment. While we're not sure if all of the methods could be usefully applied by environmental managers without very specialized training, the book provides an example of the depth of research and implementation practice that exists on this topic.

8.6   References and notes

1   Of course, there is no law that says alternatives cannot be presented in rows and objectives in columns, or that some presentation format other than rows and columns can't be used. We do it this

way, following standard decision analysis practice, but we recognize that others may have different ideas.

2   Hammond, J., Keeney, R.L., & Raiffa, H. (1998) Even swaps: a rational method for making trade-offs. *Harvard Business Review*, March-April, 3–11.

3   Stokey, E. & Zeckhauser, R. (1978) *A Primer for Policy Analysis*. Norton, New York.

4   Peters, E., Dieckmann, N.F., Dixon, A., Hibbard, J.H. & Mertz, C.K. (2007) Less is more in presenting quality information to consumers. *Medical Care Research and Review*, **64**, 169–190.

5   National Research Council. (2009) *Science and Decisions: Advancing Risk Assessment*. The National Academies Press, Washington, DC.

6   Pidgeon, N.F., Lorenzoni, I. & Poortinga, W. (2008) Climate change or nuclear power – no thanks! A quantitative study of public perceptions and risk framing in Britain. *Global Environmental Change*, **18**, 69–85.

7   Funtowitz, S. & Ravetz, J. (1997). Environmental problems, post-normal science, and extended peer communities. *Etud. Rech. Syst. Agraires Dev.* 30: 169–175.

8   Cullen, A. & Small, M. (2004) Uncertain risk: the role and limits of quantitative assessment. In: *Risk Analysis and Society* (eds T. McDaniels & M. Small). Cambridge University Press, New York.

9   Berkes, F., Colding, J. & Folke, C. (2000) Redis-covery of traditional ecological knowledge as adaptive management. *Ecological Applications*, **10**, 1251–1262.

10 Failing, L., Gregory, R. & Harstone, M. (2007) Integrating science and local knowledge in environmental science and local knowledge in environmental risk management: a decision-focused approach. *Ecological Economics*, **64**, 47–60.

11 National Research Council (NRC). Committee on Risk Characterization. (1996) *Understanding Risk: Informing Decisions in a Democratic Society*. National Academy Press, Washington, DC.

12 Gregory, R., Failing, L., Ohlson, D. & McDaniels, T. (2006) Some pitfalls of an overemphasis on sci-ence in environmental risk management deci-sions. *Journal of Risk Research*, **9**, 717–735.

13 Coglianese, C. & Marchant, G. (2004) Shifting sands: the limits of science in setting risk stand-ards. *University of Pennsylvania Law Review*, **152**, 1255–1360.

14 Turner, N.J., Gregory, R., Brooks, C., Failing, L. & Satterfield, T. (2008) From invisibility to transpar-ency: identifying the implications. *Ecology and Society*, **13**. Retrieved from http://www.ecology andsociety.org/vol13/iss2/art7/

15 Berkes, F. (1999) *Sacred Ecology: Traditional Ecological Knowledge and Resource Management*. Taylor & Francis, Philadelphia, PA.

16 Gregory, R., Failing, L. & Higgins, P. (2006) Adaptive management and environmental deci-sion making: a case study application to water use planning. *Ecological Economics*, **58**, 434–447.

17 Burgman, M., Carr, A., Godden, L., Gregory, R., McBride, M., Flander, L. & Maguire, L. (in press). Redefining expertise and improving ecological judgment. *Conservation Letters*.

18 Burgman, M. (2005) *Risks and Decisions for Conservation and Environmental Management*. Cambridge University Press, UK.

19 Not in this decision context, at any rate; any of the management agencies could decide to fund the study to increase their science knowledge base, which would contribute to other goals.

20 Burgman, M. (2004) Expert frailties in conserva-tion risk assessment and listing decisions. In: *Threatened Species Legislation: Is it Just an Act?* (eds P. Hutchings, D. Lunney & C. Dickman). pp. 20–29. Royal Zoological Society of New South Wales, Mosman, Australia.

21 Intergovernmental Panel on Climate Change (IPCC) (1995, July) *Guidance Notes For Lead Authors of the IPCC Fourth Assessment Report on Addressing Uncertainties*. Retrieved from http://www.ipcc.ch/meetings/ar4-workshops-express-meetings/uncertainty-guidance-note.pdf

22 Hora, S.C. (2007) Eliciting probabilities from experts. In: *Advances in Decision Analysis: From Foundations to Applications* (eds W. Edwards, R.F. Miles, Jr & D. von Winterfeldt). pp. 129–153. Cambridge University Press, Cambridge.

23 If you're working with biosecurity experts, you might ask 'how many incidences of white rice infes-tation were identified at Australian ports between January 1 2010 and December 31 2010?' If you're working with terrestrial ecosystems experts, a calibration question might be 'what percentage of the world's wetlands is in Canada's boreal forests?' Have participants work through a four point elici-tation procedure (see below) and demonstrate over-confidence, and talk about ways to reduce it.

24 Speirs-Bridge, A., Fidler, F., McBride, M., Flander, L., Cumming, G. & Burgman, M. (2010). Reducing overconfidence in the interval judgments of experts. *Risk Analysis*, **30**, 512–523.

25 Walshe, T. & Burgman, M. (2010) A framework for assessing and managing risks posed by emerging diseases. *Risk Analysis*, **30**, 236–249.

26 Status was characterized in various ways, including population sizes and important demographic ratios, e.g. male:female ratios, age class ratios, etc.

27 Walker, V.R. (1996) Risk characterization and the weight of evidence: adapting gatekeeping concepts from the courts. *Risk Analysis*, **16**, 793–799.

28 Clearly this is not a hypothesis that Popper would recognize, and the word 'hypothesis' was used only for want of a better term. The intent was not to develop an H0 ready for experimental testing; an impossible luxury in this context. Rather, the purpose was to help clearly articulate a 'theory' about what constitutes the main factors contributing to ongoing recruitment failure.

9

Making Trade-Offs

By this point you've set clear objectives and performance measures, identified a creative range of alternatives, estimated consequences, and provided an explicit and honest representation of uncertainty. You've eliminated unnecessary information and alternatives that are clear 'losers'. In some cases, this will already have led to a preferred solution or broadly accepted management action. But suppose you still face complicated choices – multiple objectives, many alternatives, complicated trade-offs – and there is no obvious solution. What now?

This chapter will introduce some ways of thinking about value trade-offs and present tools to help deal with them constructively. Trade-offs are an unavoidable reality in environmental management. That shouldn't be surprising: we face them all the time in our personal lives without getting too overwhelmed. We purchase a softcover book, knowing it won't wear as well as its more costly hardcover version. We buy a smaller car because we're willing to have less leg room in exchange for better environmental performance. These are conscious choices we make and if we've reflected on them carefully, we feel good about them. The only 'bad' trade-offs are the ones we make unknowingly, or without fully appreciating their implications. The goals of SDM with respect to trade-offs are:

1 To avoid unnecessary trade-offs, by iteratively developing high quality alternatives that find win-wins wherever possible.
2 To expose unavoidable trade-offs and promote constructive deliberation about them;
3 To make trade-offs explicitly and transparently, informed by a good understanding of consequences and their significance.
4 To create a basis for communicating the rationale for a decision to a broader public.

This chapter introduces what we mean by trade-offs and why they matter. It summarizes research on how people think about them and then outlines an approach for dealing constructively with trade-offs in a deliberative environment. We end, as in other chapters, with some tips for dealing with tricky but common situations involving tough trade-offs faced by groups working on environmental management decisions.

The approaches we've chosen to highlight are not the only ways to make multi-attribute choices. You'll find many other methods in the literature and we provide some references at the end of the chapter. Our focus is on methods that we believe are at once: (a) technically sound – at least enough to be sure they won't be the weak link in your analysis; b) useful in facilitating dialogue and producing insights in a deliberative environment,

Structured Decision Making: A Practical Guide to Environmental Management Choices, First Edition. R. Gregory, L. Failing, M. Harstone, G. Long, T. McDaniels, and D. Ohlson.
© 2012 R. Gregory, L. Failing, M. Harstone, G. Long, T. McDaniels, and D. Ohlson. Published 2012 by Blackwell Publishing Ltd.

and; (c) understandable and intuitively appealing to a broad range of both technical and non-technical people. Of course, these attributes are not entirely independent. The methods we present here are also largely methods that can be implemented effectively by environmental managers without a great deal of specialized training or proprietary software.

9.1 The basics

9.1.1 What are trade-offs and why do we need to make them?

Value trade-offs involve making judgments about how much you would give up on one objective in order to achieve gains on another objective. They are inevitable in a process that develops a creative range of alternatives. Of course, a creative decision process may also deliver win–wins – gains relative to the status quo for many or all objectives. But if you've done your job up to this stage, the alternatives still on the table will each deliver a different balance across the objectives and, as a result, making a choice among them involves making value-based trade-offs.

Some people don't like the term trade-offs – it has connotations of giving up something of value, perhaps in an unfair trade. If this is a concern then you can think of it as finding an acceptable balance across objectives, or simply making tough choices. The key point is that sensible decisions need to be based on the specific amounts of anticipated gains and losses, rather than on some general statement of principles or priorities. This point is important. We almost never ask for general priorities in SDM applications, whereas it's common elsewhere – public opinion surveys, for example, often include questions such as: what's more important, the economy or the environment? For nearly all resource-management decisions, such questions are meaningless – beyond understanding the broadest notion of a person's worldviews – and any specific application of their answers may be misleading (see Text box 9.1).

Text box 9.1:   How to ask a meaningless question about what matters

Decision analyst Ralph Keeney (reported in Keeney[1]) once asked a group of television and newspaper reporters interested in the clean-up of hazardous waste sites to provide importance rankings for three considerations: economic costs of site remediation, potential human life loss or sickness due to hazards from the site, and potential damage to the environment (flora and fauna). Seventy-nine out of 80 respondents ranked human life loss or sickness first – even though, as Keeney notes, no one asked about the magnitude of the estimated human fatalities or environmental damages. He then asked them to rank the importance of spending $3 billion, avoiding a mild two-day illness to 30 people, or destroying 10 square miles of mature, dense forest. Not surprisingly, their preferences completely flipped. Obviously, spending $3 billion is more significant – we could spend $3 billion and get much more for our money in terms of better health or in terms of better environment or both. The point is that people cannot think clearly about value trade-offs without some consideration of the consequences. Yet as confirmed by many newspaper headlines that summarize survey results, people are often very willing to provide answers to these meaningless questions about which of several concerns is most important, without ever asking how much of each consequence might be involved.

The real key to making sensible value-trade-offs is to focus on the choice at hand. Not generalized priorities, not questions about hypothetical willingness-to-pay, but a real choice between two or more concrete alternatives. Recall the Stave River Water-Use Planning process from Chapter 1. In this case the central choice was between Alternative 4, which provided higher revenues and greater fish benefits (albeit modest and uncertain), and Alternative 5, which provided greater access to and protection of cultural sites in the drawdown area of the reservoir. These sites had immense cultural and spiritual value to the community. Imagine if we had asked: which is more important, money or your sacred sites? Or, how much would you be willing to pay to protect these sites? Such questions would have been at best useless, and more likely downright offensive. What we did ask was: given that a decision will be made, and given that these are the best of all the alternatives we've identified, which do you prefer? Are you willing to give up the fisheries and revenue gains of Alternative 4 in order to achieve the gains for access and protection of cultural sites under Alternative 5? This question was answerable. There are of course plenty of times when it is perfectly reasonable to ask people whether they're willing to pay for something – anytime it's not hypothetical. If a $3 million program is proposed to clean up 10 hectares of contaminated soil, then it's entirely reasonable to ask a decision maker if they're willing to pay it.

9.1.2   Are there standards for addressing trade-offs in public decision making?

In this book we focus on public decision making. We think that there are different expectations with respect to accountability for public decisions. People talk easily and naturally about the quality of information in a decision process, and there are emerging standards for what constitutes the 'best available' information, as we've discussed in Chapter 8. What about standards for the value judgments that underlie difficult trade-offs? Is it possible to talk meaningfully about the quality of a value trade-off or the assessment process that produced it?

Although it would be a stretch to say that there are clear standards or best practices for the use of value judgments in decision making, we suggest that, at minimum, practitioners should strive to ensure that trade-off judgments meet five criteria:

1 *Informed.* Trade-offs should be based on good information about the extent to which alternatives achieve objectives and performance measures and how they are likely to affect the different participating stakeholders.
2 *Context-specific.* Trade-offs should be based on context-specific estimates of predicted outcomes rather than generalized priorities or worldviews.
3 *Consistent.* If Alternative A is preferred to Alternative B, and B is preferred to C, then either A should be preferred to C or a clear rationale should be provided. This transitivity should also hold for direct expressions of consequences.
4 *Stable.* Expressed trade-offs should endure beyond a single situation. In novel decision-making contexts this likely involves giving attention to both cognitive and intuitive modes of thinking and requires internal reflection about expressed values.
5 *Transparent.* The rationale for value trade-offs should be openly communicated.

This last point can raise problems in some contexts, especially for trade-offs that are highly controversial, but from a prescriptive perspective, a commitment to good governance of natural resources demands nothing less.

SDM methods challenge people to be disciplined in their thinking, not just about consequences but about values and preferences. In a public decision-making context, we think participants carry an extra burden to demonstrate that they have considered the problem from multiple perspectives, that their preferences are consistent, that trade-offs have been addressed explicitly, and that the rational underlying a choice or recommendation is transparent.

9.1.3   How people think about trade-offs

In Chapter 2, we introduced the idea that people employ two 'systems' to help them make complex choices. System 1 is the experiential or intuitive system. Fast, automatic, and effortless, the experiential system generates *impressions*. These impressions are not always explicit, nor are they necessarily intentional. System 2 on the other hand, is cognitive or analytical, and generates *judgments*. Judgments are usually explicit and always intentional, the product of a slower, more controlled and effortful reasoning process.

Intuitive decisions are affected by emotional and psychological factors that influence the accessibility of different features of the situation – how naturally, quickly, and easily these features come to mind. Loewenstein *et al.*[2], for example, discuss the importance of anticipatory emotions, such as fear or anxiety, their role as informational inputs into decision making, and the negative consequences that can result if these inputs are blocked (for example, through poor facilitation). Unfortunately, there is no reason to believe that the most accessible features are also the most relevant to a good decision, and as Nobel laureate Daniel Kahneman[3] (p. 481) notes, the 'highly accessible impressions produced by System 1 control judgment and preferences, unless modified or overridden by the deliberate operations of System 2'.

If the intuition-based System 1 sometimes overrides the cognitively-based System 2, largely because of accessibility, then it seems important to understand what it is that might make some considerations relatively more or less accessible in the context of trade-offs. For present purposes (and acknowledging the more detailed discussion in Chapter 2), we limit our focus to identifying four factors that might prominently affect individual choices and group deliberations about trade-offs.

'Framing effects' describes how our choices can vary in response to presumably irrelevant variations in how questions are posed. In a classic study, McNeill *et al.*[4] tested individuals' preferences for surgery versus radiation therapy, under two frames. In one, surgery outcomes were described in terms of survival rates (90% survival). In the other, they were described in terms of mortality rates (10% mortality). Although the descriptions are technically equivalent, the good doctors

at Harvard exhibited a significantly higher preference for recommending that their patients undergo surgery when outcomes were reported using the decision frame emphasizing survival. Decades of effort to reduce or avoid framing effects have yielded decidedly mixed results (see Kahneman and Tversky[5]).

Looking at such framing effects more generally, prospect theory[6] predicts that expressed preferences over alternatives will be (a) more sensitive to changes and differences from a reference point than to absolute levels of an effect, and that (b) anticipated losses will be valued more highly than anticipated gains. As a result, knowing that a proposed action is estimated to bring an additional $2 million into a community from one source (e.g. increased tourist revenues) may elicit less response than learning that the same action is also likely to result in a loss to the community of $1.5 million relative to the current situation – and despite the overall presumption of gain, the action is likely to be opposed. Two implications follow for the assessment of trade-offs in the context of environmental management decisions. First, the simple ratio of benefits to costs may be misleading, to the extent that losses are valued more highly than the potential gains[7]. Second, the selection of the base case, or reference condition, is critical to the valuation of many proposed environmental management actions. Consider a proposed improvement in the water quality of a river running near an urban center. If conditions under the proposed clean-up are described relative to current conditions, they would appear as a gain (relative to the status quo) and have a moderate positive value. If the base case is instead a historical condition, then the new condition may be viewed as restoring a loss, in which case the value would be higher and decision makers may be willing to spend substantially more on clean-up plans[9].

Affect refers to both the immediate perception of a thing as good or bad and to subsequent feelings or mood states, such as calmness or excitement[9]. Affective responses are powerful determinants of choice, yet decision makers or analysts cannot always accurately predict which stimuli will be perceived affectively as good or bad by participants in a decision process; although some stimuli are almost universally received as good (e.g. happy newborns) or bad (e.g. dead animals), many are subject-specific and different people will perceive them in different ways[10]. For example, some participants may perceive that information from an industry or academic source is highly credible, whereas others may immediately consider them as 'tainted' or 'bad'[11]. If a consequence is presented with pictures or descriptive stories that elicit an emotional response, people may be more strongly motivated to choose alternatives that reduce these effects.

Judgmental heuristics (or mental short-cuts) include among others (from Chapter 2) anchoring, representativeness, and overconfidence. The bottom line is that when people are faced with a complex choice, particularly one that is new or that evokes strong emotions, they are likely to adopt simplifying strategies that may or may not help them make good choices. For example, faced with complicated trade-offs over multiple objectives, people routinely simplify things by focusing on only one of the objectives (sometimes called the prominence effect). Similarly, people tend to remember information that confirms what they already believe and tend to forget information that counters their beliefs. So in the absence of structure, and when making difficult choices as part of a multi-dimensional decision, important information relevant to assessing trade-offs across alternatives may be ignored.

So where does all this leave an environmental manager working with diverse stakeholders on a problem involving tough trade-offs? It seems clear that a participant whose strong emotional response to a proposed action effectively blocks further information inputs is likely to miss something important – this can't be good. And the sensitivity of preferences to irrelevant or inadvertent changes in framing is certainly troubling. But the effects of other factors are not so clear. Is it wrong to be more sensitive to losses and gains than absolute values? Or to value losses more than

gains? Are our affective responses to various alternatives and their consequences necessarily wrong or misleading, or could they be alerting us to important aspects of a decision that haven't been addressed by carefully prepared analyses? The important message for people involved in deliberations about complex environmental problems is that both our immediate intuitions and our more considered judgments have something to offer in terms of providing useful insights, and a defensible approach to addressing trade-offs will need to encourage and incorporate inputs from both.

What does all this suggest? Some elements of good practice include:

1 Use multiple framings to guard against affective responses and arbitrary effects related to losses and gains.
2 Use multiple elicitation methods to draw on both cognitive and intuitive systems of thinking and to draw on both individual and group thinking.

9.1.4 Taboo trade-offs

Are there any 'taboo' trade-offs - things we cannot compromise on or compensate for? Definitely: but perhaps not as many as we think. As already noted in this chapter, the key is to frame difficult choices as exactly that – choices. Not as valuation exercises or generalized priority setting exercises, but as choices between alternative actions or sets of actions. By and large, the approaches we use in this book assume that trade-offs can be addressed – that poor performance on one objective can be compensated, at least in part, by better performance on another. This doesn't preclude the possibility that firm thresholds exist and there may be some minimum level of environmental performance (water quality, forest loss, species protection, etc.) beyond which further trade-offs are not acceptable, and in some cases, declarations of taboo trade-offs will be at the center of disputes about how an environmental assessment should be undertaken.

An example comes from work with the national Environmental Risk Management Authority (ERMA) of New Zealand, which is responsible for reviewing applications for the importation of biotechnologies. We were part of a group asked to assist with the development and testing of a broadly acceptable decision-making protocol that would take both tangible and intangible effects of biotechnologies into account, and to do so in a way that encourages a productive dialogue encompassing human communities (with an emphasis, in the New Zealand case, on the country's indigenous Maori population), the natural environment, health, and culture[12].

An SDM approach was introduced to create a risk framework for biotechnologies responsive to stakeholder values as an aid to ERMA decision makers. The debate centered on questions of perceived taboo trade-offs and whether Maori spiritual beliefs should be allowed to override the potential economic benefits of biotechnologies. The context already was controversial due to government handling of an earlier decision, later blocked by a government-imposed moratorium, to field trial GMO dairy cattle whose DNA had been modified to produce a human protein (in the milk of adult cows). Maori concerns were complex and fundamental; cornerstone values of Maori culture were considered to be at risk, in particular a complex of metaphysical properties (including the concepts of *mauri, wairua, tikanga*, and *whakapapa*; see Roberts *et al.*[13]) that express the potency of all beings or entities in the human and non-human world.

Our starting point was to outline a risk framework for helping the multiple parties to engage in a decision-oriented dialogue. We based our approach on the existing process followed by ERMA but modified it in several important ways (see Figure 9.1), including the explicit identification of both physical effects (common and easily measured) and metaphysical or spiritual effects (less easily measured but important).

When applications involving the use of biotechnologies are received, they will be red-flagged if they have potential cultural or technical concerns. These applications are then reviewed by a multi-party committee. If they are deemed to violate one of the essential principles, they are rejected ('no-go'). If provisionally accepted, they continue on to a full multi-attribute evaluation using objectives and performance measures that include both physical and metaphysical effects that address the question 'do the expected negative impacts to society outweigh the positive benefits'? Figure 9.1 shows only the first, information-gathering phase of a proposed four-part evaluation process; subsequent phases involve detailed deliberations with Maori and other participating stakeholders and a final project evaluation and recommendation. Monitoring and periodic reviews were also built into the decision framework, with their profile and scale dependent on the novelty and significance of the proposed project.

9.2 Working with trade-offs

The trade-offs faced in environmental management are as complex as they come, and they are made in a deliberative environment marked by multiple world views, technically intensive debates, strong emotions and difficult group dynamics. It's a tough environment, hardly the conditions of 'idealized reflection' that philosophers and decision theorists propose for making sound value-based choices. It should come as no surprise that people involved in such choices rely on both intuitive and cognitive systems for constructing value judgments and expressing their preferences. It's typical to anchor on some bits of information and ignore others, according to what is most salient or accessible. It's typical – for any stakeholder, whether government minister or local politician, First Nation elder or high school senior, academic expert or community resource user – to be influenced by framing effects, overconfidence, perceptions of losses and gains, affective responses and emotions, often in ways that aren't recognized and that we have no reason to believe would

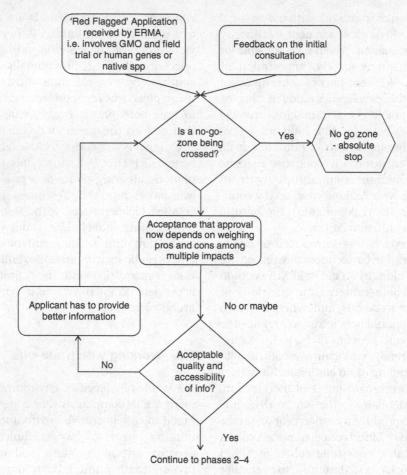

Figure 9.1 New Zealand risk framework, Phase 1 example (information gathering).

support good decisions. An over-reliance on intuitive systems is clearly troubling for decision quality. On the other hand, the 'economist' and 'decision analyst' scenarios from Chapter 1 demonstrate what happens when we overemphasize the rational and controlled judgments of the cognitive system and try to suppress the intuitive system. A solution determined by a model or formula appeals to the technocrat in all of us, but presumes that our model is appropriate and complete, which of course it can't always be.

In this section we outline an iterative approach to working through trade-offs that builds on both the prescriptive advice of decision analysis, about how best to make informed choices, and practical knowledge that stems from observing how people actually think about real-world environmental management choices. We emphasize a multi-stakeholder, multi-interest, multi-meeting decision context, although many of the same steps apply to a decision faced over the course of a single afternoon by a small group of colleagues within the same resource-management agency.

The approach consists of:

1 Eliminating dominated alternatives and insensitive measures.
2 Comparing pairs of alternatives and eliminating practically dominated ones.

3 Using multi-attribute trade-off methods to provide insight about value trade-offs.
4 Identifying areas of agreement and disagreement.

9.2.1 Eliminating dominated alternatives

As we describe in Chapter 7, participants in SDM processes iteratively develop and refine alternatives. Their consequences are estimated using some combination of predictive models and expert judgment, resulting in a consequence table. First, get rid of alternatives that are clearly 'dominated' – that is, outperformed on all performance measures by one or more of the other alternatives. Provided all aspects of interest have been captured in the set of performance measures, there is no reason why anyone should object to the removal of these inferior alternatives. At the same time, it's useful to eliminate insensitive performance measures – measures that don't vary across the set of alternatives under consideration.

Using dominance to simplify a set of alternatives is usually uncontroversial. As long as the list of objectives and measures is complete and participants accept the estimates of consequence, no one should be supporting an alternative that's dominated by another. Removing objectives and measures due to insensitivity can be a little trickier. Objectives are there because they were identified as being important, and it's not uncommon for people to object to their removal. However, removing an insensitive objective doesn't mean it isn't important in a general sense. In fact, that objective may have played a vital role in eliminating unacceptable alternatives at earlier stages of the process. But once a shortlisted set of alternatives is on the table, objectives that don't help to discriminate among them should be removed.

9.2.2 Exploring trade-offs with 'practical dominance'

Once alternatives that are 'strictly' dominated are eliminated, it's important to look for practically dominated alternatives. This involves comparing alternatives, usually two at a time, and determining whether one can be said to outperform the other on balance, in consideration of its performance on all measures. The outperformed alternative then is said to be practically dominated. This step clearly involves value judgments. For groups that have been involved in a structured decision-making process from the start – setting the objectives and performance measures, exploring preliminary alternatives, identifying the experts and methods that will provide consequence estimates – it's often surprisingly straightforward for a group to reach agreement about at least some of the key trade-offs in a decision. However, you are now clearly outside the zone of technical analysis and into a value-based discussion – facilitation skills matter. There are probably lots of ways to do this, but here's how we usually approach it.

Start by identifying an increment of change that will be considered to be a 'significant' difference in estimated consequences. For example: if the best estimate of flooded cropland is 1000 hectares under one alternative and 1150 hectares under another, is this incremental change considered to be significant? The answer is typically a function of the sensitivity of the outcome under consideration, the accuracy of measurement and models used to predict the consequences, and the value judgments of participants about what matters. In one case the difference across alternatives might have to be 20 or 30% before it would be considered significant; in other cases, it's around 5–10%. As soon as you establish this increment of change, you may find some more alternatives that are strictly dominated.

Begin with the alternatives that differ significantly on only two or three measures, if possible. When comparing, focus on the specific trade-offs – the amount of the difference in the value of the performance measure across alternatives. Consider, for example, two alternatives that are not significantly different except for two measures: alternative A delivers 100 more hectares of critical habitat for a species, whereas B 500 meters more of riparian habitat. If everyone agrees that A is preferred to B, then B is practically dominated, and can be removed from further consideration.

If an obvious winner doesn't emerge, you can check whether 'even swaps' can help you to simplify further. Even swaps is a method that expresses the effects of one measure in terms of another and can sometimes be helpful during paired comparisons – see below.

Before you eliminate the inferior alternative, check whether there's anything unique and valuable about it that could be incorporated into another alternative (assuming you're still in the refining stage).

Every time an alternative is removed, check to see if any performance measures have become unnecessary. If there is one that no longer varies across the remaining alternatives, it can be removed from further consideration.

So far, we've been describing a situation in which a group is working together and making choices through discussion. However, it's important to be on the lookout for the effects of group-think – participants may feel pressure to go along with a proposed solution because they don't want to be the spoilers of an emerging agreement. If that's a concern, it's possible to design individual exercises to test for practically dominated alternatives and then compare the individual results to see if there's any agreement on practical dominance. This, however, takes a lot more time, and typically, we do this as a group.

It's useful to test periodically for interim agreement. It isn't always necessary to work through every single pair of alternatives. Instead, it can be useful to step in and ask a group whether they're starting to see a promising solution, one that everyone might be willing to support.

There are several possible outcomes from the comparison process. Sometimes, clarity about a preferred alternative is quickly gained, and the process ends with its selection. At other times, the paired comparison process will serve as a precursor to a more detailed trade-off analysis. It eliminates alternatives that the group agrees are inferior and simplifies the set of alternatives put forward for detailed analysis. Difficult value judgments on the short-listed alternatives can then be informed by the use of more sophisticated trade-off tools and methods.

Text box 9.2:   Simplifying trade-offs through even swaps

The authors of *Smart Choices* have coined the term 'even swaps' for describing how a focus on trade-offs can simplify complex decisions[15]. The key is to think about the value placed on one objective in terms of another. Suppose that the objectives driving a decision about protecting a community from wildfires include evacuation times, management costs, tourism revenues, and forest health. Using the now-familiar consequence table, paying attention to potential even swaps focuses attention on the key trade-offs and, by using a common currency for expressing different impacts, seek to simplify decisions by deleting dominated alternatives. Suppose a review of the wildfire consequence table shows that two of the four alternatives vary on only two dimensions, with Alternative A showing lower expected evacuation times and Alternative B lower estimated costs. A focus on even swaps would ask: what reduction in evacuation times will just cancel out the cost advantage of Alternative B? This re-framing of the key trade offs question in terms of the same units (either hours or dollars, in this case) may then lead to the elimination of either A or B as a dominated alternative. Although highlighting even swaps doesn't introduce anything new – we've previously discussed the importance of simplifying the decision structure – it serves as an intuitive reminder of good structuring practice: by making the easier swaps first, and by making all swaps consistently, complex decisions can become simplified and attention focused on the value trade-offs that matter most.

Every time we write this method up, we're disappointed by how mundane it sounds. However, over the years this process has led to more tide-changing 'a-ha' moments than any other tool or method we use. Perhaps because it comes at the end of a well structured process, it's the moment when decision makers, or stakeholders working together to help inform decision makers, really get that moment of clarity and heave a big sigh of relief. It doesn't look like much, but don't underestimate just how powerful it is.

The best way to illustrate this is with an example[14]. The first case-study in section 9.3 provides a detailed example of how this multi-method approach to clarifying trade-offs can be used to aid deliberations in a multi-stakeholder context.

9.2.3 Using multi-attribute trade-off methods

If the paired comparison of alternatives has led to a decision, then an explicit quantitative weighting isn't necessary. But where an obvious solution doesn't begin to emerge, sometimes quantitative methods can help. Typically they help in three ways:

1 Improving the quality and consistency of individual judgments.
2 Moving discussions from positional to performance based.
3 Clarifying where dialogue will be most useful.

Many such methods exist to formally model values and express trade-offs. The key to success in a multi-stakeholder deliberative context is to use a method that is:

1 Technically defensible (rooted in multi-attribute utility theory).
2 Understandable to participants.
3 Quick and easy to use so that it can be applied interactively.
4 Able to produce results that help to focus group deliberations in productive areas.

In the remainder of this section, we describe an approach we use to help groups gain insight about value-based choices by using a combination of direct ranking and value modeling methods. Direct ranking assigns performance scores to the alternatives directly, and a value model calculates performance scores by assigning weights to the measures. We turn first to value modeling.

Value modeling

Recall from Chapter 2 that SDM is based on the concept of linear value modeling. Mathematically, a linear value model represents an alternative's performance score as the weighted sum of its consequences:

$$\text{Overall score} = W_1X_1 + W_2X_2 + W_3X_3 + \cdots\cdots$$

where X_1 is the score assigned to measure 1, W_1 is the weight or importance assigned to measure 1, X_2 is the score assigned to measure 2, W_2 is the weight of measure 2, and so on.

In order to calculate a meaningful overall score when individual performance measures are recorded in different units (e.g. dollars, hectares, jobs, etc.), the individual performance measure scores must first be 'normalized' and then weighted using some reputable method. Generally, the steps of quantitative value modeling involve:

1 Define the objectives and measures.
2 Identify the alternatives.
3 Assign consequences (or consequence scores) to each alternative on each measure.
4 Assign weights to the measures.
5 Normalize the consequence scores so that they can be aggregated.
6 Calculate the weighted normalized scores (sometimes called 'desirability' scores) to rank alternatives.

As we described in Chapter 2, the steps of SDM reflect a value modeling approach. You have a set of objectives, measures, alternatives, and estimated consequences. If you've followed the advice

Text box 9.3: An additive value model for siting a ski facility

Imagine that you're deciding between three possible sites for a cross-country ski facility and you need to assess the desirability of the ski conditions. You've identified three important objectives and corresponding measures to define desirability: average annual snow-fall (in cm); feasible trail length (in km); and topographical variety (m of elevation change per km^2). In this case, for each of the measures, higher numbers are better.

Three alternatives have been identified, imaginatively called Sites 1, 2, and 3. The table below summarizes the performance of each alternative on each of the identified measures.

	Site 1	Site 2	Site 3
Average snow fall (cm/y)	400	450	200
Trail length (km)	100	50	80
Topographical variety (m/km)	15	70	50

To select a preferred site, you need to address trade-offs by assigning weights to the measures. Using a reputable weighting method, suppose you assign weights as follows: average snow fall 50%, trail length 30% and topographical variety 20%.

When measures are reported in different units, they have to be *normalized* before they can be combined in a value model. Normalizing recalculates performance scores as a unitless number between 0 and 1, and in this case we want 0 to correspond to the worst value across the range of alternatives and 1 to the best. The normalized score for average snowfall for Site 1 then is:

$$\text{Normalized score} = \frac{\text{score} - \text{worst score}}{\text{best score} - \text{worst score}} = \frac{400 - 200}{450 - 200} = 0.8$$

Normalized values for all three of the measures are as follows:

	Weight	Site 1	Site 2	Site 3
Average snow fall	0.5	0.8	1	0
Trail length	0.3	1	0	0.6
Topographical variety	0.2	0	1	0.64
Weighted sum		0.7	0.7	0.31

The weighted sum for Site 1 = 0.5 (0.8) + 0.3 (1) + 0.2 (0) = 0.7.

According to your value model, you would be indifferent between Sites 1 and 2 but both are preferred to Site 3, at least from the perspective of the desirability of the ski conditions. (Other factors such as cost, community support, estimated demand, ecological implications, etc. would likely be factors in your choice as well.)

in structuring objectives and measures in Chapters 4 and 5, you've set things up so that you can use a simple linear value model to calculate the desirability of the alternatives under consideration. As you've worked through the comparisons of pairs of alternatives looking for practical dominance, you've still been using this value model, but qualitatively: You've been looking at consequences, and assigning 'importance weights' to them implicitly in order to make trade-offs. In contrast, when you use a quantitative value modeling approach, you assign the weights explicitly and use them to calculate the desirability of the alternatives.

In order to make a decision involving trade-offs, decision makers *always* assign weights. The only question is whether they do it implicitly or explicitly. If they don't do it explicitly, then they are using some kind of mental model to assign a qualitative importance to each consequence and they attempt to mentally integrate across all types of consequences to construct their preferences. This is what's going on implicitly in the practical dominance procedure described above. It's practical and not without its merits as the quality of the dialogue tends to be very high. But it is nonetheless important to realize that when choices are made without explicit assignment of weights there's some loss of rigor and transparency, and some potential for loss of quality due to group effects and lack of consistency checks.

The simple example in Text box 9.3 illustrates the math that underlies the calculation of alternatives' performance scores using a linear value model. The question then is where do the weights come from?

There are many methods for assigning weights. The key is to use one that is simple, intuitive, and reputable. Not all of them are. To be useful and defensible, the weights assigned to performance measures must consider both the inherent importance of the fundamental objective the measure represents, and, critically, the range of possible consequences across the alternatives

under consideration. Are fish more important than birds? It depends. If fish abundances are expected to change greatly and bird abundances only slightly, then the fish measure (in this case) may be more important than the bird measure.

A full review of weighting methods is beyond the scope of this book. We use swing weighting, which is described in detail in Edwards and von Winterfeld[16]. It's simple and intuitive, and generally seen to minimize some of the problems associated with other methods. The first case study in section 9.3 demonstrates its use.

An important consideration is what we do with those weights. In general, we don't advocate aggregating weights across participants, at least not in a multi-stakeholder setting. The main exception is when the decision maker needs a *system* for making choices rather than a single discrete choice. In such cases, it may make sense to work toward consensus on weights for individual measures, and then use calculated scores to rank alternatives. This can work well when decision makers share a similar value system – especially for example if they are all from the same organization. However, whenever there are multiple stakeholders with very different perspectives and values working together on controversial, high stakes decisions, we doubt that any attempt to use average weights or to reach consensus on the weights assigned to measures will be productive.

When we use quantitative value modeling in a multi-stakeholder setting, we build a value model for every participant – everyone uses the same objectives, measures, and consequence scores, but they apply their own weights, and calculate their own individual preference scores for the alternatives.

A multi-method trade-offs approach

If we use quantitative multi-attribute methods to explore trade-offs, we often adopt a multi-method approach for either of two reasons. First, there is no 'correct' approach: as we discuss in detail later in this section, all methods involve helping individuals to understand and articulate their preferences and all have strengths and pitfalls[17]. Second, using two methods and then comparing the results is a good way to make sure that quantitative methods are used to aid, rather than prescribe, decision makers' choices.

Our typical use of a multi-method approach involves direct ranking and weighting techniques. Direct ranking asks participants to rank alternatives directly, based on a review of the consequence table and a holistic consideration of the trade-offs. A weighting method (such as swing weighting) asks participants to assign explicit weights to the net change in performance measures across the alternatives, with results then used to calculate performance or desirability scores for the alternatives. Findings from both approaches can be presented to a group and used to focus deliberations on key areas of agreement and disagreement. After discussion, participants are asked to reconcile the two methods and provide a final ranking (or a judgment about a preferred alternative or set of acceptable alternatives). These steps are summarized in Table 9.1.

Both methods lend insight that help stakeholders and decision makers to understand the key trade-offs. But both have pitfalls when used on their own. Pitfalls of a direct method include:

Table 9.1 Steps in a generalized multi-method trade-offs approach.

1 The group reviews and confirms understanding of the objectives, measures, alternatives, and consequences
2 Each participant directly ranks and scores the alternatives
3 Each participant weights the performance measures; scores and ranks are then calculated for each alternative for each participant
4 Each participant reviews their individual results and examines inconsistencies across the two methods
5 The group reviews aggregated results. Areas of agreement and difference among individuals are identified and discussed
6 Each participant provides a final ranking or preferences, based on what they've learned
7 The group clarifies key areas of agreement and disagreement and provides reasons

1 It's vulnerable to cognitive biases and errors. The more complicated the decision, the more likely people are to use simplifying rules of thumb. A common problem is anchoring on a single measure – the one considered *a priori* 'most important' by that individual– and choosing the alternative that best supports that measure, without really giving adequate consideration to other measures.
2 It is vulnerable to positioning. Participants in a decision process often come to the table with a particular solution in mind. Asking them to rank the alternatives does little to facilitate a performance-based dialogue that could encourage participants to move off their positions.
3 It's vulnerable to misinterpretations of technical findings – in the Bridge River case study (discussed later in this chapter), participants initially highly weighted a 'flood' measure because they misunderstood its consequences.
4 It's sensitive to the personalities and communication skills of the people around the table. This can fail to produce socially desirable outcomes (there being no logical connection between the communication skills of participants and the social value of outcomes). There may also be procedural equity and representativeness issues, as those who are outgoing and outspoken may dominate the discussion. And ultimately, the process can be held hostage, particularly if consensus is expected, by uncompromising personalities.

Performance measure weighting methods also have potential pitfalls:

1 They are sensitive to the choice and quality of measures and will underemphasize impacts that are not well captured in the measures.
2 They can be sensitive to how measures are structured (especially whether there are multiple measures and/or whether they are structured hierarchically or not), the extent to

which non-linearities have been addressed, and the choice of a minimum significant increment of change.

3 They may be insensitive to emotions and intuition. Emotional responses often tell us something important about what's missing in our analyses[2]. Although indirect methods may facilitate agreement in the course of a meeting, such agreements may be reversed when participants return to more normal environments[18], especially if these feelings are a signal of some larger issue that was not captured in the analysis.

Decision makers don't want to – and shouldn't – hand over the decision-making authority to a formula. But we do think that methods involving weighting identified performance measures are helpful because they shift the nature of the dialogue away from positions and toward performance based or value-based deliberations. Weighting offers new ways of thinking about what matters, which yields insights and helps to expose oversights and inconsistencies.

The best way to illustrate these points is with an example. The first case study in section 9.3 provides a detailed example of how this multi-method approach to clarifying trade-offs has been used to aid deliberations in a multi-stakeholder context.

A comment on other methods

There are many other ways to explore trade-offs, and we will make no attempt here to provide a comprehensive review. For example, pair-wise comparisons of consequences can provide another approach that's rigorous and understandable; the approach is discussed in Keeney[1] and in Champ, Boyle, and Brown[19]. When managers are faced with one-off choices which involve very high stakes and significant uncertainties, there is often no substitute for formal one-on-one interviews with individual stakeholders or decision makers to review their

understanding of key consequences and to elicit value trade-offs. As always, the context for the decision-making process and the ways in which information on trade-offs will be used need to be taken into account. The methods we suggest are useful and appropriate for a large majority of environmental management problems involving multiple stakeholders, high technical complexity and significant but not catastrophic stakes. They're intended to facilitate discussion in a deliberative setting and to be accessible to an environmental manager who is working with colleagues and other stakeholders to develop a defensible management strategy.

9.2.4   Clarify areas of agreement and disagreement

We've said that consensus should not be a goal of a decision process. How then do we document areas of agreement and disagreement? And what's a decision maker to do with a non-consensus outcome?

First, define several levels of agreement. Many scales for levels of support are found in the facilitation literature. To keep things simple but flexible, we often define at least three levels:

1 Endorse = Enthusiastic support. 'This is a great solution'.
2 Accept = Support. 'Maybe not the best solution in my mind, but it is one I can support'.
3 Oppose = No support. 'I cannot support this solution'.

We define a 'consensus' decision as any alternative that is either endorsed or accepted by all parties.

Sometimes it's also helpful to elicit degrees of support for short-listed alternatives. If after discussing trade-offs you reach a point where there are two to three alternatives remaining on the table and no clear resolution, ask participants

Table 9.2 Documenting support for alternatives.

	Number who endorse	Number who accept	Number who oppose	Main reasons for support	Main reasons against
Alternative 1	III	IIIII	II	Protects aquatic habitat, wildlife and local employment	High cost
Alternative 2	IIIIII	IIII		Good balance across objectives	None
Alternative 3	II	IIIIIII	I	Supports local employment and recreation	Uncertainty about aquatic residual impacts

to indicate their degree of support for each shortlisted alternative along with a concise rationale for his/her choices. You'll end up with a table that looks something like Table 9.2.

Participants can endorse none, one, some, or all of the alternatives. Normally they won't oppose them all. Often someone will abstain, either entirely or from indicating support for a particular alternative. All of these responses: it's a good summary to take to decision makers.

Finally, ask participants to suggest ways to reach agreement. If only a few stakeholders are opposed to a particular alternative, ask 'what would it take to make this alternative acceptable to you'? Or ask supporters: given the concerns that you've heard, can you suggest a modification to this alternative that might make it acceptable to others? How far to go with this line of questioning depends on the decision context. If you truly need consensus (any kind of shared decision-making situation, such as two sovereign nations negotiating a transboundary agreement, for example) then you may spend quite a bit of time in this territory. But for most environmental management decisions, what managers and decision makers need is a good understanding of how stakeholders feel about the trade-offs, and you probably have that by now. The key point is to be explicit about the bottom line. Most of the time it's not consensus, it's a clear idea about who supports or is opposed to an alternative and why they feel this way.

9.3   Case studies

Here we present three case studies. In the first, we return to the water use planning context that we discussed in Chapter 1. This time we present one of the water use plans that went on to use formal trade-off analysis methods. We walk you through some of their deliberations, how they simplified the problem using dominance and practical dominance, and then how they applied a combination of direct and weighting methods to gain insights about values that led to clarity in the selection of a preferred alternative. This first case is lengthy. The next two are shorter and describe respectively a case when consensus was not reached but SDM succeeded in weeding out inferior alternatives, and a case using a more formulaic approach to trade-offs, with uncertain weights explored through sensitivity analysis.

9.3.1   A multi-method approach to value trade-offs on the Bridge River

As discussed in Chapter 1, BC Hydro's Water-Use Planning program is a multi-stakeholder multi-objective planning process to examine water flows at hydroelectric facilities throughout British Columbia, Canada. For one plan, involving the operation of three reservoirs at BC Hydro's Bridge River facilities, stakeholders representing the provincial treasury board, provincial and federal fish and wildlife regulators, local residents, and aboriginal communities participated in a structured decision process to examine alternative ways of operating the facilities to better balance power, fish, wildlife, water quality, and recreation interests. Over a period of 24 months, participants set objectives and measures, identified and evaluated alternatives, and addressed a difficult set of trade-offs (for a more detailed account, see Failing *et al.*[20]).

Structuring objectives, measures, and alternatives

Participants began by setting objectives and measures. Over 30 performance measures were initially defined and used in preliminary analyses. Some of these were later combined into indices, whereas others were eliminated early on because they were either insensitive to the alternatives or strongly correlated with another measure. Numerous information gaps were identified. The measures were used to prioritize proposed studies, a task that was conducted collaboratively by the committee[21]. Upon completion of the studies, the measures were refined and used to assess alternatives. The committee worked through several rounds of alternative generation and evaluation.

Location-specific objectives and performance measures were organized into six fundamental objectives. The objectives were:

1 *Flooding.* Minimize adverse effects of flooding on personal safety and property (reported as number of days/year that the most vulnerable road bridge would be inundated).
2 *Fish.* Maximize the abundance and diversity of fish in all parts of the system (reported initially by proxy measures and later by a fish index that was constructed by fisheries experts).
3 *Water supply.* Preserve access to and maintain the quality of water for domestic supply (reported by total suspended solids in tonnes/year).
4 *Wildlife habitat.* Maximize the area and productivity of wetland and riparian habitat (reported by an index that weights different types of habitat according to its benefits for wildlife).

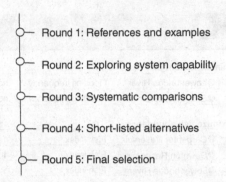

Round 1: References and examples

Round 2: Exploring system capability

Round 3: Systematic comparisons

Round 4: Short-listed alternatives

Round 5: Final selection

Figure 9.2 Five rounds of alternatives.

5 *Recreation and aesthetics.* Maximize the quality of the recreation and tourism experience (reported by the number of days of boat access).
6 *Power.* Maximize the value of power produced at the Bridge River facilities (reported by annual revenues in levelized $ million/year).

The design of measures was an intensive process. Care was taken to ensure that they could, if necessary and helpful, be used in formal preference assessment. Indices or constructed scales (described in Chapter 5) were particularly useful. For example, three proxy measures for impacts on fish in the reservoirs (littoral productivity, tributary access and stranding) were individually weighted and combined into a normalized index indicating the overall utility for fish of each operating alternative. Similarly, a wildlife habitat index was developed that estimated total habitat area, with different types of habitat weighted differently due to their different contribution to key wildlife. These context-specific constructed scales, initially novel, became familiar to participants through repeated use.

Using a value-focused approach, participants worked through several rounds of alternatives, as in Figure 9.2.

Dominated and practically dominated alternatives

The committee next worked through the process of eliminating dominated alternatives and

Objective Sub-objective	Performance measure	Units	Dir	Alt 1	Alt 2	Alt 3	Alt 4	Alt 5	Alt 6	Alt 7	Alt 8
Minimize flooding											
@Lower bridge River	Flooding frequency	no. days / year	L	1	1	0	0	0	6	7	0
@Seton Reservoir	Flooding frequency	no. days / year	L	6	6	6	6	6	12	5	6
Maximize fish abundance											
@Carpenter Reservoir	Fish index	1→100	H	69	70	41	41	29	32	25	29
@Downton Reservoir	Fish index	1→100	H	42	71	48	69	65	55	22	69
@Lower bridge River	Fish index	1→100	H	100	100	100	90	25	100	25	10
@Seton Reservoir	Fish index	1→100	H	66	66	66	66	33	55	42	10
Maximize water quality											
@Seton Reservoir	Water suspended solids	Tonnes / y	L	94	89	77	84	108	110	125	78
@Carpenter Reservoir	Water suspended solids	Tonnes / y	L	80	85	88	83	85	89	88	88
Maximize vegetated area											
@Downton Reservoir	Weighted area	Hectares	H	223	231	322	313	295	280	220	300
@Carpenter Reservoir	Weighted area	Hectares	H	759	522	758	520	602	510	500	600
@Seton Reservoir	Weighted area	Hectares	H	48	48	48	48	48	40	60	33
Maximize recreation											
@Carpenter Reservoir	Boat access	no. days / y	H	90	89	92	90	88	90	100	86
@Seton Reservoir	Boat access	no. days / y	H	119	118	122	122	124	120	122	120
Maximize power benefits											
Maximize power revenues	Revenues	Levelized $ M y	H	141	145	146	149	144	122	110	145

Figure 9.3 Early consequence table for Bridge River deliberations. The details of the dominance and practical dominance sequence have been adapted and reproduced to facilitate a more clear and concise illustration.

insensitive measures (as described in section 9.2). Figure 9.3 shows a consequence table with six fundamental objectives (some with more than one sub-objective) and eight alternatives. The sub-objectives refer to different reservoirs and rivers in the interconnected hydroelectric system.

The first two columns show the objectives and sub-objectives. The third, fourth, and fifth show performance measures, units and preferred direction respectively (H = higher numbers are preferred; L = lower numbers are preferred). Using this consequence table, a consultative committee compared each alternative in terms of its predicted consequences, facilitated by the use of an Excel spreadsheet which uses conditional color coding to highlight key trade-offs and facilitate the comparison and refinement of alternatives.

In Figure 9.3, the 'Alt 6' alternative is selected (shown as a dark grey shade). Where another alternative performs better than the Alt 6 in terms of a specific performance measure, the corresponding cell turns to a light grey. Black cells, in contrast, indicate poorer performance relative to Alt 6, and no shade indicates 'no significant difference' relative to Alt 6 for that performance measure. What constitutes a significant difference varies by performance measure, but the group decided that for most of these measures, a change of less than 10% would not be considered significant. This reflects an expectation that such error is within the level of error of the data and models used to estimate consequences. The exceptions were power and recreation measures, for which the data and models are considered quite

Objective Sub-objective	Performance measure	Units	Dir	Alt 1	Alt 2	Alt 3	Alt 4	Alt 5	Alt 7	Alt 8
Minimize flooding										
@Lower bridge River	Flooding frequency	no. days/year	L	1	1	0	0	0	7	0
@Seton Reservoir	Flooding frequency	no. days/year	L	6	6	6	6	6	5	6
Maximize fish abundance										
@Carpenter Reservoir	Fish index	1→100	H	69	70	41	41	29	25	29
@Downton Reservoir	Fish index	1→100	H	42	71	48	69	65	22	69
@Lower bridge River	Fish index	1→100	H	100	100	100	90	25	25	10
@Seton Reservoir	Fish index	1→100	H	66	66	66	66	33	42	10
Maximize water quality										
@Seton Reservoir	Water suspended solids	Tonnes/y	L	94	89	77	84	108	125	78
Maximize vegetated area										
@Downton Reservoir	Weighted area	Hectares	H	223	231	322	313	295	220	300
@Carpenter Reservoir	Weighted area	Hectares	H	759	522	758	520	602	500	600
@Seton Reservoir	Weighted area	Hectares	H	48	48	48	48	48	60	33
Maximize recreation										
@Carpenter Reservoir	Boat access	no. days/y	H	90	89	92	90	88	100	86
Maximize power benefits										
Maximize power revenues	Revenues	Levelized $ M y	H	141	145	146	149	144	110	145

Figure 9.4 Bridge River consequence table after elimination of insensitive performance measures, with Alt 7 selected.

accurate; the minimum significant increment of change here is 5%.

From Figure 9.3, it's straightforward to see that Alternative 6 is dominated by Alternative 4 – that is, Alternative 4 performs better than, or at least as well as, Alternative 6 (all the cells under Alt 4 are white or light grey). The group agreed to eliminate this Alternative from further consideration.

We now notice that two of the performance measures are not sensitive across the alternatives under consideration (water quality on Carpenter Reservoir and recreation on the Seton Reservoir) and as a result – after a confirmatory discussion – they were removed from consideration. This simplifies the situation even further to that shown in Figure 9.4.

Now we move on to consideration of practical dominance. If we compare Alt 7 with Alt 1, we see

that it is outperformed by Alt 1 on all performance measures except flooding on the Seton Reservoir, wildlife habitat on the Seton Reservoir and recreation on the Carpenter Reservoir. Deciding whether Alt 7 warrants further attention requires deliberation. One day of minor flooding on Seton was not considered very significant by participants. Further, while the recreation improvements on Carpenter are desirable, it's a seldom-used reservoir for recreational boating so this benefit wasn't judged to compensate for the poor performance on some of the other measures – in particular Alt 7 is significantly outperformed by Alt 1 on all four fish indices. Even the stakeholders most interested in improving boat access agreed that the incremental days of boat access don't offset the fish losses. A similar conclusion was reached for wildlife habitat on Seton. The group therefore agreed that Alt 7 is 'practically dominated' by Alt 1 (and in fact by Alts 1 through 4) and could be removed from further consideration. With Alt 7 gone, wildlife habitat on Seton Reservoir and recreation on Carpenter Reservoir were considered by the group to be no longer helpful in choosing alternatives, the result being shown in Figure 9.5.

In Figure 9.5, we've highlighted Alternative 3 to illustrate one of the fundamental trade-offs this group would continue to face; that between fish performance in Downton Reservoir and the wildlife habitat enhancements in Carpenter Reservoir. Recognition of this trade-off stimulated the search for new alternatives – for example, participants sought ways to modify Alternative 4 to improve its performance with respect to wildlife habitat on Carpenter.

By iteratively working through a series of paired comparisons of alternatives, the committee eliminated other alternatives that members collectively agreed were outperformed. New alternatives were developed by combining desirable elements of one alternative with desirable elements of another to create hybrid alternatives that better met multiple objectives. Ultimately, they were left with the consequence table in Figure 9.5, and here they met an impasse – which required the explicit consideration of trade-offs.

Using a multi-method approach to examine trade-offs

To help the Bridge River participants find new ways to think and talk about trade-offs, we introduced a multi-method approach to trade-offs. Participants first completed a three-step 'direct ranking' process, using the questionnaire shown in Table 9.3.

Objective	Sub-objective	Performance measure	Units	Dir	Alt 1	Alt 2	Alt 3	Alt 4	Alt 5	Alt 8
Minimize flooding										
	@Lower bridge River	Flooding frequency	no. days/year	L	1	1	0	0	0	0
	@Seton Reservoir	Flooding frequency	no. days/year	L	6	6	6	6	6	6
Maximize fish abundance										
	@Carpenter Reservoir	Fish index	1→100	H	69	70	41	41	29	29
	@Downton Reservoir	Fish index	1→100	H	42	71	48	69	65	69
	@Lower bridge River	Fish index	1→100	H	100	100	100	90	25	10
	@Seton Reservoir	Fish index	1→100	H	66	66	66	66	33	10
Maximize water quality										
	@Seton Reservoir	Water suspended solids	Tonnes/y	L	94	89	77	84	108	78
Maximize vegetated area										
	@Downton Reservoir	Weighted area	Hectares	H	223	231	322	313	295	300
	@Carpenter Reservoir	Weighted area	Hectares	H	759	522	758	520	602	600
Maximize power benefits										
	Maximize power revenues	Revenues	Levelized $ M y	H	141	145	146	149	144	145

Figure 9.5 Bridge River consequence table after consideration of practically dominated alternatives and insensitive measures.

Step 1. Based on the information in the consequence table, rank the alternatives from 1 (best) to 6 (worst). Ties are allowed.

Step 2. Assign points to the alternatives starting with 100 points for your top ranked alternative (1).

Table 9.3 Direct ranking questionnaire.

Your name				
			Rank	Points
Direct ranking				
	Alternative			
1	Alt 1			
2	Alt 2			
3	Alt 3			
4	Alt 4			
5	Alt 5			
8	Alt 8			

Step 3. Assign points to the remaining alternatives in accordance with how well they perform against the top ranked alternative (1). For example, if the number 2 alternative is almost as good as the number 1 alternative, give it close to 100 points. If it is only half as good, give it 50 points. Continue down the list in the order you ranked them, entering points either the same as or lower than the one before. You may enter zero points if you like – this would be an alternative that you feel has no value. All entries must be a whole number between 0 and 100.

Participants were then asked to complete a questionnaire to assess preferences indirectly – that is by weighting the performance measures. Although there are many ways to elicit weights, we used a swing weighting technique – a widely applicable method that delivers useful results and is relatively easy for participants to use and to

Table 9.4 Swing weighting questionnaire.

Swing weighting								
Sub-objective-level swing Weighting					Worst	Best	Rank	Points
Objective	Sub-objective	Performance measure	Units	Dir				
Minimize Flooding								
	@Lower Bridge River	Flooding Frequency	# days/year	L	1	0		
	@ Seton Reservoir	Flooding Frequency	# days/year	L	6	6		
Maximize fish abundance								
	@Carpenter Reservoir	Fish Index	1–100	H	29	70		
	@Downton Reservoir	Fish Index	1–100	H	42	71		
	@Lower Bridge River	Fish Index	1–100	H	10	100		
	@Seton Reservoir	Fish Index	1–100	H	10	66		
Maximize water quality								
	@Seton Reservoir	Water suspended solids	Tonnes/yr	L	108	77		
Maximize vegetated area								
	@Downton Reservoir	Weighted area	Hectares	H	223	322		
	@Carpenter Reservoir	Weighted area	Hectares	H	520	759		
Maximize power benefits								
Maximize power Revenue		Revenues	Levelized $ M yr	H	141	149		

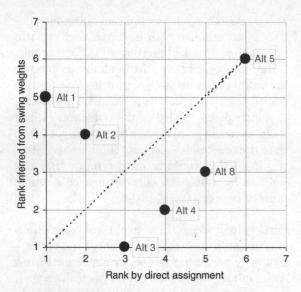

Figure 9.6 Comparison of ranks by direct and indirect methods for one participant.

understand[22]. Participants were provided with a table summarizing the worst and best scores for each measure across the range of alternatives under consideration (see Table 9.4) and asked the following four questions, in sequence.

Step 1. Imagine a hypothetical alternative that has the worst values across all of the measures in the table. Now suppose that you are able to 'swing' one (and only one) measure from its worst to best value. Which would you choose? Give it a rank of 1.

Step 2. Assign a rank of 2 to the second most important measure swing, and so on.

Step 3. Assign 100 points to the swing that ranked number 1. Assign points to the remaining swings in accordance with how important they are relative to the top ranked one.

Step 4. Review your results to make sure they reflect your priorities. That is, try to make sure that the points you've assigned reflect the relative importance in the measure swings. For example, if you assign 100 points to swing 1 and 50 points to swing 2, you are saying that achieving the swing in measure 2 is about half as important as achieving the swing in measure 1.

Participant responses to the questionnaire

Participant responses to the questionnaire were entered into a spreadsheet-based value model. For each participant, the model calculates that individual's weights for the measures and then uses them to compute performance scores and ranks for the alternatives. Essentially, it develops a personal value model for each participant. Each participant can compare his or her directly assigned ranks with the indirect ranks computed by the model. In the following sections we discuss three particularly useful ways to use the results to facilitate both useful group dialogue and individual reflection:

1 Explore the consistency of individual ranks across direct and indirect methods.
2 Explore areas of similarity and difference in weights assigned to the measures across individuals in the group.
3 Compare the ranks assigned to the alternatives across individuals in the group.

Exploring individual consistency across methods

Figure 9.6 compares the ranks assigned by one individual via the direct method with the ranks computed using the swing weighting method. Options ranked the same by both methods fall on the 45° line. Options that fall far from the 45° line are inconsistently ranked and should trigger a re-examination of that alternative by the stakeholder. For example, this illustrative stakeholder's ranks are quite consistent across the two methods except for Alternative 1. Alternative 1 is ranked as best by the direct method, but is ranked only number 5 by the weighted method. While this does not necessarily mean that the direct rank is wrong, it may indicate problems such as:

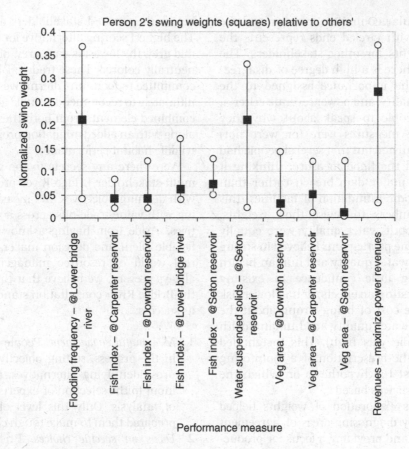

Figure 9.7 Range of weights assigned to the measures.

1 Mixing up the alternatives in the direct ranking (common when there are many alternatives).
2 Overlooking some elements of performance in the direct ranking (common when there are many performance measures).
3 Overlooking alternatives that are less controversial or less visible (common when there are polarized positions for or against other alternatives which lead them to dominate discussions).

Alternatively, the direct ranking may be a more accurate reflection of the stakeholder's values if the measures do not adequately capture some impor-

tant elements of performance. The intent of the multi-method approach is therefore not to say that one method is better than another, but to expose inconsistencies, clarify thinking, and improve the transparency and accountability of choices.

Exploring differences in weights across participants

Figure 9.7 shows the weights assigned by one participant (using the swing weighting approach) relative to the range of weights assigned by other participants (measures are shown across the bottom, with weights on the vertical axis). The squares represent weights for this particular committee member and the vertical line with circled ends represents the range of weights for other stakeholders. The figure shows there is a high degree of disagreement about the importance assigned to the flood, water quality, and power revenue criteria. We invited people to speak about why they thought these measures were (or were not) important. It turned out that several people had misunderstood the flood measure, thinking it represented a major dam breach rather than a modest periodic inundation of facilities; this led some members to revise their weights. Discussions about water quality were equally productive; some participants believed in strong links between water quality and human health effects (having little confidence in existing analyses suggesting no effects) or had lower risk tolerances (the cost of being wrong about the links between water quality and human health for these people was high). This insight led ultimately to the prescription of a monitoring program to test the hypotheses on which the existing analysis was based.

In sum, this exploration of weights helped deliberations by diagnosing areas of agreement and difference and providing a focus for productive discussion. It exposed factual errors, value differences, risk tolerances, and key uncertainties, which in some cases affected monitoring priorities.

Exploring areas of agreement and difference in ranks assigned to alternatives

Finally, once individual choices and group weights have been discussed and, in some cases, modified, trade-offs assigned by all group members can be explored. Figure 9.8 summarizes the alternatives with different levels of shading according to the scores assigned by stakeholders to each alternative by each method (whereby each person effectively distributed one point across each of the alternatives for each of the two weighting methods). Alternatives are shown

across the top and stakeholders down the side. The highest scoring alternative for each person is mid grey, the lowest is dark grey, and the rest are neutrally colored. These trade-off results led the committee to focus on Alternatives 3 and 4, and ultimately to recommend a single alternative that combined elements from both these alternatives along with an added mitigation project to improve wildlife habitat performance.

Were there any secrets to the success of the multi-stakeholder Bridge River process, starting with defining objectives and measures and ending with value trade-offs across a set of alternatives? Aside from having a knowledgeable and flexible client and decision maker, which in the real world of resource management are not always present, we believe that three aspects of the Bridge River consultation stand out as key to its success:

1 *Meaningful engagement.* People were engaged in the process, setting objectives and measures, identifying alternatives, and providing input to the selection of experts and methods of analysis. Only this level of involvement prepared them to make tough choices.
2 *Focus on specific choices.* Preferences were assessed in the context of specific choices. If we had asked participants 'which is more important, wildlife habitat or fish?' they'd have either left the room or given us the official position of the committees or agencies they represented. As noted earlier (see Text box 9.1), such general statements of priority are all but useless in a decision context; instead, we asked for and obtained preferences stated with reference to the consequences of specific trade-offs.
3 *Learning and 'constructing' preferences.* People learn, and their preferences change through the course of deliberation. Many of the participants came to the process driven by a firm desire to see improvements in the riparian wildlife habitat on Carpenter Reservoir. In the end they unanimously supported an alternative that left the majority of this drawdown

		Alt 1	Alt 2	Alt 3	Alt 4	Alt 5	Alt 8
Direct	Person1	0.13	0.25	0.19	0.31	0.06	0.06
	Person2	0.13	0.25	0.19	0.31	0.06	0.06
	Person3	0.32	0.03	0.19	0.26	0.13	0.06
	Person4	0.19	0.26	0.13	0.32	0.06	0.03
	Person5	0.19	0.26	0.13	0.32	0.06	0.03
	Person6	0.19	0.26	0.13	0.06	0.32	0.03
	Person7	0.13	0.19	0.26	0.32	0.03	0.06
	Person8	0.26	0.06	0.13	0.19	0.32	0.03
	Person9	0.32	0.13	0.19	0.06	0.26	0.03
	Person10	0.13	0.26	0.19	0.32	0.06	0.03
	Person11	0.26	0.13	0.19	0.32	0.06	0.03
	Person12	0.26	0.19	0.13	0.32	0.06	0.03

		Alt 1	Alt 2	Alt 3	Alt 4	Alt 5	Alt 8
Swing	Person1	0.11	0.14	0.23	0.23	0.13	0.16
	Person2	0.08	0.13	0.24	0.24	0.12	0.19
	Person3	0.16	0.19	0.22	0.23	0.11	0.10
	Person4	0.11	0.14	0.23	0.23	0.13	0.16
	Person5	0.08	0.13	0.24	0.24	0.12	0.19
	Person6	0.16	0.19	0.22	0.23	0.11	0.10
	Person7	0.11	0.14	0.23	0.23	0.13	0.16
	Person8	0.08	0.13	0.24	0.24	0.12	0.19
	Person9	0.16	0.19	0.22	0.23	0.11	0.10
	Person10	0.11	0.14	0.23	0.23	0.13	0.16
	Person11	0.08	0.13	0.24	0.24	0.12	0.19
	Person12	0.16	0.19	0.22	0.23	0.11	0.10

Figure 9.8 Group ranks by different methods. For two techniques, one preference point is distributed across each alternative by each person. Worst scoring alternatives have a dark shade, highest have a lighter shade, and the remainder are neutrally colored.

zone denuded – and not just grudgingly accepted it, but endorsed it as a good decision. They did this because they came to understand, after a comprehensive search for creative alternatives, that it was physically infeasible to both enhance the reservoirs and the downstream river at the same time. Only then were they prepared to make choices. Of course, the recommended alternative also included some plantings that helped, but these were far smaller in scope than originally hoped for.

These three points hold important lessons for decision makers. It is all too common to prematurely screen out alternatives based on the presumption that stakeholders will reject them. People can make surprising choices when they are truly engaged in the decision process and offered an opportunity to be meaningfully engaged and to learn.

9.3.2 Informing and bounding negotiations on the Athabasca River

In Chapter 5 we introduced how SDM methods were used to help a multi-interest stakeholder committee develop performance measures to evaluate management alternatives relating to the use of water from the Athabasca River by oil sands mining corporations in Alberta. Over the course of 18 months, we led the committee through an SDM process – defining the decision context (extraction rules that would work across the full range of forecast oil sands development into the future), identifying objectives (primarily relating to fish and river protection versus the cost to industry of curtailed withdrawal opportunities during the winter), estimating consequences, and developing alternative regulatory rules. Working with technical experts, an interactive spread-sheet-based predictive model was developed to estimate the flow impacts of numerous water management alternatives (or 'rule sets') and experts were then asked to populate a sequence of iterations of consequence tables with state-of-the-art (and ultimately peer-reviewed) analyses.

The committee worked through several rounds of alternatives; each involved significant learning about trade-offs. The first round involved developing extreme bookend alternatives that focused on the wish list of individual interests (e.g. an eco-friendly alternative, an industry-friendly alternative, etc.). The second round focused in on a narrower range of alternatives that lay on the efficiency frontier defined along two axes, with loss of fish habitat on one and industry costs for off-stream storage on the other. Using a paired comparison of alternatives

process, Round 3 ultimately identified one alternative with the most potential for mutual agreement.

General agreement was reached on a base water management alternative that covered all flow conditions except for management during low flow events (defined by the one-in-100 year event). In this case, there was disagreement about how to deal with this rare but highly consequential event. Some participants felt that because of the higher uncertainty associated with impacts to aquatic life in very low flow situations, the precautionary principle should be invoked and all consumptive use on the river should cease until natural flows increased. Others considered this to be unnecessary in light of modeling results, emphasized the associated financial impact to industry and highlighted the legal challenges that would ensue with the necessary retrenchment of existing water license rights.

The committee was unable agree on this key issue and the hoped-for full consensus agreement was not achieved. In spite of this, the process produced a great deal of clarity for the federal and provincial regulators. It clarified objectives, identified creative alternatives, and narrowed the likely solution space considerably. Insight was achieved on many aspects of water management, including flow rules for the vast majority of circumstances, implementation protocols, and priorities for a long-term adaptive-management and monitoring framework. Government agencies had a comprehensive and structured information package which they subsequently used as the basis for consultations with First Nations and the broader public.

This case demonstrates an important point. Sometimes the diverse perspectives of people involved in making a difficult choice simply can't be reconciled. This is sometimes disappointing, but is not a barrier to defensible decision making about public resources. Decision making is the job of the relevant government agencies, elected officials, or their delegates. There seems little doubt that final decisions in

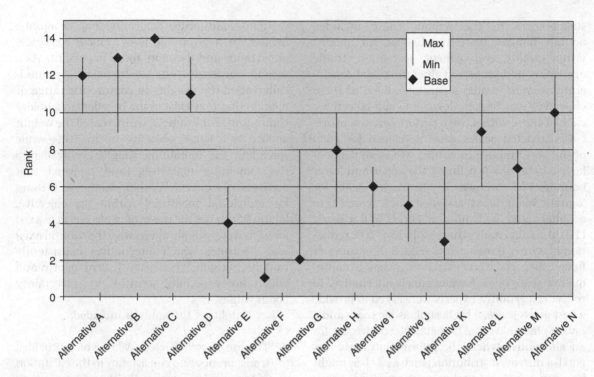

Figure 9.9 Rankings of water supply alternatives.

these cases are influenced by a host of social and political pressures and, at times, closed door negotiations. SDM will never change that. However, what we hope for is that the scope of negotiations is bounded by the outcomes of an SDM process which clearly eliminates poor alternatives and tightly bounds the areas under debate. In the case of developing a multi-objective, multi-user framework for Athabasca River water withdrawals, that's exactly what was achieved.

9.3.3 Ranking municipal water supply sources

This case study provides an example of a more formulaic approach to weighting measures and selecting alternatives. We introduce it to demonstrate the role that sensitivity analysis – on consequence estimates and on value-based weights – can play in informing trade-off decisions.

As part of its long term water supply planning process, the Greater Vancouver Water District commissioned a screening-level study to rank 14 potential source water supply sources. The study was to be based on existing data and expert judgment about their environmental and social impacts with consideration for how they might be perceived by regional stakeholders. No new field studies or consultation was included at the screening stage. The alternative water sources were diverse, ranging from expansion of three existing water supply reservoirs in the mountains immediately north of the metropolitan area, to development of distant lakes with significant long distance pipeline and pumping requirements, to the development of well-fields along a major river.

We guided a technical team to develop a set of objectives and associated measures based on a review of existing documentation of the water supply alternatives under consideration and the general concerns of regulatory agencies and stakeholders in the region. These included aquatic habitat, terrestrial habitat, air quality, visual quality, employment, recreation, traffic and noise, and property values (financial considerations were treated separately). Each of these objectives was broken down into sub-objectives, each of those with its own performance measure.

Constructed scales were developed for each of the performance measures. We were particularly careful about defining the upper and lower bounds. For example, a constructed scale for 'aquatic habitat loss' assigned a best score (1) to an alternative with no flow effects and a worst (10) to an alternative that results in a '10% reduction in winter flows or 50% reduction in summer flows'. Experts were asked to assign a consequence score to each water supply alternative. To reflect uncertainty, experts were asked to provide a best guess score, a highest plausible score and a lowest plausible score for each alternative. So, for an alternative where the information basis was good, a narrow distribution (such as 3-4-5) might be given, whereas in other cases a wider distribution (such as 2-4-9) might be given.

Once the alternatives were rated, the team addressed trade-offs across the objectives by assigning importance weights to each objective (e.g. aquatic = 30%, terrestrial = 30%, aesthetics = 10%, etc.) and then for each measure (within the aquatic objective, measures included direct habitat loss = 25%, changes in channel morphology = 20%, etc.). A value model calculated desirability scores for each alternative as the weighted sum across all the objectives and sub-objectives. To account for uncertainty, the low/median/high scores provided by experts were used to describe a probability distribution for each measure and a Monte Carlo simulation[23] produced a distribution of ranks for each option.

We recognized that the initial assignment of weights reflected only the expert team's opinion about the relative importance of each measure. But what if regional stakeholders hold a different opinion regarding the relative importance of lost forest habitat vs an altered viewscape? What if

regulatory authorities determine that the aquatic habitat for a given site is of critical ecological importance and therefore more important? As a second sensitivity test, we therefore systematically varied the weights to capture the range of opinion that may exist in the broader stakeholder community. For example, we increased the weight applied to a single objective up to 50%, while spreading the remaining weight evenly across the remaining objectives (and repeated that across all objectives). We then did the same thing for individual measures within an objective. Figure 9.9 shows the range of ranks (vertical axis) for each water supply alternative (horizontal axis) and illustrates which alternatives consistently rank highly, which consistently rank poorly, and which are especially sensitive to uncertainty about values.

Key benefits of the analysis included:

1 The use of constructed scales, which provided transparency and consistency in the elicitation of technical judgments about performance.
2 The screening of alternatives robust to uncertainty (in both technical judgments and public values).
3 Identification of the objectives and specific measures to which the rankings were most sensitive. This provided much-needed focus for subsequent data collection and for effective public consultation with regulatory agencies and the public.

9.4 Some practical considerations

9.4.1 Common pitfalls and traps

A number of errors are commonly made in working with stakeholders to develop trade-offs. These include:

1 Asking people to make tough choices without first searching for and evaluating a range of alternatives.
2 Moving too quickly to a single 'recommended' alternative, and then failing to seriously

deliberate about and record input on other alternatives. If this alternative is subsequently rejected by decision makers for any reason, the consultation has nothing to offer. Always try to provide input on a range of alternatives.

3 Trying to get people to agree on weights. This may be possible when participants all come from a single organization, or when the decision involves relatively routine, repeated choices for which efficiency is of paramount importance. But for controversial, one-off multi-stakeholder processes, this is unlikely to be anything but divisive. Further, it's not necessary. People can agree on a preferred or acceptable set of alternatives even when they have very different weights.

4 Failing to make context-specific choices. This can include assigning weights without looking at the range of possible outcomes for *this* decision, or by focusing on pre-existing priorities or world views (environment is more important than jobs). Good choices are only made when people focus on the *actual* trade-offs associated with *this* decision.

5 Over-relying on quantitative trade-off methods. Trade-off analysis methods can help groups reach agreement on informed choices if and only if objectives have been well structured, consequences have been defensibly estimated, a range of creative alternatives have been explored, and an open and trusting environment has been fostered for talking about tough trade-offs.

6 Assuming that less technical participants will not agree to or appreciate the use of quantitative methods for addressing trade-offs. In contrast, we've had many resource users and First Nation communities participate enthusiastically. Such participants are often reluctant to participant vocally in deliberations but are happy to see their views reflected on equal footing with other more vocal participants.

7 Trying too hard for consensus. Most of the time you don't need consensus, as we argued earlier, and saying that you do doesn't make it any easier to achieve (in fact, our experience suggests quite the opposite. So, be clear about what is required before you start the meeting. What you really need will depend on where you're at in the process, what your mandate is (and any associated legislative or legal requirements), who your decision maker is, and the role assigned to the group you're working with.

9.4.2 What if uncertainty is preventing consideration of trade-offs?

Participants in an environmental management planning process may be hesitant to engage in a trade-offs elicitation and alternatives selection exercise if there remains substantial uncertainty about the consequences of management actions. What might be done in this situation?

If the uncertainty is about facts:

1 Conduct sensitivity analysis to clarify just how much various aspects of performance might change under different assumptions for uncertain quantities.

2 Identify robust alternatives, if they haven't already been developed and included in the set of alternatives under consideration.

3 Develop adaptive alternatives by bundling monitoring programs with short-listed management alternatives. For example, on the Bridge River example described in this chapter, there were significant residual uncertainties. Participants ultimately agreed to a preferred alternative based on an understanding that the key uncertainties would be monitored over time, with formal timelines that would trigger institutional mechanisms established for the collaborative review of monitoring data and reconsideration of the management decision. For particularly contentious decisions, alternatives that are explicitly experimental might be developed, provided the conditions exist to facilitate learning from experimentation (see Chapter 10).

4 Carve out a set of no-regret actions that can be implemented immediately from a broader alternative, and define a plan for reducing

uncertainty on the outstanding elements over time.

If the uncertainty is about values:

1 Conduct sensitivity analysis on the weights assigned to performance measures, and test whether that changes the ranks of the alternatives. This is relevant in cases where a formulaic approach has been adopted – that is, a single set of weights has been developed for the performance measures, and used to calculate performance scores. As described in Case Study 3 above, it can be useful when actual consultation with stakeholders is not possible.

2 Conduct a critical value analysis, sometimes called a switchover analysis. This analysis tests how high (or low) an uncertain performance measure would have to go before the choice of a preferred alternative would change. For example, suppose a decision hinges on uncertainty about the magnitude of habitat losses that might be caused by an invasive pest, and that the key uncertainty is the size of the host range. Sensitivity analysis asks: 'what would be the implication of increasing the assumed host range size by (say) 10%? How much would that increase the habitat loss and how would that change the choice of preferred alternative?' In contrast a critical value analysis works backwards from the decision. It asks: 'at what value of habitat loss would the selection of a preferred alternative change? How high would the host range size have to be for that to happen? How likely is that?' Often – though not always – a critical value analysis lends insight to the decision and demonstrates that additional predictive modeling will not be useful.

9.5 Key messages

Trade-offs are an inevitable result of a decision-making process that involves multiple objectives and that generates multiple alternatives. SDM approaches seek to make trade-offs more explicit, transparent, and informed, and to do this in a consistent manner that encourages and highlights input from other participants in the decision-making process. Contrary to (some) popular beliefs, value-based decision making doesn't mean a values free for all. It's reasonable and appropriate to have some minimum expectations with respect to the quality of deliberations and the value judgments that result from them. By the time participants reach the trade-offs stage of a decision process, a sign of deliberative quality is that they should be able to demonstrate:

1 An understanding of the decision scope and context, how it is related to other decisions, why the problem matters, and for whom the consequences are most relevant.

2 An understanding of the alternatives, their relative performance, and the key trade-offs among the alternatives.

3 An understanding of key uncertainties and their impact on the performance of the alternatives.

Some rules of thumb for a productive process:

1 Focus on the choice at hand. People will be willing and able to make difficult trade-offs when they are cast as choices among alternatives.

2 Use an iterative approach that first explores trade-offs by comparing pairs of alternatives looking for dominated and practically dominated alternatives. If a clear preference doesn't emerge, it can be useful to use a more formal approach.

3 Don't try to reach agreement on weights: you won't succeed and you don't need to in order to reach agreement on an alternative.

4 To gain the most insight about difficult choices, use a multi-method approach. A combination of direct ranking and value modeling is practical and useful. The juxtaposition of the two approaches helps to engage both intuitive and cognitive modes of making choices, and encourages independent

thinking (through self-reflection) and group thinking (through structured deliberations). Documenting the results provides important insight to decision makers, while leaving the ultimate choices in their hands.

5 Above all, don't be afraid of trade-offs. We've seen many decision makers skirting the issue of trade-offs, afraid that they'll be controversial and divisive. Trade-offs are inevitable. The key to a productive discussion about trade-offs is a well-structured process leading up to it.

9.6 Suggested reading

Baron, J. (1988) *Thinking and Deciding*. Cambridge University Press, New York. Carefully works through key elements of decision making that affect how people form and evaluate trade-offs.

Bazerman, M. & Neale, M. (1992) *Negotiating Rationally*. The Free Press, New York. This introduction to 'rational thinking in negotiations' uses diverse case studies to demonstrate value-based approaches to making difficult tradeoffs, linking good negotiated solutions back to a clear expression of interests.

Brest, P. & Hamilton Krieger, L. (2010) *Problem Solving, Decision Making, and Professional Judgement: A Guide for Lawyers And Policymakers*. Oxford University Press. This book covers the basics of problem solving and professional judgment but then goes on to provide an excellent overview of how lawyers, politicians, and other key decision makers might think, feel, and be persuaded to learn about decisions.

Keeney, R.L. (2002) Common mistakes in making value trade-offs. *Operations Research*, **50**, 935–945. An excellent and case-study filled introduction for helping managers think through, as the title summarizes, how they might avoid 'common mistakes in making value trade-offs'.

Russo, J. & Schoemaker, P. (1989) *Decision Traps: The Ten Barriers to Brilliant Decision Making and How to Overcome Them*. Simon & Schuster, New York. A highly readable introduction to common decision making short-cuts and errors, including clear suggestions for how to avoid making them.

Satterfield, T. & Slovic, S. (eds) (2004) *What's Nature Worth? Narrative Expressions of Environmental Values*. University of Utah Press, Salt Lake City. An enjoyable introduction to the writings and ideas of many of today's current leaders in the new field of nature writing. These essays provide a wealth of material to help those of us trained as scientists, economists, or decision analysts understand why some people might look askance at methods and conclusions we consider sacred.

9.7 References and notes

1 Keeney, R.L. (2002) Common mistakes in making value trade-offs. *Operations Research*, **50**, 935–945.

2 Loewenstein, G., Weber, E.U., Hsee, C.K. & Welch, E.S. (2001) Risk as feelings. *Psychological Bulletin*, **127**, 267–286.

3 Kahneman, D. (2002) *Maps of Bounded Rationality: A Perspective on Intuitive Judgment and Choice*. Retrieved from http://nobelprize.org/nobel_prizes/economics/laureates/2002/kahnemann-lecture.pdf.

4 McNeil, B.J., Pauker, S.G., Sox, H.C., Jr & Tversky, A. (1982) On the elicitation of preferences for alternative therapies. *New England Journal of Medicine*, **306**, 1259–1262.

5 Kahneman, D. & Tversky, A. (eds) (2000) *Choices, Values and Frames*. Cambridge University Press, New York.

6 Kahneman, D.,& Tversky, A. (1979) Prospect theory: an analysis of decision under risk. *Econometrica*, **47**, 263–291.

7 Knetsch, J.L. (1992) Preferences and nonreversibility of indifference curves. *Journal of Economic Behavior and Organization*, **17**, 131–139.

8 Gregory, R., Lichtenstein, S. & MacGregor, D.G. (1993) The role of past states in determining reference points for policy decisions. *Organizational Behavior and Human Decision Processes*, **55**, 195–206.

9 Peters, E., Dieckmann, N.F., Västfjäll, D., Mertz, C.K., Slovic, P. & Hibbard, J.H. (2009) Bringing meaning to numbers: the impact of evaluative categories on decisions. *Journal of Experimental Psychology: Applied*, **15**, 213–227.

10 Rottenstreich, Y. & Shu, S.B. (2004) The connections between affect and decision making: nine resulting phenomena. In: *The Blackwell Handbook*

of Judgment and Decision Making (eds D. Koehler & N. Harvey). pp. 444–463. Oxford University Press, UK.

11 Loder (Loder, N. [2000] UK scientists under pressure to please. Nature, **403**, 689) reports that 30% of scientists working in government or recently privatized laboratories in Britain were asked to alter research findings so they were more closely aligned with client perspectives. For better or worse, this type of result is quickly picked up by media and widely disseminated.

12 Satterfield, T., Gregory, R. & Roberts, M. (2010) *Dealing with Differences: Policy Decision Making in the Context Of Genetically Modified Organisms.* Report to Foundation for Research Science and Technology, Wellington, New Zealand.

13 Roberts, M., Haami, B., Benton, R., Satterfield, T., Finucane, M.L., Henare, M. & Henare, M. (2004) Whakapapa as a Māori mental construct: some implications for the debate over genetic modification of organisms. *The Contemporary Pacific*, **16**, 1–28.

14 It's hard to illustrate this on the written page. We give it a try in the first case study in section 9.3. For a more compelling example, go to the website www.structureddecisionmaking.com.

15 Hammond, J., Keeney, R.L. & Raiffa, H. (1998) Even swaps: a rational method for making trade-offs. *Harvard Business Review, March-April*.

16 Edwards, W. & von Winterfeldt, D. (1987). Public values in risk debates. *Risk Analysis*, **7**, 141–158.

17 Hobbs, B. & Horn, G. (1997) Building public confidence in energy planning: a multimethod MCDM approach to demand-side planning at BC gas. *Energy Policy*, **25**, 357–375.

18 Gregory, R., Fischhoff, B. & McDaniels, T. (2005) Acceptable input: using decision analysis to guide public decisions. *Decision Analysis*, **2**, 4–16.

19 Champ, P.A., Boyle, K.J. & Brown, T.C. (2003) *A Primer On Nonmarket Valuation*. Kluwer Academic Press, Boston.

20 Failing, L., Gregory, R. & Higgins, P. (2011) *Leveling the Playing Field: Structuring Consultations to Improve Decision Making*. Manuscript submitted for publication.

21 The execution of field studies explains the long elapsed time of the project; deliberations were suspended while field studies to address key information gaps were conducted.

22 von Winterfeldt, D. & Edwards, W. (1986) *Decision Analysis and Behavioral Research*. Cambridge University Press, New York.

23 A Monte Carlo simulation involves the use of a computer simulation to review a large number of combinations of consequences; those which are most likely will be generated often whereas those which are more unlikely will be rare. Good descriptions are found in many texts, including Goodwin, P. and Wright, G. (2004), *Decision Analysis for Management Judgment*. Wiley & Sons, New Jersey.

10

Learning

The final stage of an SDM decision process focuses on learning, both to improve the quality of *future* environmental management decisions and to construct preferred alternatives and reach agreement on the *current* decision. However, learning in SDM does not only take place near the end of the process; if it did, groups would rarely (if ever) make it through the earlier steps. Learning occurs, and must occur, at every stage of the decision process. It involves learning both about facts – what is known and not known about the range of potential consequences of proposed actions – and about values – how the consequences are expected to be viewed and valued by the participants, by other potentially affected parties, and by the decision makers.

Participants in a deliberative process begin their collaborative learning with discussions about the problem framing and initial decision sketch (step 1). It is critical that they build a common understanding of what's in and what's out of scope, and why[1]. As discussed in Chapter 2, one of the key ways in which SDM methods differ from other approaches used in environmental management is in the emphasis placed on the deliberate construction of individuals' understanding and preferences. It's also critical that participants have a common understanding of the process they will use for examining the decision problem – when and how key value-based and fact-based issues will be addressed, building a common language and a set of techniques that will be used to examine these issues.

At the objective setting stage (step 2), a rigorous approach to clarifying, structuring, and articulating objectives and reporting them with unambiguous performance measures builds common understanding about what matters and a common language for expressing the effects of actions. At the alternatives stage, participants jointly explore creative new solutions, and as they explore consequences they may learn that some favored alternatives have unexpected consequences (steps 3 and 4). This link between generating alternatives and learning about what each alternative can and can't achieve is essential to creation of a decision-making framework within which people are willing to address difficult trade-offs – no one takes on tough trade-offs, whether in their personal or professional lives, unless they're sure a full range of alternatives is under review. Given the inevitable existence of a diversity of relevant information and expertise, participants in complex decisions typically spend a significant amount of time learning together about impact hypotheses and predictive modeling tools, all with the goal of building a common understanding of what is known and not known

Structured Decision Making: A Practical Guide to Environmental Management Choices, First Edition. R. Gregory, L. Failing, M. Harstone, G. Long, T. McDaniels, and D. Ohlson.

about the consequences of proposed alternatives. At the trade-offs stage (step 5), participants do some tough introspective thinking about their own values, influenced by learning about the values of others and how these are expressed through trade-offs.

So back to our initial point: learning at step 6 of the SDM process is not a new idea or a new demand on participants; it's fundamental to making informed choices that can be supported by a broad range of stakeholders. In this chapter, we focus first on practical advice for encouraging learning – specifically, what concrete steps can we take to ensure both the availability and use of better information for future decisions? In the second half of the chapter, we examine some of the conditions that are needed to foster learning more generally in deliberative processes, including an examination of what's needed to build effective adaptive-management plans.

10.1   Improving the quality of information for future decisions

The usual response to a call for improving the quality of information in environmental decision making is to monitor, for several good reasons. First, it helps to establish baseline or current conditions, key to assessing the incremental effects of management actions. Second, monitoring helps with reporting on compliance (were committed actions implemented?) and reporting on progress toward the achievement of objectives (did actions achieve what they were intended to achieve?). In this case, there may be little explicit focus on learning per se: the goal is simply to openly track and report on actual versus expected actions and results. A third reason to monitor is to inform staged or later decisions; what is learned about the effects of actions taken now can help to anticipate the effects of identical or similar actions taken at a later time or in a different but similar location.

Yet sometimes what's needed for improving the quality of information (and, in turn, the quality of management decisions) goes well beyond monitoring. For example, what is required may not be just better information about the consequences of proposed actions, but the development of an enhanced capacity to use and interpret that information. The scope of actions under consideration to encourage learning and improve information quality therefore could include many different elements:

1 Baseline data collection – against which the effects of future management actions can be compared.
2 Compliance monitoring – checking that the actions that were committed to by resource-management agencies actually were implemented.
3 Effects monitoring – were the predicted outcomes realized? Why or why not?
4 Capacity building initiatives for agencies or communities – examples include training in monitoring technologies, or post-secondary education for indigenous community members with a goal of ensuring greater technical competence in future decisions.
5 Co-management and trust building initiatives – joint agency–community monitoring teams might be established, or funding for data collection efforts could be shared between resource users and technical consultants.
6 Research – setting funds aside to examine new ideas for mitigation or habitat restoration actions which need further investigation before they might legitimately be considered as candidate actions.
7 Development of assessment techniques – these could involve new predictive modeling tools or better ways to collect data on the effectiveness of actions.

While environmental managers conventionally think of monitoring as applying to the tracking of biological variables, it may be equally important to monitor social or economic outcomes – recreational use rates, jobs produced, tourism revenues, health outcomes, political support – depending on the problem context. In cases where local or aboriginal knowledge plays an important role, there may be a need to improve gaps in local knowledge or to augment informal observations with a more rigorous observation strategy (which underscores the point that traditional knowledge is a dynamic concept; see Berkes & Folke[2]). Where competing hypotheses stem from different knowledge sources, this focus on collecting new information – and doing so in ways considered to be culturally appropriate – is particularly important.

10.1.1   Identify which uncertainties matter

From a decision-making perspective, information on sources of uncertainty has value only if it improves the ability to make informed choices, and then only if different choices would result in different outcomes with respect to the identified objectives. The task is to figure out which uncertainties, if reduced, would inform decision makers and are likely to result in improved performance.

One way to figure out which uncertainties might matter the most is to calculate the value of information (VOI). This is a fairly standard analytical approach, at least within the decision analysis world, which uses decision trees and expected value calculations to produce an estimate of the expected benefits (as a probability-weighted average) from improved information[3]. The results could be reported in dollars (if the benefits are financial) or natural units (say in expected tonnes or kilograms, if the relevant benefits are related to biomass of a managed species). In multi-attribute problems, results might be reported as an aggregate performance or utility score. Clear examples and additional detail are given in Clemen[4] and Goodwin and Wright[5]. Important insights can come from this kind of analysis – see Text box 10.1 for an example involving the use of VOI calculations to prioritize reductions in the uncertainty associated with proposed actions for reducing the spread of white-nose syndrome among bats.

A second approach to understanding and ranking different sources of uncertainty is more deliberative: assess which uncertainties hinder the ability of an individual or group to choose a preferred alternative. This might be aided by some analytical tools (such as sensitivity analysis or critical value analysis, for example) used for insight at the trade-off stage. If individuals or an advisory group are finding it difficult to select a preferred alternative because of disagreements or discomfort concerning the uncertainty of specific consequences, then these consequence estimates are strong candidates for long-term monitoring.

Consider a situation where extensive chemical contamination from a factory has led to concerns about the health of residents in a nearby agricultural center. A clean-up plan is proposed under the US Environmental Protection Agency's Superfund program, and several months of structured discussions and multi-attribute trade-off methods help a diverse consultative group to focus in on two short-listed alternatives. Uncertainties related to the predicted effectiveness of these two methods in safely removing the primary contaminants has prevented decision makers from making an informed choice. Judgments elicited from experts demonstrated that the difference was potentially meaningful, with substantially different outcomes for human health estimated under competing plausible hypotheses. Final agreement in this case could be achieved by funding a two-year set of initial clean-up and mitigation actions (e.g. determining the size of the contaminated area, checking whether groundwater supplies are contaminated, testing local agricultural produce to see if it's OK to eat and sell) and simultaneously initiating programs (monitoring and experimental) to ensure ongoing reduction of key uncertainties about the effectiveness of the methods and their impacts on health and other objectives, with the goal of making a decision within the two-year period.

Text box 10.1:   Value of information for management of white-nose syndrome in bats

White-nose syndrome (WNS) is an emergent, rapidly spreading and often fatal disease affecting cave-dwelling bat populations in the eastern United States. Bats showing evidence of WNS – for many affected animals, a white fungus on the bat's face, wings, or tail membranes – are thought to be dying (in some hibernacula, by the tens of thousands) as the result of depleted fat resources.

In summer 2009, an SDM process led by Jennifer Szymanski, Mary Parkin, Mark Armstrong (USFWS) and Mike Runge (USGS) identified six fundamental objectives of decision makers for WNS management, developed a suite of management alternatives, estimated the consequences of actions using formal expert elicitation methods, and identified a recommended alternative (for a concise summary see Szymanski, Runge, Parkin & Armstrong[6]). As is often the case, the analysis was time and information constrained. For example, further work is needed on development of quantitative, predictive models of disease dynamics as a function of management actions. Nevertheless, the SDM process facilitated learning among a diverse group of stakeholders that helped to establish a common understanding of the problem on the part of agency managers and biologists and led to a clear recommendation of a preferred short-term management alternative.

A key part of the process involved a sensitivity analysis of the expected value of information (VOI), focusing on potential reductions in uncertainty identified by experts for estimated consequences related to the fundamental objectives. In this case, experts were only asked to predict the consequences of a management alternative with respect to the specific objective within their domain of expertise. Thus, no experts were asked to rank alternatives in terms of their overall preference, which would have involved policy or value judgments. Although either group or individual face-to-face elicitations were preferred, logistic and time constraints required the elicitation process to be conducted via email. Predictions of several experts were averaged to obtain a simple composite score for each consequence.

Calculations of the expected value of perfect information (EVPI) examined the question: if uncertainty could be resolved on only one attribute, which would be the most important (i.e. which reduction of uncertainty would most increase expected performance of the alternative?). Across the six objectives, calculation of the EVPI showed significant differences: resolution of the uncertainty about one objective (the impact of management actions on cave-obligate biota besides bats) would dramatically increase expected performance, whereas the resolution of uncertainty about another objective (the number of bats directly killed) had an EPVI of 0 – which means that refining this estimate is not consequential to choosing among management actions.

The SDM analysts also asked what alternatives, if selected, would be the most informative in resolving uncertainty relevant to deciding what to do in subsequent years. This calculation, the expected value of sample information[7], led managers to make two helpful observations. First, the value of learning from observed management actions will depend on the ability to monitor these responses with some precision. Second, there are likely to be trade-offs between short-term management objectives and the longer-term value of learning. Both of these insights will figure prominently in ongoing debates about how to incorporate an adaptive-management framework as part of management efforts to halt the spread of WNS.

10.1.2 Develop and bundle monitoring programs

Monitoring often sounds like a good idea as part of an environmental management initiative, but unless monitoring is carefully designed and implemented it may fail to achieve its anticipated purpose. A decision focus highlights concerns when designing monitoring programs to inform environmental management decisions: start early on to think about monitoring needs, always link monitoring indicators to objectives or performance measures, establish clear hypotheses, and ensure that effects can be detected.

Start early

Uncertainty arises at many stages of the decision process. Keeping in mind the practical bent of this book, we suggest that a useful approach is to develop a 'monitoring bin' at an early stage and to add candidates for monitoring throughout each step of the SDM process.

For example, uncertainty arises when performance measures are being proposed – in the context of an environmental management decision, examples include the selection of a preferred performance measure (e.g. what measure of air emissions is most appropriate?) and the designated methods for calculating it (where and how often should measurements be taken?; with what degree of precision?). While it may be possible to reduce these uncertainties through dialogue or analysis prior to the detailed assessment of consequences, some uncertainties will persist. These are candidates for longer term monitoring, and it's useful to put them formally and visibly into a monitoring bin and to tell participants that their significance will be addressed as part of step 6. Uncertainty related to performance measures also arises as the result of iteratively generating alternatives. For example, creative or unexpected new alternatives often are proposed in the latter stages of a decision process and their costs and benefits may not be captured fully by the previously agreed upon performance measures. It may not be possible (e.g. due to time or resource constraints) to cycle back and select new measures or to build new models, so estimates typically are made using expert judgments. When uncertainty in these estimates is high and the issue is thought to be significant, then these costs or benefits get added to the bin and become candidates for future monitoring.

In some circumstances we also have found it useful to develop an 'implementation bin'. This occurs in cases where the underlying concerns and interests of stakeholders relate as much to 'what happens on the ground' as to 'who is supposed to do what, and when'. While it is often outside the formal scope to delve too far into the mechanics of implementation, crafting a set of implementation recommendations as a part of the SDM process can provide useful and important learning for resource managers and decision makers.

Link monitoring indicators to objectives and measures

As part of an SDM process, an attempt should always be made to link monitoring indicators to objectives or performance measures. This is critical for ensuring that the information from monitoring will be useful to decision makers, whether as part of a current or future environmental management process. Fisheries applications, for example, commonly have an objective related to the abundance of fish, perhaps with fish biomass as a performance measure. Monitoring programs might attempt to distinguish between competing hypotheses about what causes changes in fish abundance, so in addition to monitoring fundamental outcomes of concern (fish abundance) they could also monitor potential explanatory variables – e.g. changes in food supply, water temperature, and so on. These monitoring indicators are distinct from the fundamental objectives, so in this sense they are not performance measures even though they can be mapped to what matters – water temperature, food supply, etc. may help to explain observed changes in biomass (the performance measure) or in fish abundance (the fundamental objective).

Establish clear hypotheses

Many monitoring and adaptive-management initiatives assume that just by implementing actions and measuring outcomes, something useful will be learned for management. This casual approach to monitoring and to adaptive management, albeit widespread, is both naïve and wasteful. The single most important thing managers can do to avoid this trap is to proactively define clear hypotheses linking an action to an outcome: make a prediction and then design a monitoring program to generate evidence to help you determine whether it's true.

From a technical perspective, the principles of adaptive management (see section 10.3) emphasize the importance of developing and testing *alternative* hypotheses and designing monitoring programs accordingly. From an SDM process perspective, we encourage hypothesis development (and detailing candidate statistical approaces to draw inferences from results) to be done collaboratively, including both technical experts and representatives from key interested parties. Equally competent participants often will have different opinions, so this is a situation where a little proactive engagement can save a lot of frustration (and a lot of time) later on.

Examine the ability to detect an effect

In large ecological and social systems, monitoring is difficult and not always successful. The design of programs that can actually detect changes attributable to a management action requires special expertise and should not be based on just common sense. It will involve some consideration of the expected natural variability in the monitoring data and the detectable effect size. Monitoring effectiveness also will depend on the duration and intensity of monitoring – which often depends on institutional stability and the existence of a secure, sufficient source of funding over time. It is not uncommon for technical experts or committees to evaluate several alternative monitoring designs before selecting one or two to present to decision makers. Alternative designs might reflect different levels of financial investment, different geographic scope, or different durations; they should clearly state implications for the expected quality of information and, hence, different probabilities of realizing the long-term benefits of management initiatives.

The uncomfortable reality is that there are many cases, especially in working with complex or large ecological systems, when uncertainty cannot be reduced significantly with any amount of monitoring. On one hand, the obvious advice is not to waste your money – if uncertainty cannot meaningfully be reduced, then monitoring programs should be abandoned and the money redirected. However, monitoring is sometimes viewed as a critical foundation for transparency and accountability in resource management: choosing not to monitor is sometimes simply not an option. In this case, we emphasize that if expert advice suggests that uncertainty cannot meaningfully be reduced, then resource managers should make sure decision makers know it. They may choose to monitor anyway, as a means of demonstrating a commitment to transparency and accountability. As always, our role is not to judge the rightness or wrongness of such choices – just to illuminate the trade-offs and seek to ensure that decision makers have access to the insights required for making an informed choice.

10.1.3 Evaluate programs

At the most basic level, managers will need to evaluate (a) the cost of monitoring programs, (b) the probability of gaining useful information (sufficiently reliable for use in decision making) in a relevant time period, and (c) the importance of the information to linked decisions – how likely is it that resolution of this uncertainty would result in beneficial changes to management actions or significant improvements in outcomes[8]. What exactly is done should depend on the stakes (how significant are the consequences of being wrong?; how costly are the proposed programs?) and the available resources. The importance of information to future decisions could be calculated as a

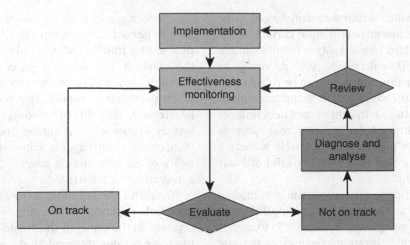

Figure 10.1 Generic implementation, monitoring, and review cycle.

formal VOI, as discussed earlier. In deliberative settings, however, we usually choose to turn this evaluation into a multi-attribute problem – an example of a decision within a decision – and work with participants to define performance measures for cost, usefulness, and importance to management actions or outcomes, typically estimating costs in dollars and developing constructed scales for usefulness and importance.

Once experts assign scores to the programs and the deliberative group weights the different measures, then it's straightforward to calculate overall performance scores for each proposed monitoring program. This is the classic strategy development problem that we discussed in Chapter 7: rather than just scoring monitoring actions or programs and picking off the top scorers up to a budget cap, it's usually helpful to also consider a few other design criteria such as spreading programs over time, across geographic areas, or across different objectives. Further, some monitoring programs may have been bundled with specific management actions as part of reaching agreement on a particular action. So this evaluation process, although important, is rarely tidy.

The good news is that by this stage, a group involved in an SDM process will have the game figured out. They should be able to quickly set objectives and evaluation criteria for monitoring programs, work together to develop scoring systems for usefulness and importance, and agree on who has the expertise to assign them. They will mix and match individual monitoring actions into logical packages or programs that make sense together, put them all in a consequence table, and come up with a collaborative way to examine the trade-offs.

10.1.4 Establish triggers and mechanisms for review

New information is of little use if management systems and institutional mechanisms aren't in place for responding to it. Consider Figure 10.1, which shows how implementation and monitoring fit into a classic management review cycle. This neat and (we hope) intuitively appealing diagram hides quite a number of difficult choices that need to be addressed by a consultative committee, a resource manager, or a designated decision maker.

Consider the 'diagnose and analyse' step – that is, the review of monitoring results and determination of whether or not implementation is 'on track' – as part of the design of the monitoring programs themselves. A first question to consider is whether explicit reference points or thresholds have been established for use in the evaluation of monitoring results. Are results considered to be on track if for some environmental variable – say, mean annual bird nest density – results show a 5% decrease? 10% decrease? 50% decrease? In some cases, existing standards can be relied upon (e.g. the IUCN thresholds for determining endangered species status). In other cases benchmarks from monitoring similar systems may provide useful reference points. Yet only rarely is there a true bright line that indicates any kind of clear response threshold.

It's equally important to specify the institutional structure for incorporating monitoring information into the management cycle as it is to design effective monitoring programs themselves. In some cases the regulatory or resource-management agency charged with implementing a monitoring program will have an established and trusted record of carrying through on its responsibilities; in other cases this positive expectation may not hold and, in far too may instances, monitoring may be left to an unnamed agency or consultant or simply designated as something to be addressed at a future time. An additional consideration is that participating stakeholders may recognize that they have invested significant effort as part of an SDM process and, as a result, they may want (and feel entitled to) an ongoing role. This can include a seat at the table when evaluating monitoring results or when decisions are made, to review monitoring protocols or, especially, when it's time to revisit the management actions.

10.2   Deliberative learning

We noted at the start of this chapter that learning encompasses both values and facts. It also involves both personal learning and group or collaborative learning. For example, an individual may recognize part-way through a deliberative SDM process that something previously felt to be important really isn't, at least not in this decision context and not over the expected range of consequence. At other times, participants may report that, based on learning that has occurred over the days or weeks a group has worked together, they've gained a new perspective on what objectives matter that they simply did not have prior to the start of the decision process. Sometimes learning also affects process considerations, as when participants report they have gained (or lost) trust in the ability of a resource-management agency to stay with an agreed-upon monitoring plan, or when participants acknowledge the credibility of an information source that previously was doubted or mistrusted.

Successful deliberation as part of an SDM approach depends on learning, which in turn depends on the ability of those leading the process to create an environment that fosters dialogue, questioning, and self-reflection. In this section we discuss several often-neglected conditions, largely within the control of those leading an SDM group, which can greatly help to facilitate learning. Before we get there we'll also introduce three leading contributors to deliberative learning: joint fact-finding and peer review, values clarification, and adaptive learning.

The three types of learning correspond roughly to the single, double, and triple loop learning models first identified in the early 1970s by social learning theorists[9] and recently expanded in the context of environmental and resource management[10]. Incremental or *single-loop learning* focuses on the identification of strategies (e.g. use of modified gill nets) for resolving specific issues (such as increasing catch). *Double-loop learning* occurs when underlying values or worldviews are challenged, with the result that fundamental changes may be made in management processes and stakeholder behavior. *Triple loop learning* focuses on governance, such as long-term efforts to improve decision-making capacity and quality over time.

10.2.1   Types of deliberative learning

Learning through joint fact-finding and peer review

Participants in a deliberative process work toward a common understanding of what is known and not known about the consequences of proposed

Text box 10.2:   White sturgeon recovery planning

In Chapter 8, we introduced a case study involving the exploration of competing hypotheses about the causes of white sturgeon recruitment (reproduction) failure in the Upper Columbia River. We described how participants examined the evidential basis for competing hypotheses using an impacts pathway structure to reduce verbal ambiguity.

Attention then shifted to the implications for management decisions. After hearing the evidence for and against each hypothesis, the participating experts were assigned three tasks:

1 Give your best estimate, based on the available evidence, of the relative importance of the hypothesis as a contributing factor to recruitment failure.
2 State your degree of confidence in the hypothesis.
3 Assess the feasibility of reducing the uncertainty around the hypothesis through further research, within a reasonable timeline.

Figure 10.2 illustrates how these judgments were communicated and debated, in this case showing the ratings for anonymous expert 'X.' The circles in the chart represent hypotheses 1 to 6.

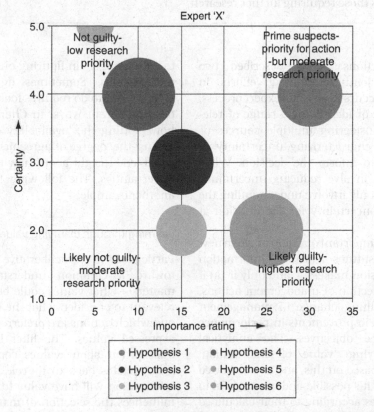

Figure 10.2 A visual tool for examining management decisions in light of recovery hypotheses.

The further to the right on the x-axis, the more important the expert feels the hypothesis is as a cause of recruitment failure, based on available information. The higher up on the y-axis, the more certain the expert is about the importance rating. Finally, the bigger the circle, the more practical it is to undertake more research to reduce the uncertainties surrounding the hypothesis.

The location of the circles on the chart give an indication of the most appropriate kinds of action to be taken for each hypothesis. Those in the top right of the chart are the 'prime suspects' – that is, they are considered 'certainly important'. Given the time constraints, an appropriate response might be, therefore, to work on developing near-term mitigation actions that will help reverse the recruitment failure (if such possibilities indeed exist), and building the technical (and political) case to promote their implementation. Those on the bottom right might be important, but there is high uncertainty about them – if, as in the case here, such hypotheses are also considered amenable to further research, then a tightly focused research program aimed at clarifying the uncertainties around this issue might be the way to go. In the top left are the 'not guilty' parties – hypotheses that should be considered lowest priority for follow-up unless new information comes to bear. Finally, those in the bottom left have a chance of being important, but all else being equal are probably of lower research priority than those to their right. Participants found these distinctions to be helpful in distinguishing between areas requiring urgent action and near-term funding versus those requiring further research.

management actions. We've described two approaches to joint fact-finding efforts in Chapter 8 – one occurs in a formal expert process, the other as part of identifying a range of relevant experts, considering multiple sources of knowledge, and characterizing uncertainty in ways that help to inform the decision. While these efforts can involve reducing uncertainty, they more commonly involve understanding the implications of uncertainty for the decision at hand.

This raises the important question of when new information constitutes learning. Information has value to decision makers if, and only if (a) it influences the selection of management actions, and (b) the resulting change in management actions results in improvements in performance on the identified objectives. The analytical methods underlying 'value of information' calculations are based on this, and as we showed in Text box 10.1, it's possible and often useful to rank uncertainties according to their calculated value of information. However, it's important to remember that these calculations are only a tool

to aid managers in figuring out what to do with uncertainties. Sometimes figuring out what makes sense to do requires looking at uncertainties in different ways. In Chapter 6 we showed how plotting the 'availability of data/evidence' against the 'degree of agreement among experts' could lend insight about how to deal with those uncertainties. The following text box provides another example.

Learning through clarifying values

Participants in the deliberative process also work toward a common understanding of what matters – what values could be affected and are relevant to consider, what the key trade-offs are, and which actions are preferred – in light of the expressed values. The filter for whether new information about values constitutes learning again relates back to its role in the decision – information will have value to the extent that it influences the selection of management actions which, in turn, lead to improvements in performance on the identified objectives. In practice, what this means is that new information about values must iteratively be reflected in the hierarchy of objectives, the definition of performance measures, and the relative weights that are assigned.

Learning about values involves group learning – through listening to and understanding the preferences expressed by others. Sources of group learning include those individuals who are directly involved in the consultation or decision process as well as others who may not be in the room but have an interest in the outcome of the decision. Learning about values also involves individual learning – each participant explores his or her own values, a process that involves personal reflection on preferences, priorities, and risk tolerances. It is the combination of dialogue and introspection within decision-focused structuring that largely facilitates this learning as part of SDM approaches – in particular, the relentless clarification of means and ends, an ongoing assault on linguistic ambiguity, and careful attention to the properties of good objectives and performance measures.

Is learning itself a fundamental objective? We think not. Instead, it is a means to an end – the end being better performance on the specified set of fundamental objectives. From a practical standpoint, however, it's sometimes useful to identify learning as one objective of an environmental management plan because this designation will lead to more informed decision making. Having an explicit objective can help lead to an agreement among participants on what constitutes learning and, in turn, to the clarification of performance measures and the selection of actions designed to improve knowledge.

As part of several multi-stakeholder SDM processes we've therefore sometimes found it helpful to develop an explicit 'learning scale', which helps participants to make explicit judgments about how much learning is anticipated as the result of a proposed action. One example is the following five-point constructed scale, used to assess the quality of anticipated learning as part of the comparison of different management

Text box 10.3:   Learning about values and objectives for Cultus Lake sockeye salmon

A good example of learning about and clarifying values – what matters, and what information should be collected to help reduce uncertainty and create improved environmental management alternatives – comes from the structured deliberations that occurred as part of creating a species recovery plan for Cultus Lake sockeye salmon (described in Chapter 7). In this case, the 'obvious' fundamental objective of conserving the species turned out to be not so obvious. A full understanding of what constituted 'conservation' involved four sub-objectives, related to the probability of meeting recovery plan objectives (a legal responsibility of the federal regulatory agency), the magnitude of expected salmon returns over the next 10 years (a conservation concern of the commercial fishing industry), the probability of extirpation by 2036 (reflecting concern about long-term viability of the species), and the percent of the total fish population that were enhanced (reflecting a concern about genetic diversity of the species). This multidimensional definition of conservation was not articulated before deliberations began, and defining it required a willingness to learn and to broaden the range of concerns and measures considered to be important on the part of all participants – about their own values, about concerns important to others, and about what information could be collected efficiently and used to compare multiple management alternatives. More details are provided in Gregory and Long[11].

initiatives[12]. In this case, learning was important to both scientists and local residents, so the scale reflected the benefits of learning from conventional experimental designs as well as learning that would enhance knowledge among the resident resource users.

1 *Poor:* The opportunity to learn is negligible, because of inadequate resources and low commitment to monitoring, or because the actions to be undertaken are already well understood.
2 *Fair:* Learning will occur but there are serious concerns about the ability to interpret and use findings, either as a result of undetectable effect size or poor monitoring information quality.
3 *Good:* Learning is anticipated that will be usable in decision making, although important gaps (related to endpoints, options or reliability of information) are likely to be present.
4 *Very good:* Learning is anticipated that will directly be used by decision makers; gaps may remain but their influence on information quality or the range of endpoints or options addressed should be minor.
5 *Excellent:* Significant opportunities are expected for enhancing both scientists' and local resource users' knowledge with reliable and relevant new information.

We re-emphasize that learning is clearly a means to an end. But if the desire to learn and the definition of learning influences what you do and how you do it, then it can be useful to have a learning objective and an explicit agreement about how it will be measured. In this case, there were important trade-offs about learning. As one example, experimental designs proposed for the flow releases on the river would preclude habitat work because the resulting changes would make it impossible to interpret results from the experiments (i.e. the habitat initiatives would be a confounding factor). Because this habitat work was seen as potentially important, both for its environmental benefits and for its community capacity building benefits, there was value in constructing an explicit objective and scale for learning so that the learning–habitat trade-off could be discussed and carefully evaluated.

Adaptive learning

A deliberative group builds capacity, over time, to work together through complex or controversial issues. This begins at the initial problem bounding and objectives identification phase, when participants learn that others view the environmental management problem in ways that are both similar to, and different from, their own. As analysts and facilitators, we often observe situations where the initial closed reaction of some participants to dialogue and input from other members (you know the stance: eyes down, arms folded across the chest) magically begins to change – quickly or slowly, over the course of a few hours or a few meetings – as they become interested in learning and start to integrate and incorporate the perspectives of other participants: recognizing new dimensions to the problem (timing constraints, legislative mandates), adding new objectives (including social, cultural, or political concerns), or changing their view of what constitutes the 'best available information' as they recognize the knowledge contributions of other stakeholders (community residents or resource users). The accompanying response we often hear from participants is one of surprise: their understanding of the resource-management opportunity turns out to be less complete than they thought, and their own perspective turns out to be more flexible and adaptive than previously considered.

This introduction of adaptive learning through deliberation also arises as part of the consideration of sub-topics within the context of a larger environmental management issue. There may be a big problem, e.g. developing a clean-up plan for a contaminated site, or a multi-user water allocation plan for a region, that is difficult and controversial and already has required several months of dedicated effort as part of a SDM process. When a sub-problem arises, participants find that a vocabulary and problem structuring approach and high level of comfort with inter-participant dialogue already are established – a decision-aiding framework for addressing multidimensional problems already has been agreed upon. Without extensive debate or argument, the group can just get to work: setting objectives, developing performance criteria, identifying alternatives, estimating consequences, and addressing the relevant value trade-offs. The success of this type of decision-within-a-decision adaptive learning also relies strongly on the development of trust, both in other participants in the organizing and structuring capabilities of the SDM approach.

In other cases, participants are asked if they want to commit to intentional long-term learning that will improve the information base available for future decisions – in other words, to a long-term adaptive-management plan. Adaptive management involves reducing decision-relevant uncertainties, but in many cases (particularly if extensive experimental trials are involved) the proposed initiatives will require many years or even decades to provide clear information. This raises significant governance issues for a deliberative learning-based process: how will a multiyear learning process be funded, who will be responsible for incorporating results, and will the deliberative environment (e.g. an inter-agency technical working group, or a multi-stakeholder consultative committee) still be in existence when results become available? We discuss these and related issues in section 10.3.

10.2.2   Conditions for learning

We've said that collaborative learning is critical to sound decision making – and we've said that it happens in SDM. Why? What are the conditions that help to facilitate collaborative learning?

Meaningful participation

We believe that a key benefit, and also a responsibility, of an SDM approach is to provide a deliberative framework for meaningful participation in resource management, and that this will lead to learning and ultimately to better decisions. Having experts denote key aspects of the problem from their perspective, and then conducting analyses to address these issues, provides neither learning nor participation on the part of other stakeholders – nor does simply asking participants to voice their goals or concerns through small-group or town-hall meetings and then creating long lists of issues which are passed along to managers. For example, Durham and Brown[13] (p. 462), after interviewing stakeholders active in United States watershed planning initiatives, reported that 'fewer than 50% noted that participation was useful in clarifying the issues'.

The numerous cases of failed or only partially successful community-based environmental planning and risk initiatives speak to this same underlying problem. In the United States, recent examples include the USEPAs national estuary clean-up program, the US Department of Energy's attempts to store high-level nuclear wastes underground, or the US Department of Agriculture Forest Service's attempts to adopt a more ecologically responsible 'ecosystem management' approach. Contrast this with the findings of an independent review of British Columbia's water use planning process – which used a structured decision-making approach at 23 facilities province-wide – in which a survey of participants reported that over 88% felt they had learned about options and trade-offs, 93% had learned about others' interests and values, and 84% reported they had learned about their own interests and values[14].

Environmental management decisions involve multiple objectives and often require unfamiliar choices (few people routinely think about the pros and cons of different air emission controls or biosecurity planning for imported plants). Trade-offs frequently are made across very unlike outcomes, as occurs when a decision to protect ecological services may result in a loss of recreational or employment opportunities; and consequences usually are uncertain, which requires that choices need to recognize a distribution of possible outcomes. Each of these considerations can become more tractable as a result of the learning that's possible through participation in structured discussions.

Allowing emotions

Another key to effective resource-management deliberations is to recognize that learning is, at times, emotionally demanding because it brings up difficult feelings or controversial moral and ethical concerns. Examples include management actions – increasing capacity in a kraft-paper mill, or increasing mining or natural gas production near to an urban center – that are expected to decrease the (statistical) health of vulnerable populations (e.g. young or old individuals) in return for larger industry profits, near-term employment opportunities, or government tax revenues. These choices are sufficiently controversial that no matter what is decided, some people are likely to be angry or frustrated[15]. This does not imply that highly emotional issues should be avoided as part of group deliberations, instead, it means that facilitators, in response, need to make use of techniques that seek to establish a safe yet productive environment where effective deliberations and learning can take place.

One common response to highly emotional issues is that people may adopt simplistic decision rules that discourage any balancing of preferences across multiple objectives or avoid the loss of any valued resource, however small. This type of concern has been referred to by Baron and Spranca[16] as a 'protected' value; the term 'taboo trade-offs' also is used in this context[17]. The use of SDM or other decision-analytic approaches can help to address such refusals and, in some cases, encourage learning and more productive deliberations (while defusing confrontations) by providing explicit, transparent ways to express emotion-laden concerns[18]. Constructed performance measures, described in Chapter 5, provide one means to enhance the responsiveness of the decision process and to incorporate strong emotions associated with potential deaths or injuries, mistrust or perceptions of mismanagement, or worries about the loss of typically 'invisible' objectives such as social status or community image[19]. Knowing that these objectives and measures can be weighted differently by individuals,

and that standards exist for the construction of such measures, may help to reassure stakeholders and help to foster a desire on their part to learn about the nature of others' concerns.

The ability of SDM approaches to simultaneously handle different metrics also can help to discourage facilitators from the usual response of calling for calm or rationality. From an SDM, insight-based perspective, encouraging deliberative participants to stay calm may exclude important information, in that a negative or angry response may itself carry insights about how individuals perceive a choice and why they would refuse to support a proposed alternative. Working with different metrics also helps to address emotionally difficult management questions that arise from what the political scientist Philip Tetlock refers to as 'transactions that transgress the spheres of justice'[20]. Such concerns typically are the result of a request to express something of value (such as the intergenerational transmission of knowledge) in terms of a fundamentally different metric (such as dollars). The ability of SDM processes to express concerns in either natural, proxy, or constructed measures can help to avoid such problems. These different measures also will help to move past broad, general statements of moral or ethical belief – for example, using vague terms such as precautionary or sustainable, which lack predictive ability – and, instead, to help participants in a deliberative process to focus on the realistic and more specific choices that confront the resource manager: how much precaution, applied to what objectives or species, over what time periods, and at what cost?

Allowing mistakes

SDM approaches emphasize intentional learning, which occurs through both successes and mistakes. Learning from mistakes requires explicit attention to several things, including judgmental factors (were we overconfident, and hence overstepped the limits of our expertise?), the range of outcome feedback to which we're attentive (did we only look for confirming evidence and thus ignore some important information?), and systemic motivational forces (which in most cases tend to reward successful outcomes and penalize what are seen as failures[21]).

Only recently have environmental managers and decision makers begun to recognize the usefulness of methods or techniques – experimental management approaches, for example – that explicitly permit mistakes and failures as a mode of learning. The challenge is for participants in a decision process – especially where the context is novel, complex and consequential – to routinely examine their own motivations and risk tolerance: is the natural desire to avoid mistakes leading to an over-reliance on conservative or habitual methods of response? In our experience, a high level of underlying trust is a prerequisite to a willingness among diverse stakeholders to try something new and unconventional. A good way to build that trust is through iterative deliberations and the design of initiatives that allow for mistakes. This already is commonly done; for example, large-scale habitat restoration efforts on a large river might begin with several smaller trials, conducted in the controlled environment of a laboratory or in suitable tributaries. The role of an SDM approach would be to ensure that the benefits and costs of various alternatives involving trials – more or fewer, larger or smaller scale, of shorter or longer duration – are clearly identified and compared, along with a careful analysis of the learning that is anticipated and both the benefits and risks that are associated with likely failures or mistakes.

Encouraging iteration

The majority of environmental management processes are conducted under time and resource constraints; it is rare to hear from managers that they consider either the duration or resources allocated to a planning or evaluation process to be sufficient. Remembering our 'practical' focus, calls for iteration may therefore fall on deaf ears: managers may agree with the concept, but find themselves perpetually up against a wall in terms

of the need to deliver quickly results to senior staff or to decision makers. What does this mean for our emphasis on encouraging iteration as part of step 6 of the SDM process?

We draw a firm conclusion when it comes to iteration: in our experience, there almost always are significant benefits to careful iteration as part of an environmental management decision-making process. In previous chapters we have highlighted the difficulties that individuals experience in coming up with a comprehensive list of objectives[22], an informative set of performance measures, or a sufficiently broad group of alternatives. Many of these difficulties are compounded in the context of a multi-issue, multi-interest environmental management setting, where there is a need to integrate across the usual disciplinary boundaries: engineers are asked to understand why changes in water flows might adversely influence hunting access or the conduct of traditional ceremonies by an Aboriginal population, and community residents are asked to understand why setting aside land for a new park might reduce government royalties from timber harvests and adversely affect future funding for local schools. The earlier discussions also highlighted a variety of behavioral considerations that also underscore the need for iteration: in particular, individuals tend to be overconfident about the quality and relevance of their knowledge.

There is no instant fix for these problematic issues: simply telling a group that they need to work 'extra hard' on coming up with good objectives or a full set of alternatives will not automatically lead to desired results because some information that will be helpful is not yet available – it only comes from later steps in the decision-making process. Because each aspect of an SDM process is intended to inform both subsequent and preceding steps – thinking about performance measures helps to refine the understanding of objectives, or thinking about consequences helps to create better and more responsive alternatives – iteration is an essential component of a well-designed and well-implemented decision process. Nor does iteration necessarily require a substantial time commitment. In our experience, substantial gains can be associated with quick check-ins with participants (e.g. at the start of each of a series of meetings) and with creating an environment where iteration is encouraged, so that participants learn that it is appropriate and helpful for them to rethink previous discussions and to begin a consultative meeting or a technical working group session by stating: 'I've been rethinking something we talked about at our last meeting . . .'

10.3 Adaptive management revisited

Over the past three decades, adaptive-management (AM) approaches have been developed to address the uncertainty inherent in the function of natural ecosystems and to encourage learning over time with respect to both ecosystem behavior and the effectiveness of management actions[23,24]. As we described in Chapter 2, we view the frameworks for SDM and for AM as highly complementary.

As decision-oriented practitioners, we've moved toward using SDM as the overarching framework for an environmental management process and adapt the steps to ensure that a rigorous approach to identifying and testing hypotheses can be accommodated. We've found this to be a practical and useful way to integrate the science and learning objectives of adaptive management with the decision-making objectives of SDM. As others have noted, AM can be viewed as a special case of SDM for iterative decisions. Our perspective is that most substantive environmental management decisions are iterative. There are precious few that will not be reviewed at some point in the future, and for which learning about key uncertainties is not a key priority. So iteration and learning are built in to the SDM process. Further, we suggest that an SDM approach helps to address many of the implementation problems that typically are associated with AM, and that the explicit and *a priori* emphasis on

defining how subsequent decisions will be made – who will make them, using what objectives and measures as evaluation criteria, and through what analytical and deliberative processes – can enhance the likelihood of successful AM implementation.

10.3.1 The evolution of adaptive management

Early representations of AM as a continuous learning cycle looked much like that shown in Figure 10.3(a). The emphasis is on the design and implementation of management actions (often as experimental trials) and the associated monitoring. Important tasks such as defining the decision context, setting objectives and defining measures are included under the problem assessment (sometimes called the problem structuring) stage, but little guidance is typically provided about how to do this. Guidance about the 'evaluate' stage is similarly sketchy, which becomes problematic when there are trade-offs across objectives. There is broad agreement that the 'adjust' stage is where many AM initiatives ultimately falter, as the ability to translate monitoring results into a decision is complicated by inconclusive science and/or the diverse sometimes changing values of stakeholders and decision makers. In short, early attempts at AM seemed to fail as a result of a lack of attention to the institutional processes for planning, decision-making, and implementation[25,26].

More recent representations of AM (see Figure 10[b]), such as that provided in the USDOIs 'technical guide' to adaptive management, recognize the need for a more distinct set-up phase (i.e. planning) and iterative phase (i.e. implementation). In our opinion this framework enables a closer linkage between the ecological goals of adaptive management and the multidimensional goals and decision-making realities of resource-management agencies, including processes for engaging stakeholders and clearly identifying their objectives, risk tolerances, and trade-offs[27]. Within this framework, 'adaptive management is framed within the context of SDM, with an

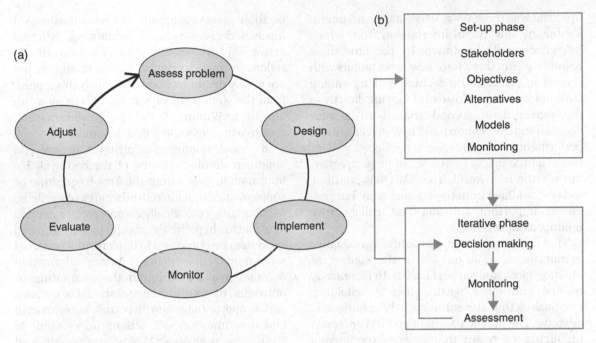

(a)

(b)

Figure 10.3 The evolution of frameworks for adaptive management (source: Williams *et al.*[28]).

emphasis on uncertainty about resource responses to management actions and the value of reducing that uncertainty to improve management'[28] (p. vii).

Our own experience is that the SDM process is sufficient for accommodating a sophisticated approach to implementation, monitoring and adjustment – in other words, what usually is described as 'the AM cycle.' Under an SDM approach, alternatives are developed for the specified decision context, based on stakeholders' fundamental objectives; these may include but are not limited to experimental and monitoring alternatives. Their consequences are estimated using explicit hypotheses based on best available information and models, with uncertainties examined in detail. If uncertainty is low or there are other reasons that result in a clear preference for a single alternative, then it may be selected, implemented and monitored. If the choice among multiple alternatives is complicated by uncertainty and the stakes are high (and a

variety of technical requirements are met such as the ability to detect an effect), then an experimental approach may be preferred. Different experimental designs will be possible, each with their own trade-offs (e.g. cost and time versus quality of information, risks to one species versus another, etc.). This may trigger another round of evaluation before final experimental designs are selected. Alternatives are then implemented and monitored and a review process is triggered.

Upon review, it is not uncommon for there to be changes in the institutional context surrounding who decides and how, and there may be new perspectives about which objectives are relevant. In our experience even the idea of a single set-up phase followed by an iterative phase, in which different trials are implemented, may be overturned. Along with the ecologist Paul Higgins, we worked with a multi-stakeholder group to define and reach agreement on a 10- to 12-year series of experimental trials designed to help restore a damaged coastal ecosystem. The first trial was three years long – a bare minimum for having any useful information from which inferences could be drawn. By the time these results were in, there were new expectations with respect to collaborative decision making among stakeholders and so, instead of moving directly to the agreed-upon second trial, both a new decision-making context and new objectives and performance measures were developed. All of this is a little messy from a scientific perspective, but it's the real world. The AM trials continue today – modified from the original plan, but providing important learning and collaborative management benefits[29].

This decision-focused perspective on adaptive management, as it unfolds in the context of building ideas and support from both technically trained and other participating stakeholders, emphasizes that the entire cycle is iterative and adaptive. The practical focus of an SDM approach highlights the reality that for any experimental trial that requires significant resources or a significant amount of time to complete, there needs to be an evolving discussion and exchange of both factual and values information among the people who are involved. This will necessarily involve iteration and learning, over all six SDM steps, because key aspects of the decision framework will continue to be discussed and revisited and altered based on the results of analyses and deliberations. The recognition of uncertainty, about the stability and role of institutions and about the influences from the external world that may influence the results of any AM trial, also plays an important role. The world does not stop or cease to change while an AM trial is going on: AM practitioners ignore these influences, and the associated lessons for the design of successful AM trials, at their peril.

10.3.2   Types of adaptive management

A distinction is often made between two primary types of adaptive management, 'passive' and 'active'. Although both approaches are valuable, their implications for the resource-management decision process are quite different: passive AM involves decision-focused monitoring, whereas active AM involves decision-focused experimentation. In our experience, both AM approaches are widely misunderstood and widely misapplied, from the standpoint of serving effectively as an aid to facilitating learning by stakeholders, resource managers, and decision makers.

In theory, managers applying a passive AM approach should make use of the best available information, either from the area in question or supplemented by data from other areas considered to be roughly ecologically comparable, to develop alternative hypotheses about the anticipated ecosystem response and to implement a preferred set of management actions. As new information becomes available through the monitoring of outcomes, the existing data set should be updated and, if appropriate, new hypotheses constructed and new management actions undertaken. *In practice*, many passive AM programs are criticized as being little more than expensive exercises in 'trial and error' learning. We believe that if examined closely, failure can always be traced to failure to adhere to at least one of the core requirements listed above.

The feature that distinguishes active AM is that the implementation phase incorporates actions intended to deliberately perturb the system. Management interventions may take place at different (yet comparable) locations or at the same location over time. For example, the effects of different water flows in a managed system may be expressed across three similar-sized reservoirs or, at a single facility, water levels may be set at one height for several years and then changed, with the results compared in terms of explicit, pre-set variables[30]. The goal is to develop statistically testable information about the system of concern through targeted manipulations. However, these clear and sensible goals of an active adaptive-management approach are beset by several potentially serious problems. First, it may be difficult or impossible to evaluate results in the absence of contamination from outside or external sources. For example, the comparison of growth rates of riparian vegetation in response to different water flows may be affected by variations in precipitation or temperature, changes in use (e.g. incursions by hikers or cows), or any number of other factors. A second problem is that what scientists perceive to be a 'short period of time' may be viewed quite differently by the management agencies in charge of the area, by community residents, or by local politicians – whose timeframes rarely exceed the date of the next election. And a third problem is that the proposed comparisons may result in clear winners and losers; in this event, it is likely that those who were affected by the less beneficial treatment will complain to the decision makers and possibly seek compensation or pursue legal action.

Regardless of which approach is taken, passive or active, there are challenges to be faced if the goal is to explicitly link learning to managers' decision-making practices. Given the cost and long-term commitment typically required for successful AM, it's useful to consider some decision-focused criteria for guiding the decisions about whether and when adaptive-management approaches might be useful.

10.3.3   Conditions for learning through adaptive management

Adaptive-management processes are not a one-size-fits all proposition: they couldn't be, in light of the diversity of scale and content of the different resource-management problems they are designed to address. Some applications of AM, for example, have been relatively simple and small scale. For example, the British Columbia Forest Service conducted experiments throughout the 1990s to evaluate a range of forest-harvesting techniques; these studies posed no threat to individual species or to the larger ecosystem and the results were not intended to be useful to other locations[31]. Other AM applications, such as the large-scale adaptive-management experiments on the Columbia River Basin[32] or the work by Walters, Gunderson, and Holling[33] on the Florida Everglades, have affected multiple interests – fishers, Native Americans,

farmers, industry, recreationists, and local governments – and held the potential to affect significantly a variety of endangered species. Not surprisingly, the benefits of an adaptive-management approach have been widely debated, with serious questions raised about the feasibility of adopting an AM approach specifically when it is often most needed – on multi-issue, multi-stakeholder environmental management problems characterized by complex ecological and human interactions that are subject to significant uncertainty.

Although the range of potential problems and institutional settings is large, five main questions (adapted from Gregory et al.[34]) cover the more critical decision-focused issues and can help to determine, a priori, if an adaptive approach to resource management is likely to be effective as an aid to learning and better resource management.

1 Do the uncertainties affect the choice of future management actions? – and specifically, in consideration of all the objectives that a management action must respond to (economic, environmental, and social), is it plausible that a reduction in the targeted uncertainty will result in a change of management action?
2 Does the spatial and temporal scale allow for formal and effective monitoring or experimentation? Will we be able to draw meaningful inferences in a timeframe that's relevant for management? Can we get all the parties who need to be involved to collaborate?
3 Given the multiple objectives of stakeholders and potential risks, is experimentation really an option? Will perturbation of the system over long time periods be acceptable to a broad base of stakeholders?
4 Is there sufficient institutional capacity to support an AM program? This refers not only to the resourcing required for monitoring and review, but to building in review cycles to institutionalize the capacity to learn and to respond to new information over time. It also involves giving careful attention to the

processes that are required to build a shared understanding of both the selected management approach and the results among diverse stakeholders, recognizing that the people engaged in these activities may not be the same: a period of several years may pass before results becomes available to address the questions raised by an advisory committee of stakeholders.

5 Are there alternative ways to reduce the relevant uncertainties? For example, careful elicitation of experts' viewpoints or a review of practices in a comparable location may provide sufficient new knowledge that additional experimental trials are not needed.

These questions should be carefully considered before a decision is made to enhance learning and reduce uncertainty through a time-consuming and often costly adaptive-management initiative. As we pointed out in Chapter 2, it's important to recognize that conducting an AM experiment is itself a multiobjective choice, involving value-based trade-offs[34]. When choosing among alternative management actions, sometimes uncertainty severely hinders our ability to choose a preferred one – and in this case, one or more of the alternatives may involve experimentation. But it's only one way to build alternatives, and a value-based choice still may be needed between implementing an AM experiment and implementing other management actions. Neither analysts nor decision makers should be drawn into the appealing nature of AM as a construct, despite the intuitive attraction of a 'learning by doing' proposal.

For example, many resource-management initiatives are driven by the desire for learning about a specific aspect of ecosystem behavior under uncertainty – for example, the effect of climate change on the quality or abundance of lichens that serve as a food for migrating caribou. Yet the choice of a management practice is never based on only a single objective. Once 'experimentation' is selected there are still value-based trade-offs to make among experimental designs. The promise of an AM approach is that

shorter-term learning will improve longer-term management possibilities[35]. Yet acceptance of this logic still leaves resource managers and decision makers with tough choices: longer experiments may deliver higher quality information but they defer potential on-ground benefits, some experimental designs impose more stress on the resources under consideration than will others, and the selection of which alternatives merit exploration (i.e. which alternatives make the short list for experimentation) is itself a value-based choice, especially when there are multiple objectives. The bottom line is that the SDM process is designed to facilitate learning, and it should be able to accommodate the careful examination of decision-relevant AM initiatives – involving either experiments or monitoring – through the rigorous comparison of a broad range of alternatives, some of which may in the end prove to be better (albeit perhaps more mundane) choices than any of the AM-based options.

10.4 Key messages

SDM methods promote and emphasize the importance of learning as part of decision-making processes. This chapter describes how an emphasis on learning can improve the quality of current and future environmental management decisions. We examine how learning can improve the quality of information used in estimating the consequences of proposed actions, including helping managers and stakeholders to understand which sources of uncertainty might matter the most (and therefore be most deserving of further resources). The discussion also covers issues of capacity building, of linking monitoring to the uncertainties that affected decisions, and the need to institutionalize trigger mechanisms for reviews of ongoing plans. Special attention is given to the promise and perils of deliberative learning, including both fact-finding opportunities and methods for helping participants to clarify their values in the face of novel decision problems and new perspectives. Learning through adoption of an adaptive-management approach is examined in the context of an SDM perspective, with emphasis on the need for clearly defined management objectives and hypotheses that recognize problems associated with the future detection and interpretation of the consequences of management actions.

10.5 Suggested reading

Gregory, R., Ohlson, D. & Arvai, J. (2006) Deconstructing adaptive management: criteria for applications to environmental management. *Ecological Applications*, **16**, 2411–2425. Applies the logic of SDM methods to develop criteria for the successful application of adaptive-management principles to environmental management problems.

Holling, C.S. (1978) *Adaptive Environmental Assessment and Management*. Wiley, Chichester. Environmental impact assessments have not been the same since Holling pulled together this edited selection of papers on adaptive management, linking environmental planning to the dynamics of human and ecological systems. It was path-breaking when first published and it is still worth a careful read (or a fifth re-reading).

Ludwig, D. (2001) The era of management is over. *Ecosystems*, **4**, 758–764. Makes the argument that many of the most important environmental issues are sufficiently complicated that the only hope for solutions lies in establishing an adaptive process, responsive to the objectives of resource users and management agencies, and to maintain an ongoing dialogue among interested parties.

Runge, M., Converse, S. & Lyons, J. (2011) Which uncertainty? Using expert elicitation and expected value of information to design an adaptive program. *Biological Conservation*, **144**, 1214–1223. Using expert judgment elicitations, the authors demonstrate the benefits of identifying which sources of uncertainty are most important to development of an effective resource-management strategy for whooping cranes.

Walters, C. (1997) Challenges in adaptive management of riparian and coastal ecosystems. *Conservation Ecology*, **1**, 1. Retrieved from http://www.ecology andsociety.org/vol1/iss2/art1/. A provocative and insightful review of adaptive-management applications to fisheries problems, informed by the author's extensive experience, clear thinking, and strong opinions.

Williams, B.K., Szaro, R.C. & Shapiro, C.D. (2007) *Adaptive Management: The U S Department of the Interior Technical Guide*. US Department of the Interior, Washington, DC. Retrieved from http://www.doi.gov/initiatives/AdaptiveManagement/TechGuide.pdf. Summarizes methods and experience of the US Department of the Interior in developing and applying adaptive-management approaches for addressing uncertainty in environmental management, based on insights from structured decision-making methods.

10.6 References and notes

1 Ludwig, D. (2001) The era of management is over. *Ecosystems*, **4**, 758–764.

2 Berkes, F. & Folke, C. (2002) Back to the future: ecosystem dynamics and local knowledge. In: *Panarchy: Understanding Transformations in Systems of Humans and Nature* (eds L.H. Gunderson & C.S. Holling). pp. 121–146. Island Press, Washington, DC.

3 Runge, M., Converse, S. & Lyons, J. (2011) Which uncertainty? Using expert elicitation and expected value of information to design an adaptive program. *Biological Conservation*, **144**, 1214–1223.

4 Clemen, R.T. (2004) *Making Hard Decisions: An Introduction to Decision Analysis*, 4th edn. Duxbury, Belmont.

5 Goodwin, P. & Wright, G. (2004) *Decision Analysis for Management Judgment*, 3rd edn. Wiley, Chichester.

6 Szymanski, J.A., Runge, M.C., Parkin, M.J. & Armstrong, M. (2009, October) *White-Nose Syndrome Management: Report on Structured Decision Making Initiative*. Retrieved from http://www.fws.gov/northeast/pdf/WNS_SDM_Report_Final_14Oct09.pdf

7 Yokota, F. & Thompson, K.M. (2004) Value of information analysis in environmental health risk management decisions: past, present, and future. *Risk Analysis*, **24**, 635–650.

8 Lyons, J., Runge, M., Laskowski, H. & Kendall, W. (2008) Monitoring in the context of structured decision-making and adaptive management. *The Journal of Wildlife Management*, **72**, 1683–1692.

9   Argyris, C. & Schön, D.A. (1974) *Theory in Practice: Increasing Professional Effectiveness.* Jossey-Bass, San Francisco.

10 Gunderson, L.H., Holling, C.S. & Light, S.S. (1995) *Barriers and Bridges to Renewal of Ecosystems and Institutions.* Columbia University Press, New York.

11 Gregory, R. & Long, G. (2009) Using structured decision making to help implement a precautionary approach to endangered species management. *Risk Analysis*, **29**, 518–532.

12 Failing, L., Gregory, R. & Higgins, P. (2011) Decision-focused consultations to help implement adaptive management plans. Manuscript submitted for publication.

13 Durham, L. & Brown, K. (1999) Assessing public participation in US watershed planning initiatives. *Society and Natural Resources*, **12**, 455–467.

14 Dovetail Consulting (2005) R*eview of B.C. Hydro's Water Use Plan Collaborative Process.* Vancouver, British Columbia.

15 Lichtenstein, S., Gregory, R. & Irwin, J. (2007) What's bad is easy: taboo values, affect, and cognition. *Judgment and Decision Making*, **2**, 169–188. Retrieved from http://journal.sjdm.org/jdm7314.pdf

16 Baron, J. & Spranca, M. (1997) Protected values. *Organizational Behavior and Human Decision Processes*, **70**, 1–16.

17 Fiske, A.P. & Tetlock, P.E. (1997) Taboo trade-offs: reactions to transactions that transgress spheres of justice. *Political Psychology*, **18**, 255–297.

18 In most SDM applications where such taboo trade-offs arise, decision-making authority rests with elected officials or a regulatory body. The task facing the SDM practitioner is to ensure that decision makers have the best possible information about the values of potentially affected stakeholders and implications for their support or opposition, along with an understandable presentation of the relevant facts and key sources of uncertainty.

19 Turner, N.J., Gregory, R., Brooks, C., Failing, L. & Satterfield, T. (2008) From invisibility to transparency: identifying the implications. *Ecology and Society*, **13**. Retrieved from http://www.ecologyandsociety.org/vol13/iss2/art7/

20 Tetlock, P.E., Peterson, R.S. & Lerner, J.S. (1996) Revising the value pluralism model: incorporating social content and context postulates. In: *The Psychology of Values: The Ontario Symposium on Personality and Social Psychology* (eds C. Seligman, J.M. Olson & M.P. Zanna), Vol. 8. pp. 25–51.: Erlbaum, Mahwah.

21 Schoemaker, P.J.H. & Gunther, R.E. (2006) The wisdom of deliberate mistakes. *Harvard Business Review.* Retrieved from http://hbr.org/2006/06/the-wisdom-of-deliberate-mistakes/ar/1

22 Bond, S.D., Carlson, K.A. & Keeney, R.L. (2008) Generating objectives: can decision makers articulate what they want? *Management Science*, **54**, 56–70.

23 Holling, C.S. (1978) *Adaptive Environmental Assessment and Management.* Wiley, Chichester.

24 Walters, C. (1986) *Adaptive Management of Renewable Resources.* MacMillan, New York.

25 Folke, C., Carpenter, S., Walker, B., Scheffer, M., Elmqvist, T., Gunderson, L. & Holling, C.S. (2004) Regime shifts, resilience, and biodiversity in ecosystem management. *Annual Review of Ecology, Evolution, and Systematics*, **35**, 557–581.

26 Walters, C. (1997) Challenges in adaptive management of riparian and coastal ecosystems. *Conservation Ecology*, **1**, 1. Retrieved from http://www.ecologyandsociety.org/vol1/iss2/art1/

27 National Research Council (2005) Decision Making for the Environment: Social and Behavioral Science Research Priorities. National Academies Press, Washington DC.

28 Williams, B.K., Szaro, R.C. & Shapiro, C.D. (2007) *Adaptive Management: The US Department of the Interior Technical Guide.* US Department of the Interior, Washington, DC. Retrieved from http://www.doi.gov/initiatives/AdaptiveManagement/TechGuide.pdf

29 Bradford, M., Higgins, P., Korman, J. & Sneep, J. (in press). Does more water mean more fish? Test of an experimental flow release in the Bridge River, British Columbia. *Freshwater Biology*.

30 Gregory, R., Failing, L. & Higgins, P. (2006) Adaptive management and environmental decision making: a case study application to water use planning. *Ecological Economics*, **58**, 434–447.

31 Taylor, B., Kresmater, L. & Ellis, R. (1997) Adaptive management of forests in BC. Report for the Forest Practices Branch, Strategic Policy Section, BC Ministry of Forests. Queens' Printer, Victoria BC.

32 Lee, K. (1993) *Compass and Gyroscope: Integrating Science and Politics for the Environment.* Island Press, Washington, DC.

33 Walters, C., Gunderson, L. & Holling, C.S. (1992) Experimental policies for water management in the Everglades. *Ecological Applications,* **2**, 189–202.

34 Gregory, R., Failing, L. & Higgins, P. (2006). Adaptive management and environmental decision making: A case-study application to water use planning. *Ecological Economics* 58: 434–447.

11

Reality Check: Implementation

The previous chapters have provided numerous suggestions about how to proceed with selecting and implementing an SDM approach to an environmental management decision. We've sought to provide a wide range of case studies and relevant techniques, so that any readers with experience in resource management will recognize at least some of the situations we've posed and exclaim: that's it, I've been there, and I've done that (or, I wish I had or hadn't done that). But even if you now feel that you've learned some additional methods or concerns to complement your existing knowledge of SDM, or if you are new to this and now feel there might be some benefit in trying an SDM approach, there are a few more issues to raise so that the lessons of this book can improve the likelihood that SDM will, in fact, prove to be a useful aid to your environmental management decisions.

Hence the title: a reality check, with a focus on implementation challenges. We begin by reviewing several of the concerns that are often raised when SDM approaches are introduced as part of resource-management planning. We then address some unglamorous but important issues associated with organizing and working with advisory and technical committees as part of multi-stakeholder consultations. We also discuss ways for integrating an SDM process with larger public processes; this linkage is often essential for the successful communication and implementation of even the most superbly conducted analyses and deliberations. This still leaves thorny issues associated with (at times) finding ways to actively involve decision makers in the analysis itself or bringing the results of an SDM analysis to their attention, keeping in mind that not all 'decision makers' are particularly skilled at making good decisions. We then review a typical SDM work plan, covering meeting tasks and timing issues, and show how this model was adapted to work well as part of a real-life case study. Where did the work plan hold up? Where and when was it important to make adjustments? Finally in this chapter, we want to demonstrate the range of SDM applications and, after all this talk of methods and techniques, to present several final case studies that return us to the roots of SDM – as a sensible, consistent decision structuring approach intended to provide insights for improving environmental management choices.

11.1 Addressing concerns about using structured decision making approaches

As noted in Chapter 2, multi-attribute methods based in decision analysis (including SDM) are not used as often by environmental managers as

Structured Decision Making: A Practical Guide to Environmental Management Choices, First Edition. R. Gregory, L. Failing, M. Harstone, G. Long, T. McDaniels, and D. Ohlson.
© 2012 R. Gregory, L. Failing, M. Harstone, G. Long, T. McDaniels, and D. Ohlson. Published 2012 by Blackwell Publishing Ltd.

are a variety of other economic, ecological, and risk-based methods. The reasons are diverse:

1 Historical; since decision making has received scant attention as a distinct body of study or practice and decision-analytic methods have been in widespread use for less than 50 years.
2 Disciplinary; since a reliance on their training in the natural sciences (e.g. ecology, forestry, wildlife ecology, oceanography, or conservation biology), economic analyses, or dispute resolution approaches still guides most resource-managers.
3 Evolutionary, since citizen involvement in environmental decision making and interest in deliberative processes is a relatively recent initiative, far more common today than only 20 or 30 years ago.

As a result, any mid-level environmental manager interested in trying some of the methods described in this book, as part of a partial or full-blown SDM application, is likely to hear something less than ringing initial support from senior managers or elected decision makers. This first section of our reality check addresses seven concerns that often are brought up as reasons to not try elements of an SDM approach.

Issue 1: Structured decision making sounds expensive

It is true that the initial phases of an SDM process can require relatively more resources than a multi-criteria, risk analysis, or contingent valuation process. The reason for this, as noted in earlier chapters, is that SDM places greater emphasis on problem structuring – bounding the problem, identifying key stakeholders, helping them to define objectives and articulate effective performance measures – and (consistent with the constructed preferences paradigm discussed in Chapter 2) on making sure that deliberations about these foundational steps take place among participants. However, problem structuring is not expensive (compared with gathering biological

data, for example) and by helping to identify the types of anticipated impacts for which data are required or additional research and study are needed, SDM can ultimately conserve resources and reduce managers' costs. A good decision structuring process is thus likely to reduce costs or, at worst, to shift costs (e.g. from baseline biological studies to the assessment of stakeholder objectives and management options) rather than increase them. This is particularly true for some of the more contentious aspects of evaluation, such as making defensible trade-offs across different considerations. In its review of a decade of SDM application to procurement decisions, for example, BC Hydro concluded that 'the SDM-informed ranking and weighting system ... has resulted in decreased costs'[1](p. 5).

Issue 2: Structured decision making will take too long

There are several responses to this assertion. First, SDM approaches are flexible, and can be designed to fit different timeframes. Examples discussed throughout this book include SDM processes that took place over less than one day, three days, one month, and two years. In addition, even a more lengthy SDM process can quickly identify activities that everyone agrees on and actions relating to these 'low-hanging fruit' can be initiated quickly, while more difficult decisions remain under evaluation.

Second, it is probably true that the first stages of an SDM process might well take longer than those of many alternative methods. However, the decision structuring phase creates a road map and a common understanding of the process, and builds a common vocabulary among participants. These help to streamline later steps. We stated in Chapter 1 that the six core steps of an SDM approach serve an important organizing function, in that they provide a logical sequence for addressing environmental problems: defining the decision context sets the stage for discussing stakeholder concerns and relevant values, objectives are defined before generating alternatives, and performance measures are clarified before value trade-offs are addressed. To those with no previous experience in using SDM, this attention to objectives, measures, and creating alternatives can sound like a recipe for a never-ending procedure. It's therefore important to remember that tools such as consequence matrices provide summary reviews of alternatives in ways that quickly delete non-helpful objectives (i.e. those which fail to distinguish among decision options) and omit dominated alternatives (i.e. those which are inferior, or nearly so, on all dimensions of value) from further discussion. These structuring qualities save time and resources. They also can save some other things even more precious, which include the credibility and transparency of the deliberations and the patience and enthusiasm of the participating stakeholders.

Issue 3: Structured decision making is too quantitative

SDM approaches make no secret about their use of quantitative analysis. The core of SDM, however, is its reliance on value-focused thinking and decision-structuring methods. Analytical techniques are used as a means for stimulating deliberation or dialogue – which, in turn, are a means to constructing informed values, understanding facts, and providing insight to decision makers. When an SDM approach is done well, its more quantitative aspects are transparent and fully in the service of answering questions that naturally arise in the course of a multi-stakeholder dialogue: how is information to be collected, how are uncertainty and experts' confidence to be portrayed, how is the importance of different objectives to be considered, and how are trade-offs to be made across the different value-based dimensions of alternatives? SDM is not always quantitative: in many cases, the problem structuring aspects of an SDM approach can provide substantial decision-aiding help to stakeholders and decision makers without the need to conduct any quantitative analyses at all. And if some of the analyses conducted as part of an SDM process are highly quantitative, then

it's the job of the facilitator/analyst to explain why these specific methods were necessary and to articulate the reasoning behind their interpretation. In our experience, participants are readily able to understand even the most complex probabilistic or statistical analyses so long as the explanation is clear.

A related issue is the concern that an SDM evaluation will be all about numbers and, as a result, ignore important qualitative or philosophical concerns – for example, helping managers to come up with initiatives that are sustainable or adaptive. Yet as we've discussed in many of the previous chapters, SDM approaches often are used to generate plans that are (for example) precautionary, robust, resilient, or adaptive. The practical nature of SDM methods can be at odds with those individuals who like to stay 'in the clouds' and wax eloquent about abstract principles, yet run from attempts to make these concrete or implementable. From our perspective, if this is the root of the criticism, then we plead guilty. In our experience, often the only way to achieve principled action is to conduct – and clearly explain – appropriate quantifiable analyses.

Issue 4: Structured decision making is an art

There is no doubt that a successful SDM process involves good chemistry among participants and subtle choices on the part of the facilitator or analyst. Experience therefore helps, personality does come into play, and all of us who have worked as consultants to environmental managers know that some participants (and some clients) are much easier to work with than others. Does this mean that being good at the practice of SDM is an 'art'? Perhaps – but no more so than being good at the job of a physician, teacher, midwife, firefighter, or chef. To do any of these jobs well requires a combination of technical skills, emotional intelligence, and facilitation skills, along with an ability to not only know the rules and procedures but also when to bend them and do things a little differently, or at least when and how to do familiar things in unfamiliar ways or out of the usual order. Being effective also requires an ability to simultaneously pay attention to both the big stuff (in the case of SDM, things like developing a concise set of relevant objectives and gathering high-quality data about consequences) and the smaller things, for example picking up on a distracted participant and asking whether they have something to contribute, or recognizing that a decision maker may need a little delicate pushing and probing to reveal an important but not obvious objective. In most ways, this concern therefore strikes us as rather naive: if being (in part) an art implies an attention to people and thoughtfulness and knowing something about when to deviate from the usual rules, then this is simply a description of business as usual for most resource managers – and why would anyone expect effective decision making to be any different?

Issue 5: Structured decision making under values scientific input

Good science is an essential element of a successful SDM process. But there are two important caveats. First, other sources exist for obtaining credible information about consequences. Second, although science clearly has an important role to play in defining objectives, performance measures, and the consequences of actions, science alone does not make decisions[2]. Choices about everything – from which facts are relevant and how to represent or analyze them, to which policies or actions to implement – are made not by science but by people's values. If the question at hand involves the evaluation of a proposal to improve a local fishery, for example, answers to questions about fish biology will come in part from the scientific community. However, other sources of knowledge, including the results of studies conducted on another continent or the experiences of long-term resource users and residents of nearby communities, may also make valuable contributions to the knowledge base. Drawing from local

and community sources of knowledge also can have significant process benefits, for example in building trust and credibility. Yet the same questions of information relevance, quality, and reliability will apply to all sources, and systems of culturally appropriate peer review will need to be incorporated as part of the analytic and deliberative process.

Issue 6: Structured decision making over values public input

A critical feature of a structured approach to environmental management is that it relies heavily on participating stakeholders and decision makers to assess the *relative* value or importance of effects, in terms of a range of different metrics. This is true whether participants are technically trained experts or a mix of resource managers and community residents. We recognize that the individuals and stakeholder groups who participate in a SDM process, for example as participants on an advisory committee, are seldom representative (in a formal sense) of the public at large. Decision makers therefore need to use common sense and good judgment when drawing wider conclusions based on the results of a small-group SDM process: the goal is neither to accept the results as strictly indicative of the larger population nor to ignore the results of such processes and, instead, to make choices on the basis of closely guarded political or interest-based considerations. SDM approaches seek to provide insight to decision makers about the reasons why support or opposition to different strategies is likely, and thus even choices that remain strongly influenced by political considerations stand a reasonable chance of being better informed and more transparent. Yet it should be clear that, except in the relatively rare cases where a formal commitment to shared decision making exists, final decisions are left to the responsible managers or officials; thus any SDM process needs to ensure that there are clear expectations about who will actually decide on what initiatives go forward.

Issue 7: Structured decision making will be ignored by decision makers

Like any evaluative process, SDM will sometimes be 'ignored' by decision makers – in the sense that a 'political' decision is made anyway, under influence of public pressure, industry lobbyists, or bilateral negotiations. But even when this happens, SDM has usually succeeded in narrowing the scope of options under consideration and highlighted key issues that need to be talked about.

Sometimes SDM will be uncomfortable for decision makers. For example, they may shy away from being so explicit about difficult trade-offs. A pro-active SDM analyst can anticipate these challenges by involving decision makers at key steps of the process, such as setting objectives and creating a short list of alternatives and reviewing trade-offs, thereby increasing the probability and extent to which SDM is directly useful. If a convincing analysis can be done, particularly one that shows status-quo management actions to be inferior to other alternatives, and if there is broad stakeholder support for implementation of these other options, then it is a rare (and probably short-lived) manager or politician who won't decide to at least make some changes to current policies.

11.2 Organizing participants

For environmental managers, nearly every problem will benefit from the perspective or involvement of other parties. In some cases this might be limited to an 'internal' group: colleagues, other departments within an organization, other ministries within government, and so on. In some cases, a multi-agency group might be convened – including representatives of different levels of government or different government and regulatory agencies, but stopping short of full public engagement. In other cases the relevant parties might include diverse stakeholders from key resource user groups and industry, interested and perhaps influential members of nearby communities and the public, and (in relevant cases) aboriginal communities or non-governmental organizations. These groups may work in relative isolation or their every move may be scrutinized by senior managers, politicians, or the media. In any of these cases the group or committee may be charged with either making a management decision or making a recommendation to a decision maker (e.g. an elected official) or decision-making body (e.g. a regulator). Different participants will have different levels of knowledge of the problem specifics and different types of training. Usually important information is known ahead of time, but key bits are nearly always missing. Sometimes time and resources are scarce, but not always.

In this section, we discuss common challenges that arise in four different settings – working with a multiple stakeholder committee, establishing a technical working group, engaging in dialogue and analysis with members of a group, and integrating the results of a small-group SDM process with broader public input – as part of using SDM for aiding environmental-management decisions.

11.2.1 Multi-stakeholder advisory committees

A multi-stakeholder committee will typically involve 10–30 people who are brought together to represent a variety of different points of view in the context of a pending resource-management issue: should the boundaries of a state park be enlarged; how can a dependable water supply be established for a new town site; what requirements should be placed on industry to protect an endangered species; what is the best way to transport electricity from a generation site to its market? The purpose of having different people in the room is to develop alternatives that characterize their different perspectives and to ultimately increase the probability that any agreed-to actions will be implemented. The values side of this is easy to see: different people are likely to have different reasons for either favoring or opposing an action, and capturing these reasons through carefully defined objectives and performance measures provides a good start on a resource-management decision. The facts side is equally important: different people are likely to have access to different sources or types of information and thus can contribute to a more comprehensive and responsive accounting of the anticipated consequences of any proposed actions. Process considerations are also often essential: unless the right people are on the advisory committee, not only are important objectives or consequences likely to be left out, but the recommended actions are far less likely to be implemented quickly or with broad support.

Ideally, the participating members of an advisory committee would, as a group, represent all key interests related to the problem or opportunity. In reality, it is easy for some perspectives to be omitted, just by virtue of peoples' availability: elderly and special-needs individuals, parents of young children, people who work 10 or 12 hours per day. It's then up to the leader to change meeting times (evening rather than day, or weekends?) or to provide special assistance (e.g. daycare so parents with younger children can attend meetings?) or special incentives (a free lunch or dinner after the meeting?). If important perspectives are left out then it's important to add additional people later on, or at minimum to find a way to incorporate their views. Sometimes the problem is the opposite one, in that an invited member of the committee doesn't attend or has to be removed for failing to respect meeting ground rules. Often the responsible manager and analyst have control over who is on a committee, and both selection decisions and add/drop decisions can be made in a straightforward fashion, whenever possible in consultation with other committee members. Sometimes the analyst has no say over the attending stakeholders, because of the history or nature of the problem (e.g. an existing committee already is formed, as in the Nicola Valley case study presented in this chapter).

11.2.2 Expert and technical committees

A different setting for application of SDM techniques involves working with an expert or technical committee. There are two main ways in which technical committees arise. One is as an offshoot of a larger, multi-stakeholder initiative. Here the issue is how best to organize a group with diverse talents and personalities so as to take full advantage of their knowledge about the identity and consequences of potential actions. In a practical sense this often means either forming a technical or expert committee with a subset of participants (e.g. those who know the most about the issue in question) or bringing in outside consultants or presenters (with specialized knowledge of a topic). In the context of a multi-stakeholder process, it is essential that a clear basis be established for discussing and deciding on the criteria for defining expertise.

As one example of how *not* to do this, one author was asked in the early 1990s to help out with an ongoing, multi-interest group in Alaska addressing issues relating to exploration of potentially valuable offshore oil sites. An important topic was how oil exploration could impact aquatic life, with a special interest in walrus (a species important for food as well as ceremonial uses by local indigenous populations). In its limited wisdom, the United States government client had earlier brought in 'experts' from outside the region who had strong western-science credentials, yet completely ignored local expertise in the form of Inuit hunters who had 40 or more years of experience with walrus. Not surprisingly, the testimonies of these outside experts were largely ignored by local residents and a significant rift had been created between the client and members of the consultative group. Our first recommendation was to incorporate local knowledge in all future discussions of consequence identification and estimation. In contrast, as part of developing a new water plan for the managed Alouette River[3], several resident members of the multi-interest advisory committee volunteered to work closely with selected outside technical experts on the development of improved performance measures and consequence estimates, with presentations later made to the entire committee so that questions could be asked and any remaining concerns addressed.

The other way in which SDM approaches are used with experts is when they ask for help in order to more clearly define their objectives, identify hypotheses relating to the topic of interest, understand the uncertainty associated with the outcomes of management actions, or address trade-offs across different policy alternatives. Several examples of this type of SDM application already have been provided: the hypothesis identification assistance provided to the joint United States/Canada technical group examining recruitment failure in Upper Columbia River white sturgeon (see Chapters 8 and 10), the portfolio development work with the three-agency management board on a recovery plan for endangered Atlantic salmon (see Chapter 7), the decision-aiding assistance on management plans to reduce catastrophic losses among bats due to white-nose syndrome (also Chapter 7), and the strategy development work with a multi-interest technical group charged with developing new wildlife management strategies (see Chapter 5).

Not every initiative designed to make use of SDM with technical or expert committees goes smoothly, however, for a variety of reasons. The most common pitfall in working with expert groups using SDM approaches is the rift between 'objective' science and 'subjective' decision making. There are three main ways that this gap might take form, with consequences for an SDM process ranging from a short-term awkwardness to an outright disaster:

1 Experts might bristle at the idea that their science is not as objective as they might think it is. Adding to this is the feeling among some scientists that methods fundamental to approaches based in decision analysis, such as the use of formal techniques to examine experts' consequence estimates (discussed in Chapter 8), are not only unnecessary but convey a mistrust of science or scientists, as if the deconstruction of experts' assumptions will somehow undermine (rather than strengthen) the credibility of their science.

2 Experts may protest when the analyst states that no one can make 'science-based decisions'. This often arises when scientists are asked for value judgments disguised as technical judgments: a common example is when an expert panel is assigned the task of setting targets or thresholds, which was noted in Chapter 4 as a recipe for disaster.

3 Experts might not want to address social science investigations with the same rigor as biological science – despite the SDM claim that social, economic, or cultural effects should enjoy the same resources and attention and should be subject to the same kinds of quality standards. Gregory and Keeney[4] (p. 1602) note that, as a consequence, it can be difficult for natural scientists to ask for help with something as presumably commonplace as simply 'making decisions' due to 'the misperception that social science techniques are largely common sense (and, therefore, not meriting careful study) rather than, as is more correctly the case, reflective of common sense'.

11.2.3 Challenges in working with groups

A third issue – one that cuts across both multi-stakeholder advisory and technical contexts – is that many resource managers have little knowledge of what might be different about working with groups as compared with working with individuals. Often, there is a hopeful assumption that group wisdom will help to overcome the frailties of individuals' judgments. Yet research on the topic suggests that little in the way of wisdom comes automatically with people getting together as a group. On the plus side, helpful insights can derive from averaging across diverse judgments. Surowieki[5] refers to this as 'the wisdom of crowds', but it arises only if several conditions are met: the group reflects a diversity of opinions and skills, members' judgments are independent and not easily influenced, the participants are truthful, and appropriate techniques are used to aggregate individual judgments. These conditions will rarely be met without skilled input on the part of a facilitator who is able to encourage candor and independence among participants. In the absence of this input, likely tendencies of group judgmental processes include groupthink, whereby pressures to conform push decision-making groups to converge quickly on an inferior outcome[6], or a reluctance to focus sufficiently on information that isn't shared among all members (termed the 'common knowledge effect' by Gigone & Hastie[7]).

There are straightforward responses to some of these concerns. For example, when eliciting objectives from more than one individual, it's usually helpful to ask each person to write down their objectives first, prior to moving on to a group discussion, so that participants don't anchor on the ideas presented by others and thus succumb to groupthink. Probing for objectives and measures, for example by asking members of a group to consider others' perspectives or to generate bookend alternatives (see Chapters 4 and 7), can help to overcome the disadvantages of the common knowledge effect. Yet group dynamics remain somewhat mysterious; for example, there is some evidence that deliberation can pull groups toward their more extreme views (rather than toward a middle or average ground) because differences across group members are magnified[8]. Common sense approaches such as devil's advocacy, whereby plans or suggestions are openly debated with the group and constructive critiques are encouraged, can be very helpful. Asking participants in a group to consider the perspective of other stakeholders also can be useful to stimulate and broaden their thinking: a horse riding enthusiast who makes use of riverside trails might be asked to think about the concerns of hikers or cross-country skiers, or an executive with a forest company might be asked about the potential concerns of suppliers, staff, or customers.

Yet any attempt to minimize biases and improve the quality of group thinking takes interest, insight, and effort, and not every resource manager will have the temperament, knowledge, or energy to work effectively with groups.

Of course, it's also true that people who are part of a group are also individuals: members of an advisory committee who serve as representatives of a resource management agency, industry, or non-governmental organization also are individuals with unique personalities, interests, and experiences. In such cases, it is important for the SDM facilitator to be clear about what perspective he or she is seeking. Our typical advice is that it is most useful to ask for individuals to function within a group, first and foremost, on behalf of their constituency, in order to make sure that the main considerations of the agency or group that they represent are expressed. Yet we also want to hear from them as individuals, bringing to the table their own perspective and concerns. This dual focus tends to ensure a more comprehensive listing of objectives and measures, informed by the professional status of the individual as well as by their position in the community: a local fire chief might also be a parent and an avid sport hunter, and if the issue under consideration involves possible changes to wildlife habitat then each of these perspectives can be valuable as part of an SDM evaluation approach.

11.2.4   Integrating structured decision making with larger public processes

An SDM approach emphasizes depth rather than breadth: a small number of people are involved in an intensive examination of values, information, and management alternatives. At times, however, this intensive evaluation needs to be matched with a parallel effort that seeks to broadly disseminate information and, in some cases, to incorporate

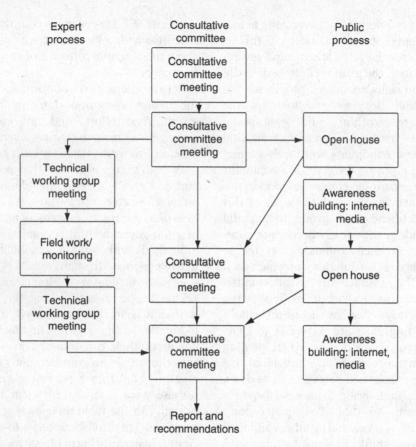

Figure 11.1 Illustrative parallel public and technical processes as part of SDM.

feedback from larger public processes or (in the case of technical advisory committees) peer reviews. Figure 11.1 illustrates a typical situation. In this example, a multi-stakeholder advisory consultative committee (CC) of 25 members schedules four meetings over three months, with the intention of providing a summary report to decision makers at the end of the process. To be successful, two parallel processes are also initiated. As shown on the left-hand side, a technical working group is formed after the second meeting of the CC, tasked with coming up with defensible estimates of the effects of potential management alternatives on a subset of the specified objectives (for example: employment effects, wildlife impacts, or cultural implications). This working group conducts field studies or modeling and reports back to the CC at the start of their fourth session. Meanwhile another process (again, linked to the CC deliberations) is undertaken to provide information about the ongoing SDM deliberations to the larger community, using a variety of public meetings, internet websites, and media releases. This includes an open house after the first CC session, which informs the discussion of objectives and initial alternatives, and again after the third CC session, with the information provided as feedback helping to inform the trade-offs session during the final CC meeting. These three parallel processes – the CC deliberations, the TWG findings, and the Public Information sessions – are all summarized as part of the report that goes forth to decision makers.

At the end of the day, an SDM process typically makes a recommendation to a decision maker. Ultimately a decision will occur – a permit will be issued (or not), conditions will be put on a license, or a management plan will be approved – and this outcome will be announced, quietly or with great fanfare, by a senior manager or elected official. From our perspective, this gap argues for a strong dose of reality when integrating SDM with larger public processes. It also suggests that practitioners of SDM should be aware of the expectations that might be built on the part of group participants. Despite the focus on 'making' good decisions, SDM processes rarely make decisions but, instead, usually serve in an advisory capacity: they provide insight to decision makers. Many potentially great decisions have failed to be enacted because of political considerations, much to the chagrin of the participating stakeholders. In some cases, history has proved the non-conforming decision makers to be right: they sensed something in the wind – an emerging shift in public values, or a change in legal frameworks – that wasn't adequately captured in the analysis (this may explain why they were successful politicians in the first place). In other cases, history has shown that the decision makers would have made a far better choice had they listened more carefully to the structured opinions of their constituents. The bottom line, well worth remembering, is that SDM approaches typically unfold in the context of, and may be subservient to, other larger public, legal, or political processes.

11.2.5 Case study: linking structured decision making to public consultation: Musqueam Legacy Trust

Background

In almost any environmental planning context, there is a possibility for tension and misunderstanding to be created between an in-depth decision-aiding process that involves a small, albeit representative, group of stakeholders and the larger group of residents and resource users who comprise 'the public'. Although approaches such as SDM might be used to encourage well-structured input from smaller groups, in most cases little methodological rigor is applied to engaging the public. As a result, key sectors of the community are alienated, choose not to participate, or shrink at the prospect of endless meetings at the same time that keenly interested, 'professional citizens' appear at all town-hall meetings and often dominate the entire process. Information open houses barely scratch the surface of engagement and opinion surveys rarely offer substantive insight into key aspects of the decision-making process, such as creating responsive alternatives or making defensible trade-offs that characterize wise, long-term resource-management decisions. Yet accountable decision makers – from small city councils to the office of state governors or provincial cabinet ministers – may not have faith in recommendations if they feel that broader public values are not well-represented. In a practical sense, this means that the results of advisory or stakeholder committees, no matter how well done, could be undermined or ignored; at minimum, a door is open to interest groups to subvert the process.

Structured decision making application

An example comes from work done by William Trousdale and his colleagues at EcoPlan International, based on principles and methods drawn from SDM, that involved input from both a representative advisory committee and the general community of the Musqueam Indian Band in Vancouver, Canada. As one of the Four Host Nations for the 2010 Winter Olympic Games, Musqueam signed onto a 2010 Legacies Agreement that created a special $17 million Legacy Trust Fund established for the benefit of present and future members of the Band. EcoPlan set up a SDM process and a clear schedule (Figure 11.2) to help the Musqueam Band decide how best to allocate this fund, working in the community at three levels: public engagement (Musqueam community members) was integrated

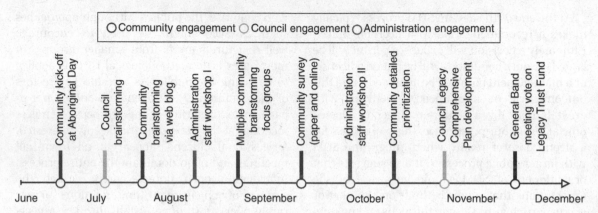

Figure 11.2 Legacy Trust fund engagement process timeline.

with small group stakeholder deliberation (Musqueam Band Administration) and meetings also were held throughout the SDM process with the decision-making organization (Musqueam Band Council).

Initial sessions with the Musqueam Band administration and community members led to the generation of over 500 ideas. Using 11 community objectives as evaluation criteria, interviews and surveys were developed and used at stakeholder meetings and modified for broad community input. The results were used to rank alternatives in terms of the leading objectives, thus creating a clear list of priority projects. The inclusiveness of the process also had the benefit of informing, educating and building trust around the results. This became clear when the final set of projects was put to a community vote at a Musqueam Indian Band General Assembly meeting; despite a history of community–Band Council tensions, the priority projects were put to a vote and passed by an unprecedented 2-to-1 margin. Further, with the confidence of the community behind them, Musqueam Band Council began implementing projects immediately and within six months work had begun on over 75% of the projects.

The Musqueam Indian Band experience suggests that this combination of working intensively with a smaller group but increasing transparency and buy-in by simultaneously involving a much larger group can increase the confidence of community leaders and decision makers, particularly when more difficult or controversial choices are required. The use of SDM methods and techniques also helped to address problems such as inconsistent attendance at meetings and concerns about the transparency of the overall planning process. Tools such as influence diagrams, consequence tables, and interactive decision models were used at both stakeholder small-group meetings and public open houses. In this and other projects, a variety of new engagement methods – including decision support software to aid direct participant input and the use of social media (e.g. Twitter, Facebook, blogging, vlogging, etc.) – are proving to be helpful for getting ideas out and input in. Integrating these tools into an overarching SDM framework that clearly shows *how* and *why* they are incorporated into the process is critical, and helps to avoid the 'engagement as an end in itself' syndrome that often occurs with less-structured processes.

11.3 A 'typical' structured decision making work plan

As a starting point, there really is no 'typical' SDM work plan beyond what is suggested by the six-step process presented in Chapter 1 (Figure 1.1): clarify the decision context, define objectives and measures, develop alternatives, estimate

Table 11.1 Typical plan for Consultative Committee (CC) meetings.

No.	Focus	Meeting tasks
1	Decision context and process	Review terms of reference and roles Summarize and understand the decision context, including scope/bounds and key issues Confirm work plan Discuss needs for Technical Working Groups (TWG) and their roles in supporting the CC
2	Building common understanding	Develop a decision sketch Provide technical presentations Develop and review initial influence diagrams
3	Objectives, measures, alternatives	Develop objectives and measures Develop preliminary alternatives
4	Information needs (identify and prioritize)	Review possible approaches to populating a consequence table – data, modeling, expert judgment Prioritize information gaps and candidate studies Identify technical review needs and process
5	Round 1 alternatives and consequences	Review key outcomes from studies Review modeling methods (peer review may have occurred in TWG) Review Round 1 alternatives and consequences Identify opportunities and trade-offs Refine alternatives
6	Round 2 alternatives and consequences	Review Round 2 alternatives and consequences Explore trade-offs: identify areas of agreement and difference and refine alternatives
7	Round 3 alternatives and consequences	Review Round 3 alternatives and consequences. Explore trade-offs: document areas of agreement and difference Identify key uncertainties; review monitoring needs
8	Monitoring program and review period	Confirm monitoring priorities Confirm review period and monitoring process Confirm recommendations to be submitted to decision makers
9	Sign off	CC reviews and signs off on recommendations and report

consequences, evaluate trade-offs and select a preferred alternative, and then implement, monitor, and review. Why is this? There are two main reasons: the context within which decisions are to be framed and choices made and information about consequences collected will differ widely across situations, and the people who are involved – their affiliations and knowledge, what they care about, and what processes they might view as valid inputs – will vary significantly across these different problems and contexts. Thus our response to presenting a typical SDM work plan is to provide a general schedule based on tasks and then to describe three quite different case studies, emphasizing ways in which they are different and similar.

First we turn to the general case, summarized in Table 11.1. This table shows the steps of SDM completed in nine meetings. This might be appropriate for a fairly in-depth consultation on a complex management decision. If budgets for the involvement of a stakeholder committee are limited or the decision context is less complex, it may take fewer meetings. The schedule assumes that quite a bit of analytical work occurs outside of meeting time in response to input from the CC at previous meetings. 'Complexity' refers to the technical details of the system – the geographic scope for example, or the number of alternatives under consideration, the availability of information for estimating consequences, the degree of uncertainty, and so on – or it may refer to the people side of the equation – it will take more meetings to build common understanding and support for a way forward in groups where trust is low for example. In the Nicola case that follows, a similar scope of work was conducted in five meetings. It is important to note that, in nearly all cases, some analysis and discussion with stakeholders typically needs to occur between meetings in order to bring new material back to the committee for its review.

How can closure be recognized? In some cases it will be when all participants agree on the same preferred management strategy, which might include a mix of near- and long-term actions as well as proposals for adapting as new information becomes available in the future. Sometimes it will be when participants agree to disagree and, as a result, voice support for two (or more) different options; in this case, information about the reasons for these disagreements should be documented and passed on to decision makers. In other cases it will be when participants agree that further progress toward selection of a management plan cannot be made until additional information has become available, which might mean that a technical working group is formed (with a specific charge) or that meetings of the consultative committee are suspended for a period of time, until the required information is in hand.

What closure in an SDM process does not mean is that a weak and temporary settlement has been achieved, or that consensus has been gained because key issues have been forgotten or hidden away under the consultative table. In the SDM universe closure requires, at minimum, that the environmental management issue has been structured sufficiently well that it is clear, to decision makers as well as stakeholders, how the problem is defined and what the key concerns are, how this information will be used to generate responsive alternatives, and how trade-offs will be addressed to come up with recommendations about preferred management plans. From the perspective of a consultative committee, whose members have donated time and energy to a resource-management planning process, closure

also requires that something be done: at minimum, that decision makers be informed about what stakeholders believe and recommend or, in some cases, that managers actually choose among different actions.

11.4   Example case studies

In this section we summarize three contrasting case studies, each relying on the decision structuring, deliberation, and assessment properties of a SDM approach to create useful and compelling insights for resource managers and decision makers. The first is a regional water planning study for an arid region facing conflicts among different water users. The second is a comparison of alternative systems for providing clean water to rural communities in east Africa. The third is an industry-based evaluation of different alternatives for purchasing transmission line poles.

11.4.1   Case study: community water planning, Nicola Valley, British Columbia

Background

The Nicola Valley is nestled in the south central interior of British Columbia. Situated between two mountain ranges, the area is arid, very hot during summer months, and subject to chronic water scarcity and drought. Major sources of water include Nicola Lake, the largest body of water in the basin (comprising an area of about 2500 hectares), and a concrete dam built in the 1980s to provide storage for agricultural irrigation projects and fisheries during low flow periods of the year. However, because of challenges associated with excavating a high point at the outlet of the lake, full storage benefits had not been realized. Forecasts for the future emphasize sustained and worsening periods of drought due to climate change and the effects from an invasive beetle infestation on local forests, and increasing pressures on water resources associated with rapid population growth, changes in agricultural land use, and meeting in-stream fishery requirements of threatened salmon runs (in 2004 the Nicola River was ranked the most endangered river in the province). These factors led the government to effectively stop issuing any new water licenses for the area, which in turn has impacted development and growth opportunities and threatened to further pit water users against one another. Ranchers and farmers, the principal water users, had a particular interest in spearheading the search for solutions prior to possible government intervention and possible introduction of tough new regulations that could threaten the viability of their businesses through reductions in licensed water rights or pending future authorizations.

These issues brought the community together to search for solutions to the problem of how limited water supplies should be conserved and divided among competing water interests. The end result was a community planning process, led by a diverse and broadly representative Multi-Stakeholder Committee (MSC) of about 50 people from different sectors within the community including ranchers and farmers, government agencies, First Nations, industry, residents, and special interest groups. The community-driven process faltered, however, when it came time to address trade-offs and to make choices on the myriad of regulatory, economic, and voluntary options to be included as part of a water-use plan. The MSC met about 25 times during the first three years of the process. Over CDN $300 000 was spent on technical studies; this research helped characterize and to a certain extent emphasize what many community residents already knew; there simply was not enough water to meet the needs of all users during dry periods and that (given increasing demands and changes in the hydrological cycle) conditions were worsening. Community members became frustrated by the lack of progress and, after discussion, the MSC agreed to try a SDM approach to reframe the collected information, address trade-offs across alternatives, and look for areas of agreement among participants.

Structured decision making application

The overall purpose of the community water plan was to 'ensure that the future water supply will be divided equitably among all water users balancing the community's social, economic, traditional and ecological values'. Key issues included:

1 Insufficient water for both irrigation and fish during summer and early fall low flows.
2 Growing tension with how water is managed for agriculture and recreation.
3 Competition for scarce water supplies between new land developments and existing surface water license holders.
4 Inadequate groundwater controls or regulations, in place, which further threaten base flows in streams.
5 Security for farmers and ranchers to access enough water in times of drought.
6 Poor water quality from past and present land use practices.

Preliminary reconnaissance discussions with community members highlighted that much of the initial work completed by the MSC focused on process and data collection. Very little effort and planning had been geared to clarifying the decision context or analyzing possible options and policy instruments that might ultimately make up a draft water-management plan. These factors – along with a limited budget and the difficulties of trying to schedule meetings with a 50 member committee – were taken into account in the development of a SDM work plan intended to provide a draft water-management strategy after only five meetings and within a condensed 11 month period. The proposed process consisted of the following seven steps, summarized in Figure 11.3.

Step 1: project scoping. Review and confirm the decision context for synthesizing a plan, confirm a work plan, confirm the available information resources, confirm the roles and responsibilities of the MSC, SC, sub-committees, and the greater public.

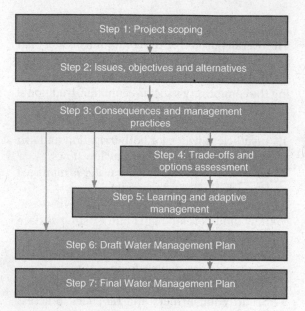

Figure 11.3 Steps in the Nicola Valley water-management process.

Table 11.2 Primary tasks conducted during each of five Nicola water-plan meetings.

Multi-stakeholder Committee meeting	Tasks
No. 1	Clarify decision context and work plan
No. 2	Review existing information
No. 3	Undertake a detailed water supply options assessment
No. 4	Review and identify a preliminary set of possible policy instruments and recommendations for the draft Water-Use Management Plan
No. 5	Reach agreement (documenting any areas of disagreement) on a final suite of policy options (recommendations) and develop a strategy for implementing the plan

Step 2: issues, objectives, and alternatives. This step involved several tasks: (i) confirm the objectives that will be used for creating evaluation criteria; (ii) compile and assess the existing information that addresses these concerns; (iii) develop an initial menu of the available options or alternatives to meet these objectives; and (iv) undertake a first categorization of these potential water-plan components according to the following split: must do[9], must not do[10], or could do[11]. Key evaluation criteria including financial costs, effectiveness, distribution of net benefits, and public acceptability.

Step 3: consequences and management practices. Screen and assess those 'must do' options identified in Step 2. It is during this step that more detailed cost estimates and implementation characteristics (including any monitoring requirements) could be assessed in further detail before inclusion into the Draft Management Plan.

Step 4: trade-offs and options assessment. Undertake a series of technical and value-based evaluations to uncover trade-offs and find the right balance across the competing objectives. Tasks include the identification of performance measures, the grouping of the components of alternatives in sensible ways, and their evaluation.

Step 5: learning and adaptive management. Develop a monitoring program to address critical uncertainties and inform future water-management decisions.

Step 6: draft Water-Use Management Plan. Draft the plan in a report.

Meetings of the MSC were arranged for only a half-day, in order to accommodate the volunteer nature of participation by many members. Given the community-driven nature of the process, timing for step 7, submission of the final plan, took place over a 12-month period to accommodate political considerations (e.g. a provincial election), coordination with external stakeholders, and to build awareness and support in the community before moving forward with the plan. Agreement on the schedule was established at the first of the SDM sessions, along with a brief overview of the approach and a brief review of key water issues and plan objectives. A breakdown of the tasks carried out during each half-day of the five MSC meetings is summarized in Table 11.2.

Step 7: final Water-Use Management Plan. Making progress on this SDM-based assessment required the facilitator/analyst to closely coordinate activities of the full MSC and a technical subgroup. At the fifth and final MSC meeting committee members reached agreement on 37 policy recommendations; these served as the foundation of a draft official Water-Use Management Plan, provided the basis for broader community consultations that led to its finalization, and became the basis for acceptance of the first community-driven water-management plan in the province.

11.4.2   Case study: selecting water filters in rural Africa[12]

Background

More than one billion people in the world still lack access to clean drinking water. Although a number of different technical solutions exist, and although a variety of aid organizations have a keen interest in enhancing the health of individuals living in developing countries, significant problems have been encountered as part of attempts to identify effective and sustainable solutions for providing clean water to families. Much of this interest has focused on rural communities in sub-Saharan Africa, where despite the presence of some of the worlds' largest lakes (e.g. Lake Victoria) a combination of low average annual rainfall and extreme poverty have resulted in tens of millions of people drinking and cooking with contaminated water and, as a result, exposing themselves and their families to water-borne diseases that include cholera, typhoid, and shigellosis. The resulting high mortality rates also are associated with high costs to the economies and social structures of these countries.

Structured decision making application

As part of a comparison of alternative systems for providing clean water to rural communities in Africa, Joe Arvai and Kristianna Post[12] employed a SDM approach that focused on the comparison of various point of use (POU) water treatment systems that could be used to provide clean water at the household level in rural Tanzania. In consultation with both in-country partners and experts in international development and microbiology, the study began by identifying five alternative POU systems that were both widely available and technically feasible (i.e. they required neither electricity nor batteries to operate). Two small villages in Tanzania were selected as study locations, with three SDM workshops held in each village.

Arvai and Post faced several severe constraints in the conduct of this research. First, although SDM methods have been used extensively in more developed western countries, there have been few applications in developing countries where participants face extreme poverty and researchers face significant political, cultural, and language barriers (also in this case Post, a graduate student at the time, undertook the unusual step of learning conversational Swahili in order to facilitate direct communication). Second, in-country advisors stated that, in light of local employment and resource constraints, the maximum duration for a SDM workshop should not exceed a half-day (3.5 hours). Third, travel restrictions meant that the workshops all were held over the course of two days, which means there was little time for feedback or for adjusting workshop protocols.

Each of the three workshops conducted in the communities began with a 30-minute session, introducing both study and community participants and reviewing project objectives. This introduction was followed by a 60-minute interactive demonstration covering five different POU methods, including demonstrations and opportunities for participants to test each of the POU methods themselves. The analysts then elicited objectives associated with using POU treatments for domestic water use: after an interactive discussion, the five leading objectives were identified as taste, odor, the amount of time required to disinfect the water, health and safety (efficacy), and ease of use. Because most of the participants in one of the communities had not worked with outside

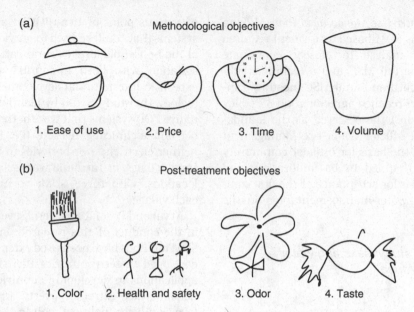

(a)   Methodological objectives

1. Ease of use   2. Price   3. Time   4. Volume

(b)   Post-treatment objectives

1. Color   2. Health and safety   3. Odor   4. Taste

Figure 11.4 Objectives and measures for comparing water treatment methods in Tanzania.

researchers before, and in order to stimulate conversations about what was important to themselves (rather than to the researchers), community members were encouraged to draw pictures on sketchpads that characterized their objectives and concerns and to raise any questions they might have concerning the operation and effectiveness of the different POU treatments.

Participants were then asked to evaluate post-treatment water samples from each of the available POU systems in terms of their ability to meet the four stated objectives, with each sample scored on a scale of 0–5. To make this scoring system more user friendly, participants placed the desired number of tokens in cups (in one village) or were asked to show a corresponding number of fingers (in the other village). Finally, trade-offs were addressed using a lexicographic decision rule based on attribute ranks, with participants discussing each of the POU systems on an attribute-by-attribute basis until settling on a preferred option (effectively, the alternative exhibiting practical dominance).

In both villages, the same POU system was preferred by the community-based participants.

Although it was known that this method would not be effective when considering all possible local water sources, the emphasis on selecting a water treatment method consistent with the objectives of community residents – rather than consistent with the technical studies of outside experts – was felt to provide important advantages in terms of encouraging use of the POU system over the long run (thus addressing a serious problem, in that previous follow-up studies have shown that the use of water treatment systems often quickly declines over time). Despite the obvious disadvantages of a highly time-constrained assessment process (e.g. requiring condensing steps 5–7 from Table 11.1), workshop participants agreed to work through a SDM process, including completion of a consequence table, and articulated their views without difficulty at each step of the SDM process (e.g. deliberations about objectives, measurement criteria, trade-offs, etc.). Finally, the values-based SDM process itself was sufficiently flexible to assist participants in the expression of their concerns: for example (see Figure 11.4), price was measured at one of the villages in terms of cashew nuts, which many people in the community harvest and sell at market; the visual marker for taste was a piece of candy; and health and safety was shown using a picture of a family.

11.4.3   Case study: industry purchase of transmission line poles[13]

Background

SDM methods are often used, in government and industry, as part of efforts to implement triple bottom line (TBL) accounting systems that address environmental, social, and cultural considerations as well as economic or financial concerns. One example comes from BC Hydro, which has won international awards for its environmentally-minded business practices and whose CEO has stated 'triple bottom line decision making is how we do business'. Guidelines for the conduct of TBL at BC Hydro are based on SDM approaches, and come into play within BC Hydro whenever the scale of investment decisions is sufficient to bring in related goals of business sustainability, regulatory approval, or public acceptance. Although an SDM approach is not the only way to construct and implement a TBL framework, it provides both an accessible philosophy and an appropriate set of techniques.

Structured decision making application

In the case of procuring electrical transmission line poles, utilities typically face a choice among various wood and metal options. In British Columbia all utility poles initially were made of cedar, but since 1965 pine poles have been used due to a shortage of appropriate cedar stock. In 2008 BC Hydro began a strategic review of its pole procurement. It was recognized that a variety of financial, social, and environmental issues impacted the procurement process, and in light of high interest levels from stakeholders BC Hydro sought to ensure that procurement was done in a systematic, rigorous, and transparent manner. The corporation's Strategic Sourcing department, which (following emerging industry standards)

had already begun to incorporate triple bottom line considerations into its purchasing processes, chose an SDM approach to guide this important procurement decision and sought the advice of a trained on-staff SDM practitioner to help lead the process.

It was quickly determined that, in light of the province's substantial forests, wooden poles were preferred for economic, social, and environmental reasons over other options. Although for several decades pine poles have been cheaper to purchase, they are also more susceptible to rot and require more extensive chemical treatment than other species (including cedar) to prolong their life. Even after treatment, the service life of pine poles is 30–35 years, which is only about one-half that of cedar. In addition, the infestation of pine beetles in BC forests threatens the long-term supply of pine and complicates reliable testing for pole quality.

SDM techniques highlighted relevant objectives and used these to distinguish a range of alternatives and to clarify key value trade-offs. Discussions among stakeholders emphasized comparisons between the lower upfront, short-term costs but also the lower quality of pine poles and the improved quality performance, reduced lifetime costs, and increased recycling ability of cedar poles. Key elements of the SDM-based analyses and deliberations included:

1   Acknowledging the widely divergent values of group members.
2   Highlighting the multiple dimensions of value associated with the decision: economic considerations associated with lifetime costs, environmental concerns associated with chemical treatments and recycling, and social issues associated with locations of supplies and the provision of poles.
3   Exploring stakeholders' trade-offs through different values elicitation techniques.

After completion of the SDM analysis, BC Hydro elected to shift its purchasing towards achieving a long-term, reliable supply of cedar poles. This switch from pine to cedar is estimated to save approximately $110 million over the next 60 years.

11.5 Key messages

Reservations are often voiced about undertaking an SDM process – concerns about the required level of effort and costs, concerns about a presumed facility with quantitative analysis, and concerns about the relative roles of scientific and stakeholder input. We address each of these concerns and introduce several case studies, quite different in their context and level of effort, to demonstrate the range, strengths, and limitations of SDM approaches. We also examine several important practical issues that need to be dealt with in order to run an effective SDM process: these include organizing participants, structuring advisory committees, and integrating the results of an in-depth SDM approach with the findings of other public policy processes.

11.6 Suggested reading

Chess, C. & Purcell, K. (1999) Public participation and the environment: do we know what works? *Environmental Science and Technology*, **33**, 2685–2692. A succinct review of the early literature on what works and what doesn't as part of public participatory processes dealing with environmental management.

Dietz, T. & Stern, P. (2008) *Public Participation in Environmental Assessment and Decision Making*. National Research Council, Washington, DC. A recent overview of public participation methods and success in environmental assessment and decision making, with contributions from many leading researchers and practitioners.

Keeney, R.L. (1992) *Value-Focused Thinking: A Path to Creative Decisionmaking*. Harvard University, Cambridge. Yes, we've noted this book before, but for entertaining case-study reading about creative ways to improve decisions it's hard to beat Chapter 13 of this book, where Keeney reviews how he has used value-focused thinking to help choose a life-partner, select a job, buy a car, rent an apartment, and name his son.

Wilsdon, J. & Willis, R. (2005) *See-Through Science: Why Public Engagement Needs to Move Upstream* (Pamphlet). Retrieved from http://www.demos.co.uk/ publications/paddlingupstream. A passionate and well-argued plea from experienced decision-aiding consultants for incorporating public engagement in the early stages of risk and environmental management debates.

Several websites also provide additional information on case studies demonstrating the use of SDM techniques. Readers are encouraged to check out websites of the US Fish and Wildlife Service, British Columbia Hydro Authority, the Australian Centre for Excellence in Risk Analysis (ACERA), and Compass Resource Management (www.structureddecision-making.org).

11.7 References and notes

1   BC Hydro (n.d.) *Triple Bottom Line and Structured Decision-making: A Case Study of BC Hydro* (Report No. GDS09-198). Retrieved from http://corostrand-berg.com/wp-content/uploads/files/Industry-Canada-SDM-Case-Study-Oct15.pdf

2   Gregory, R., Failing, L., Ohlson, D. & McDaniels, T. (2006) Some pitfalls of an overemphasis on science in environmental risk management decisions. *Journal of Risk Research*, **9**, 717–735.

3   Gregory, R., McDaniels, T. & Fields, D. (2001) Decision aiding, not dispute resolution: creating insights through structured environmental decisions. *Journal of Policy Analysis and Management*, **20**, 415–432.

4   Gregory, R.S. & Keeney, R.L. (2002) Making smarter environmental management decisions. *Journal of the American Water Resources Association*, **38**, 1601–1612.

5   Surowiecki, J. (2004) *The Wisdom of Crowds: Why the Many are Smarter than the Few*. Random House, New York.

6   Janis, I.L. (1982) *Groupthink: Psychological Studies of Policy Decisions and Fiascoes*. Houghton Mifflin, Boston.

7   Gigone, D. & Hastie, R. (1993) The common knowledge effect: information sharing and group judgment. *Journal of Personality and Social Psychology*, **65**, 959–974.

8   Sunstein, C.R. (2002) The law of group polarization. *Journal of Political Philosophy*, **10**, 175–195.

9 Must Do – or *Best Management Practices* – these are the 'low-hanging fruit' options that would likely be done no matter what. They are either low cost options with high benefits and no significant or controversial trade-offs, or plan components that are required because of some legal or regulatory constraint.

10 Must Not Do options. These are options that are readily identifiable as infeasible, either because of timeline or budget constraints, major costs or other trade-offs, or because there is an obviously better option for achieving the same results (i.e. they are dominated).

11 Could Do Options – these are options that are expected to have benefits but also possibly signifi-

cant costs, and for which the MSC will need to make choices that involve tough trade-offs. They will need more detailed evaluation.

12 This case study is based on Arvai, J.L. & Post, K. (in press) Risk management in a developing country context: structured decision making for point-of-use water treatment in rural Tanzania. *Risk Analysis*.

13 This case study draws on the discussion in the document *Triple Bottom Line and Structured Decision Making: A Case Study of BC Hydro*, available on the BC Hydro website and primarily written by Basil Stumborg, Senior Business Strategy Advisor, BC Hydro and Coro Strandberg, Principal, Strandberg Consulting.

12

Conclusion

We're all used to making decisions informally, whether on the basis of well-established patterns or by following the advice of friends, family members, or colleagues. Most of us make it through the day's typical decisions without a crippling degree of effort: we spend money for groceries, we decide on commuting routes and vacation options, we vote for preferred candidates in elections, and so on, without having to invest much time or thought. Often we do fine, occasionally we mess up. Either way, the stakes are typically low and we're usually accountable to no one but ourselves.

Environmental management decisions are different from these personal decisions, in important ways: the stakes are high, the context unfamiliar, and the outcomes of our choices will affect people other than ourselves. As a result, the trade-offs that we face in environmental decision making can be emotionally and morally challenging. Because these are decisions about public resources, most of the time we need to work collaboratively and reach agreement with others. We all have some experience in making group decisions, as part of a family unit or at work or in our community, and are familiar with the extra level of effort and very different set of skills that entails.

In recent years, both social expectations and legal requirements about who needs to be involved in environmental management choices have changed. Decisions that used to be made solely by industry or politicians or government agencies – how to balance energy production and fish protection at a dam site, where to site a road or locate a resort development, how much to spend on clean-up of contaminated wastes or improving domestic water supplies – now include input from a whole host of interested and capable people. Our goal with this book has been to help people approach complex environmental problems as choices among alternative actions, to provide tools and ideas that will support both the reasoning process and the dialogue process that underlie these choices, and ultimately to contribute to more defensible and transparent decision making.

In the preceding chapters, we've tried to synthesize what decades of research and practice have taught us about what constitutes best practice in decision making. It's a large field, and we've covered a small, but hopefully widely applicable part of it. In this brief concluding chapter, we pull together some of the main arguments we've made throughout the book.

We begin by observing that conventional approaches to decision making often don't work well in complex environmental management situations, largely because of the interplay of facts and values. Many environmental managers and decision makers rely heavily on science and the public often calls for 'science-based' decisions. But choices are fundamentally value-based: science

Structured Decision Making: A Practical Guide to Environmental Management Choices, First Edition. R. Gregory, L. Failing, M. Harstone, G. Long, T. McDaniels, and D. Ohlson.

can inform but will never make decisions. On the other hand, consensus-based approaches provide insufficient support for learning about what might work in complex systems, and insufficient space for people to explore creative ideas that might threaten a fragile consensus. What's lost with quantitative valuation approaches – whether rooted in cost benefit analysis or multi-criteria methods – is the emphasis on making sound decisions. If you're a manager, you need solutions – creative solutions that are directly responsive to the problem at hand and the perceptions of stakeholders – not a precise valuation of losses and gains or a formula-driven ranking.

What's different about an SDM approach is that it *focuses on the management decision*. Everything that's done is in the service of making a better, more informed, more defensible decision (or set of decisions, or recommendations for decisions).

In everyday terms, we think of SDM as an organized, inclusive and transparent approach to building a common understanding of complex problems and generating and evaluating creative alternatives. It's founded on the idea that good decisions are based on an in-depth understanding of both values and consequences. Designed with groups in mind, it pays special attention to the challenges and pitfalls that can trap people working together on emotionally charged and technically intensive problems – mental shortcuts and biases, groupthink, positioning, and a host of difficult group dynamics and communication challenges.

Although there are many kinds of environmental management problems and decision making contexts, SDM inevitably involves addressing six key steps: (1) define and bound the decision context; (2) clarify what matters; (3) understand what alternatives are possible; (4) think about their consequences and uncertainties in terms of what matters; (5) balance the pros and cons of different alternatives, and; (6) follow up on the decision to see what can be learned and implemented as part of future decisions. What exactly is done at each step, to what level of rigor and complexity, will depend on the nature of the

decision, the stakes and the resources and timeline available.

For a resource manager or decision maker worried about following a defensible process that will withstand close scrutiny, SDM methods provide a way to organize what can feel like an overwhelming surge of input about too many issues from too many people. For resource users and community residents, worried about how they might potentially be affected by the actions of government or industry, SDM offers a standard or guideline about what they can and should expect from a quality decision-making process. It announces recognizable best practices that are transparent and defensible but also intuitively appealing to ordinary people.

For each of the decision steps, a variety of methods have been developed to help collect and describe information and to assist in making comparisons among possible environmental management alternatives. Some of these techniques are straightforward and involve different ways to organize and present information to build common understanding among participants regardless of their technical background. Others are complex – they involve sophisticated quantitative methods and require special expertise to implement.

The trick is to link the amount of thought and effort that goes into addressing the problem to the importance of the issue and the extent to which the choice will be scrutinized. And at each step along the way, consideration is given to the quality of available information and to uncertainty: if there are things we want to know but don't, how can better information be obtained and is it worth the time and effort required to procure it?

If there is one message you take from this book, we hope it is that environmental managers need to start thinking like decision makers, and shift from a 'study' culture to a 'decision-making' culture. This doesn't mean you should stop studying; but it might change what you study and how. If you're a manager, or work in an organization that develops management plans or policies, nearly every problem you face is a decision or one that is linked to a decision. When you frame your problem as a decision – a choice with multiple objectives and alternative courses of action – it changes your point of entry into the problem and, consequently, everything else that you do: the make-up of the project team, the allocation of resources, the collection of information, the focus of technical analyses, the timing and content of public consultation, and the characterization of the outcome.

In our experience, all resource-management decisions benefit from time spent completing a quick decision sketch – a brief run-through of the key elements of the pending environmental decision, based on the six core SDM steps. Not all decisions require the same level of rigor: the thought, effort, data, and quantification that goes into each step therefore will and should vary. Yet all decisions will benefit from clear thinking about the nature of the problem, the fundamental objectives of participants, and a realistic range of alternatives – if this phase is done well, some problems may need little or no further quantitative analysis to arrive at a broadly supported decision.

The identification of objectives – concise and (as a set) complete statements of what matters to people in a decision – encourages people to focus on and express what's important to them when evaluating potential management solutions to an environmental issue. In doing so, we attempt to unhook people from their presumed preferred solutions – quite literally, to unfold their arms (and simultaneously open their minds): in an SDM process, the first stage of deliberations focus on what matters rather than what should be done. Aside from being fundamental to good decisions, developing a list of objectives is also a highly collaborative exercise. Groups previously paralyzed by mistrust and anger are tasked with doing something productive and non-threatening – without passing judgment, at this stage, on the relative merits or importance of the expressed objectives. The result can often encourage closer connections among participants and help to create a collaborative foundation for shared decision making.

At times, developing objectives can appear either to be so trite a task as to hardly warrant the time, or so overwhelmingly difficult as to be impossible. Neither perspective is accurate. Objectives lay the foundation for engaging stakeholders in the decision process and for creating and evaluating alternatives. They matter and you should take whatever time is needed to get a good and agreed-upon set of objectives. But, critically, objectives for decision making are not 'targets' and they are not weighted – at least not at this early stage. If you follow the guidance in this book, defining a mutually agreeable set of objectives is eminently achievable, even in very contentious situations.

Performance measures (Chapter 5) are used to assess and report the performance of the alternatives on the objectives. As with other aspects of decision structuring, performance measure development is introduced as a subjective exercise, to be undertaken by the decision participants who will potentially be affected by them. Development of good performance measures requires both thought and discussion.

Happily there are well-recognized, easy-to-use practices for building a useful set of objectives and performance measures along with well-accepted standards or properties for doing this defensibly.

During the course of a decision-making process, performance measures are often best developed iteratively: start simple, and add sophistication as needed (but only when needed). If you only have half an hour to sketch out a choice, go for the key considerations and the simplest, most operational five point rating scales you can come up with. If you have a bit more time and better resources, you can afford to emphasize a broader set of factors or come up with more nuanced scales. It's important to keep in mind that what's essential is to communicate *what matters*, not what can readily be measured. If you're dealing with values that are easy to identify, for example using widely understood natural or proxy measures, consider yourself lucky. If your problem involves hard to quantify or intangible values that are affected differently by the alternatives under consideration, then you'll need to find ways of communicating these values, and to clarify how they might be affected by alternatives, to stakeholders and to decision makers. With this in mind, we discuss the use of constructed performance measures, which often combined qualitative and narrative descriptions with quantitative indices, as a way to express important but hard-to-quantify values.

We suggest that SDM supports the goal of leveling the playing field in a decision process. Performance measures contribute to this goal in two ways. First, they allow hard-to-quantify things – such as maintaining a traditional lifestyle, or enhancing visual quality – to be put on the table and consistently evaluated alongside more conventional objectives such as revenues and habitat. Second, they synthesize a large volume of technical material into a summary format so that everyone in a multi-stakeholder group can understand and agree on the critical aspects of performance and how they will be communicated. As a result every participant, regardless of their technical background, can participate meaningfully in identifying, evaluating, and choosing alternatives, even in very technically intensive decisions.

Uncertainty arises in many ways as part of nearly every environmental decision process. Linguistic uncertainty – the vagueness, ambiguity, and underspecificity that leads to communication failures – is often as important a source of problems in resource decision making as is the epistemic uncertainty that arises due to a lack of knowledge (e.g. concerning measurement methods, predictions of physical effects, or external factors such as climate change and political will). The structuring tools and practice guidelines outlined in this book – objectives hierarchies, influence diagrams, properties of good performance measures, and so on – leave little room for ambiguity, and this alone is an important contribution to addressing uncertainty in decision making.

Uncertainty about consequences will be a key factor in many environmental management decisions, so we argue that today's managers must be conversant in the language of probability.

Yet the fundamental goal of further analysis is always to understand uncertainty and its implications for decisions better. Reducing specified sources of uncertainty may be important, but these reductions may not always be possible – at least within the constrained decision setting – and they always need to be linked to anticipated improvements in the quality of resource-management decisions. Organizing and structuring key uncertainties will lend insight to the decision – here we emphasize the importance of the decision sketch to help you avoid falling down the rabbit hole of detailed technical analysis at the expense of other aspects of the problem. Be specific with describing consequences and probabilities, but recognize that the assignment of probabilities is not always appropriate.

We emphasize the need to explore the risk tolerance of decision makers and provide some ways to think about that. In today's world, best practice with respect to characterizing uncertainty includes characterizing the degree of disagreement among experts. A good decision process will look for ways to improve the quality of information, including the quality of judgments under uncertainty, but sometimes, the most important way to address uncertainty is not to try to reduce or analyze it, but to find alternatives that work effectively despite it. We introduce and discuss some of the pros and cons of three leading management approaches – precautionary, adaptive, and robust – that explicitly seek to deal with uncertainty.

Why do we need alternatives? Lots of reasons, but a key one is that people will make tough trade-offs only when they are sure that a creative range of alternatives has been explored – you don't buy the first flight you find if it's unexpectedly expensive or long – you look for more options before you're willing to make a tough choice. A good environmental decision process uses the specified objectives and performance measures to generate and iteratively evaluate a broad range of value-focused alternatives. Iteratively generating and evaluating alternatives allows everyone involved to gain the same level of understanding about what works and why. As part of a multi-objective, multi-stakeholder SDM process, developing an environmental management alternative for consideration doesn't mean it is recommended by, or even acceptable to, those who put forward the idea. It can be very useful to explore alternatives that explicitly examine controversial approaches, or that explore the consequences (all the consequences) of one party's initially entrenched 'position'. Our experience is that after undertaking a thorough exploration of alternatives people will often accept and endorse trade-offs that they previously found unacceptable. Extensive research in decision theory (about how people should make decisions) and behavioral decision making (about how people really do make decisions) shows unequivocally that whether or not an alternative is acceptable to people depends on comparisons among the alternatives.

Chapter 8 covers the estimation and presentation of the consequences of alternatives, arguing that their presentation must allow consistent comparisons of relative performance across alternatives and expose key value-based trade-offs and uncertainties. Society's understanding of what constitutes 'best available information' is evolving. This profoundly influences how consequences are estimated, and how we define experts and expertise. Participants in a deliberative process work together to build an understanding of what is known and not known about the consequences of management actions. It's a kind of joint fact-finding process, except that the facts aren't just out there: there's no 'right' information for use in decision making. Sometimes the group will make use of experts – who may be scientists, resource users or residents from the local area, or elders representing an indigenous community – whose judgments will be elicited as part of a formal process that seeks to level the playing field across different sources of knowledge. Assumptions underlying predictions of effects will be identified, along with each individual's degree of confidence in their assessments. As part of a deliberative SDM process, explicit choices will be made by participants concerning different

methods for assessing consequences and characterizing their uncertainty.

With all this in hand, we are finally ready to take on the difficult and often controversial (but only rarely show-stopping) topic of trade-offs. The trade-offs faced in environmental management are as complex as they come. In many cases the performance of alternatives will be compared using wildly different objectives and measures, some expressed using precise numbers (e.g. dollars, numbers, percentages) and others in qualitative or semi-qualitative terms (e.g. constructed scales with narrative descriptions). Further, discussions typically involve species and habitats or resource uses about which many stakeholders may feel passionate, angry, or fearful. The deliberations will be taking place in collaboration with people whom you may not know or trust or like. Participants may be inundated with technical information and feel they're given insufficient time to process it. And in all likelihood, managers and other expert or public stakeholders will be part of an advisory group that is the source of both creative brilliance and startling dysfunction.

We discuss and show examples of ways that the structuring and deliberative aspects of SDM support an efficient and constructive exploration of trade-offs – for example by refining the objectives and performance measures and simplifying the decision through tests of relevance and dominance. If fancy quantitative tools won't help, then don't use them. Iterative refinement of objectives, alternatives, and performance measures and the use of dominance and practical dominance will get you most of the way, most of the time. If heavier artillery is required, we describe in some detail how managers and other SDM practitioners can make use of defensible multi-attribute trade-off analysis methods to aid decisions. Such methods should help narrow alternatives, focus discussion, and diagnose areas of agreement and disagreement among participants. They facilitate both the careful consideration of relevant technical material and focused, constructive, value-based discussion. A key message is that we think there are standards for making value-based trade-offs in public decision making – such judgments should be informed, context-specific, consistent, and transparent.

Sometimes the exploration of trade-offs leads directly to a clear solution. Sometimes it doesn't. Even when a preferred environmental management alternative isn't clear, the structured exploration of trade-offs and careful documentation of areas of stakeholder agreement and disagreement will, at minimum, inform decision makers and help to identify a more broadly acceptable set of recommendations or management actions. The goal, as always with SDM approaches, is to understand and represent the decision context so as to lend insight to decision participants, resource managers, and decision makers.

We believe that a commitment to collaborative learning is a critical requirement for successful environmental decision making. It is one of the things that fundamentally sets SDM apart as a framework for decision making. Individuals participating in SDM will need to be prepared to listen, to learn, to explore competing hypotheses, and to build a common understanding of what constitutes the best available information for identifying alternatives and assessing consequences. This forms the basis for working collaboratively on solutions. Learning, monitoring, and iteration are covered in Chapter 10. The discussion highlights the role that SDM approaches can play in ensuring that learning occurs throughout the environmental management process, including learning about the scope and nature of the problem, learning about the relevant facts, and learning about the value choices that need to be made. We give special attention to principles and application of adaptive approaches to environmental decisions, emphasizing that care needs to be taken to ensure that the context for adaptive management is carefully defined and that the recommended passive or active approaches have been carefully constructed and evaluated.

Finally, in Chapter 11, we tackle some of the barriers to undertaking an SDM process and discuss several common concerns and fears that are voiced about heading down the SDM path. For example, we recognize that the emphasis on carefully defining objectives and performance criteria, estimating the consequences of proposed actions, and creating a set of alternatives means that SDM approaches can sound expensive or time consuming. In reality, an SDM process is as scaleable as you can conceive; some approaches warrant the dedication of many months of effort, including careful data collection and eliciting probabilistic judgments from invited experts, whereas other SDM processes unfold in a day-long or afternoon session with a few people and a flipchart. Nor is an SDM approach especially complicated: the logic of the six steps is no more demanding, and (in our opinion) far more intuitive, than other environmental management processes. And in our experience, taking time to systematically work your way through an environmental management problem nearly always is faster than scrambling to a finish without due process, only to be sent back to the starting line over and over again.

The case studies included in this reality-check chapter also address ways in which an SDM process seeks to incorporate information and knowledge from a variety of different sources. An SDM process should be guided by the principle that if something is important, then a way should be found to estimate the consequences of alternatives on that concern – using whatever qualifies as the best available information to do so. We also address issues about the quantitative nature of SDM. It's true that analysis or quantitative measures are often adapted as a means for informing and summarizing choices, but the core of an SDM process is its reliance on value-focused thinking and providing insight to decision makers. Numbers are therefore just one ingredient, as are the stakeholders' stories that might accompany different levels of a constructed performance measure. In many cases the decision-structuring qualities of an SDM approach lead to a clear and early preference for one of the alternatives under consideration. We also often hear worries concerning the subjective nature of the key decision elements included as part of an SDM approach. We agree again that SDM is a subjective process – but so are all others, and as part of an SDM process we seek to address the value-based elements of management decisions in ways that are systematic, consistent, and transparent.

We hope that the practical advice we offer in this book is sufficiently helpful to encourage you to take a stab at applying an SDM approach to your environmental management issues. Perhaps consider starting with a relatively minor and well-defined issue, and take on tougher problems once you have gained both experience and confidence. At the simplest level, it's a matter of getting in the habit of asking a few key questions: What are the objectives and why do these matter to people? What alternatives exist and what are their relative strengths and weaknesses? What are the key trade-offs and uncertainties? As with everything, practice makes all the difference, which goes as much for stakeholders and decision makers as it does for analysts and facilitators.

In the preface, we note that our goal in writing this book is to reset expectations for what constitutes defensible decisions in the management of environmental resources. We hope we have persuaded you there are such things as best practices in environmental decision making. We hope that we've also convinced you that this arena is interesting for all the same reasons that it's important and challenging: environmental management problems are also human problems, so they combine the inherent complexities and uncertainty of the natural environment with the dazzling and frustrating behaviors of people. Finally, we hope that the principles and case studies we include in this book may encourage you to learn some new things and to try on a new perspective or two. After all, what's the alternative?

Index

Printed in the United States
By Bookmasters